1806355702

AF598565

Biotechnology: Enhancing Research on Tropical Crops in Africa

Edited by

G. Thottappilly, L.M. Monti, D.R. Mohan Raj and A.W. Moore

Technical Centre for Agricultural and Rural Cooperation (CTA)

International Institute of Tropical Agriculture (IITA)

Technical Centre for Agricultural and Rural Cooperation (CTA)

Postbus 380, 6700 AJ Wageningen, The Netherlands
Tel: (31) 8380 60400; telex: (44) 30169 CTA NL; fax: (31) 8380 24262

CTA was established in 1983 under the Lomé Convention. Its mandate is to help the African, Caribbean and Pacific countries which comprise the ACP group achieve greater food security by providing them with better access to scientific and technical information on all issues related to agricultural and rural development. Working in close cooperation with ACP and EEC countries and with international, regional and national institutions, CTA fulfills its mandate through a range of activities, including seminars, studies, publications and support to ACP documentation centres.

International Institute of Tropical Agriculture (IITA)

Oyo Road, PMB 5320, Ibadan, Nigeria
Tel: (234-22) 400-300 to 319; telex: 31417 TROPIB NG; fax (INMARSAT): 874 1772276

The goal of IITA is to increase the productivity of key food crops and to develop sustainable agricultural systems that can replace bush fallow or slash-and-burn cultivation in the humid and subhumid tropics of Africa. Crop improvement programmes focus on cassava, maize, plantain, cowpea, soybean and yam. Research findings are shared through international cooperation programmes, which include training, information, and germplasm exchange activities. IITA was founded in 1967. The Federal Government of Nigeria provided a land grant of 1000 hectares at Ibadan for a headquarters and experimental farm site, and the Rockefeller and Ford Foundations provided financial support. IITA is governed by an international Board of Trustees. The staff includes nearly 200 scientists and professional staff from about 40 countries, who work at the Ibadan campus and at selected locations in many countries in sub-Saharan Africa.

IITA is a member of a system of non-profit, international agricultural research centres supported by the Consultative Group on International Agricultural Research (CGIAR). Established in 1971, the CGIAR is an association of about 50 countries, international and regional organizations and private foundations. The purpose of the research effort is to improve the quantity and quality of food production in developing countries. The World Bank, the Food and Agriculture Organization of the United Nations (FAO) and the United Nations Development Programme (UNDP) are co-sponsors of this effort.

ISBN 978 131 090 1

Citation: Thottappilly, G., L. Monti, D.R. Mohan Raj, and A.W. Moore (eds) 1992. Biotechnology: Enhancing Research on Tropical Crops in Africa. CTA/IITA co-publication. IITA, Ibadan, Nigeria. 376 pp.

Prepared for publication by:
Sayce Publishing, Exeter, United Kingdom

Printed by:
Ebenezer Baylis & Son Ltd, United Kingdom

Contents

Preface

In what ways can biotechnology be used to help improve agriculture in Africa? This question was considered by some 120 scientists from 24 countries (including 16 African countries), who attended a workshop on "Biotechnology: Enhancing Research on Tropical Crops in Africa" held at IITA headquarters in Ibadan, Nigeria, from 26 to 30 November 1990. The workshop was co-sponsored by the Government of Italy, the Technical Centre for Agricultural and Rural Cooperation (CTA), the Food and Agriculture Organization of the United Nations (FAO) and the International Institute of Tropical Agriculture (IITA). It coincided with the formal opening of the new Biotechnology Laboratory at IITA, funded mainly by the Government of Italy. This volume brings together the papers presented at that workshop, as subsequently revised during a systematic process of scientific and editorial review.

Three opening addresses were given. E.E. Okon, Director General of Nigeria's Federal Ministry of Science and Technology, said that plant biotechnology, based primarily on advances in cell and molecular biology, is radically changing agricultural research. He expressed the hope that Nigeria would be able to join the rest of the world in tapping and reaping the benefits of biotechnology. The Italian Ambassador to Nigeria, S. Rastrelli, stressed the need for cooperation between developed and developing countries in agricultural research. The North cannot live without the South, he said, and "poverty is a plague for the whole world, famine equally a shame for anyone who can do something to avoid suffering and does nothing." Representing IITA, Dr K.S. Fischer, Deputy Director General for Research, said that IITA was trying to link biotechnology with ongoing agricultural research in order to address the twin challenge of feeding an expanding population and protecting the natural resource base. IITA has established the biotechnology research unit mainly to help attain those objectives that cannot be easily realized by conventional breeding within its crop improvement programmes. IITA would also aim to strengthen and broaden collaboration between advanced laboratories or institutions and those responsible for agricultural research in Africa, so that new skills in science, now being so actively pursued in commercialized crops, would also be applied to the most pressing problems in African crops.

In his keynote address, L.M. Monti, Professor of Cytogenetics at the University of Naples, Italy and Special Adviser to IITA on biotechnology, stressed the decisive role that advances in biotechnology could play in agriculture, with their ability to directly modify plants, animals and agricultural processes to meet new needs. Biotechnology brings together scientists from many disciplines, some of whom may not have studied plants in depth previously. Molecular biologists, biochemists, physiologists, and virologists work with plant geneticists, breeders, pathologists, and entomologists to establish research teams with the capacity to tackle what is clearly a revolution in agricultural sciences. Professor Monti pointed out, however, that the combination of so many disciplines in a research effort may constitute a barrier for developing countries, many of which lack adequate scientific infrastructures. International centres, such as IITA, should act as bridges between advanced research centres in the developed world and national programmes in developing countries.

After the presentation of 48 papers, in eight technical sessions, the workshop participants drafted and adopted a set of recommendations; these recommendations are given near the end of this volume. The order in which the papers are presented in this volume differs, in some instances, from their grouping during the workshop, which was governed by time considerations and the intended, rather than the known, content of the papers. During the review process, it became evident that it would be more logical to group the papers into seven sections, as reflected in the contents list. The first section, in view of its breadth, includes two keynote papers; the other sections each include one keynote paper.

In view of the multidisciplinary and interdependent nature of plant biotechnology, we sought the cooperation of scientists from numerous organizations around the world in carrying out the peer review of the papers in this volume. Although the names of these scientists appear in the acknowledgements overleaf, we would like to thank them here for their willing participation in this demanding task.

We hope that this volume is the better for all the effort expended on it by authors, reviewers and editors. Any imperfections that remain, however, are the responsibility of the authors and editors.

G. Thottappilly, L.M. Monti, D.R. Mohan Raj and A.W. Moore
November 1992

Foreword

Biotechnology is a rapidly developing area of contemporary science. It can bring new ideas, improved tools and novel approaches to the solution of some persistent, seemingly intractable problems in food crop production. Given the pressing need to enhance and stabilize food production in Africa in response to mounting population pressures and increasing poverty, there is an urgent need to explore novel technologies that will break traditional barriers.

Sensing the need for an international initiative in this regard, the Government of Italy, the Technical Centre for Agricultural and Rural Cooperation (CTA), the Food and Agriculture Organization of the United Nations (FAO) and the International Institute of Tropical Agriculture (IITA) co-sponsored a 1-week workshop in November 1990 at IITA, Ibadan, Nigeria to examine how increased use of plant biotechnology could enhance research on tropical crops to meet Africa's needs. This volume contains the papers presented at that workshop, as modified by further review and discussion between scientists, for the benefit of all who are working towards similar goals.

We hope that this initiative will contribute towards the knowledge required by scientists in Africa and elsewhere to explore the avenues available to them in plant biotechnology, both conventional and modern. We hope also that the analyses of the need for biotechnology research in crops of importance to Africa will encourage scientists everywhere to apply their talents to working for food self-sufficiency in Africa.

D. Assoumou Mba
Director, CTA

Lukas Brader
Director General, IITA

Acknowledgements

We wish to thank the Government of Italy, CTA, FAO and the Rockefeller Foundation for providing the financial assistance which enabled speakers from a large number of countries to attend the conference. The attendance of some participants was sponsored by the US Agency for International Development (USAID), the Overseas Development Administration (ODA) of the UK, the International Development Research Centre (IDRC) of Canada, the British Council and the International Fund for Agricultural Research (IFAR).

Our thanks go also to Dr Ken S. Fischer, then Deputy Director General for Research at IITA and now Deputy Director General for Research at the International Rice Research Institute (IRRI), Manila, Philippines, for his interest and wholehearted support during preparations for the workshop. We thank the contributors and reviewers for their cooperative and interactive role in preparing these papers for publication.

The reviewers included M.O. Akoroda, M.N. Anishetty, R. Asiedu, D.C. Baulcombe, M.P. Bokanga, P.D.S. Calicari, M.A. Chase, I. Cubitt, E.D. Earle, E.E. Ene-Obong, E.B. Esan, C.A. Fatokun, E. Filippone, E.A. Frison, D. Furtek, S. Gleddie, M.M. Goodman, R.I. Hamilton, J. Hammond, B.D. Harrison, R.J. Hayes, G.G. Henshaw, T. Hohn, M.A. Hughes, L.E.N. Jackai, R. Jambunathan, R.L. Jarret, G. Kahl, A Karp, A. Van Kammen, K.K. Kartha, W.A. Keller, P. McGrath, R. McMuller, B.L. Møller, L.M. Monti, T.N. Murashige, G.O. Myers, N.Q. Ng, S.Y.C. Ng, C.E.A. Okezie, S.N.C. Okonkwo, N. Olembo, S. Padulosi, P. Perrino, I.T.D. Petty, J.E. Poulton, R.C. Pratt, F. Quak, R. Rao, Sir Ralph Riley, D.J. Robinson, C.A. Ryan, F. Saccardo, S. Sarkar, R.J. Schnell, O.P. Sehgal, R.J. Shepherd, P.H. Sisco, G. Staritsky, C.W. Stuber, J. Toll, I.K. Vasil, H.J. Vetten, D. Vuylsteke, K. Waithaka, J.G. Wutoh, N.D. Young, and M. Zaitlain.

A special word of thanks should go also to O.M. Aiyelokun, who input the disk files and kept track of the papers as they went through several rounds of corrections.

Lastly, we are grateful to IITA for the time and resources it invested in this volume and to CTA for co-sponsoring its publication.

G. Thottappilly
Convenor, Organizing Committee

1.1

The role of biotechnology in agricultural research[1]

L.M. Monti

Department of Agronomy and Plant Genetics,University of Naples, Via Università 100, 80055, Portici, Italy

Abstract

This paper outlines the progress made by new biotechnologies in agriculture. There are many applications for biotechnology in micropropagation for commercial uses and in overcoming interspecific barriers. In vitro production of haploids is now performed routinely in many laboratories and several crop varieties have already been bred in this way. Somaclonal variation, gametoclonal variation, and direct selection in vitro have also been used to produce new cultivars. Somatic embryogenesis has been used in several crops, opening the way to the development of so-called "artificial seeds". Meiotic mutants that form unreduced gametes have revolutionized the breeding of potato and other crops, allowing crosses between species with different levels of ploidy. Selection techniques have made considerable progress using restriction fragment length polymorphism (RFLP). Transformation and regeneration systems are now available for about 20 crops, making them amenable to genetic engineering. In several species, transformed germplasm derived from somatic fusion or gene transfer is already being used in field trials. New immunological techniques are being used for diagnostic assays in agriculture. This paper identifies the priorities for biotechnology research in the near future.

The growing human population and the concomitant increase in use of natural resources are generating a series of negative effects on ecosystems, such as pollution, loss of genetic diversity, soil fertility decline, climatic changes, deforestation, and desertification. Agriculture is asked to satisfy two apparently contradictory needs — to become more productive and at the same time more sustainable; that is, to supply the food needed without depleting

1 Contribution No. 63 from the Research Centre for Vegetable Breeding, National Research Council, Via Università 133, 80055 Portici, Italy.

renewable resources. Breeding plants that produce higher yields of better quality but do not adversely affect the ecosystem can be achieved only through a very broad scientific input.

The New Biotechnologies

New biotechnologies could play a decisive role in agriculture because of their ability to directly modify plants, animals, and agricultural processes in response to new needs. What were seen as promising technologies a few years ago have already produced new varieties. This means that today these biotechnologies are being rushed from research to development. The concern that biotechnology is only high-risk basic research should be reviewed in the light of the results that have been obtained from applied research.

New biotechnologies should be seen not only as a means of solving problems when traditional techniques have failed, but also as a way of generating a better understanding of crop plants through the cooperation of scientists from different disciplines, who for the first time are basing the model for their studies on plants. Molecular biologists, biochemists, physiologists, and virologists, working with plant geneticists, breeders, pathologists, and entomologists, can form research teams capable of bringing about a revolution in the agricultural sciences. However, the need for a combination of several disciplines may constitute a barrier to the worldwide application of biotechnology in agriculture because it depends on a solid scientific infrastructure not always present in developing countries. For this reason, biotechnology research is being conducted mainly in developed countries.

Biotechnology represents the latest front in the ongoing scientific progress of this century. However, its increasing importance, at least in plant improvement, should not obscure the fact that traditional plant breeding, based on hybridization followed by selection and evaluation of a large population in the field, accounts for over 50% of the global increase in agricultural productivity. Not only have particularly important new genotypes been bred in Asia through the so-called Green Revolution but, worldwide, new varieties have been bred in response to the changing needs of agriculture.

Improvement of agricultural productivity is today based on research progress in universities and research centres (public and private), and in national and international organizations. These institutions complement each other, constituting an integrated system.

Problems related to the impact of new technology on developing countries and the Consultative Group on International Agricultural Research (CGIAR) system have been discussed in several fora and were highlighted in 1989 in Canberra at the seminar on "Agricultural Biotechnology: Opportunities for International Development." I would like to emphasize that there could be several fields of action for biotechnology in tropical and subtropical crops and that the international agricultural research centres (IARCs) should act as bridges between advanced centres in developed countries and national programmes in developing countries, in order to transfer new technologies and apply them to local crops.

Current Role of Biotechnology in Crop Improvement

Some of the applications of biotechnology that are currently being used in crop production are shown in Table 1. The first of these, in vitro culture of meristems and buds, is now widely used for the micropropagation of many species for commercial purposes. It is also used for

Table 1 Biotechnology in crop production

Technology	Application
Meristem and bud culture	micropropagation for commercial purposes, genetic conservation, and exchange of material
Zygotic embryo culture	interspecific crosses
Anther and microspore culture	haploid production
Cell and tissue culture	in vitro selection, somaclonal variation, somatic embryogenesis, artificial seeds
Chromosome engineering	2*n* gametes for interspecific crosses
Protoplast culture	fusion for somatic hybridization
Genetic engineering	gene transfer
Molecular markers (RFLPs)	aid to breeding programmes
Monoclonal antibodies	diagnosis of plant diseases

germplasm conservation of vegetatively propagated species and for the exchange of virus-free material.

In vitro culture of zygotic embryos has enabled us to overcome barriers to a number of interspecific crosses from zygotic failure. Several interspecific hybrids (Table 2) have been

Table 2 Hybrids in legume species obtained through embryo and ovule culture between naturally incompatible species

Hybrid	Techniques[a]
Arachis hypogea x *A.* spp.	E, O>E
Glycine max x *G. tomentella*	E
Glycine tomentella x *G. canescens*	NE, O
Lens culinaris x *L. nigricans*	O>E
Phaseolus coccineus x *P.* spp.	E
Phaseolus vulgaris x *P.* spp.	E
Vigna radiata x *V. angularis*	E
Vigna umbellulata x *V. angularis*	E
Lathyrus clymenum x *L. articulatus*	E
Lotus spp. x *L.* spp.	NE
Medicago sativa x *M.* spp.	O>E
Melilotus officinalis x *M. alba*	E
Ornithopus spp x *O.* spp.	NE
Trifolium ambiguum x *T.* spp.	E, NE
Trifolium pratense x *T. sarosiense*	E
Trifolium repens x *T.* spp.	E, NE

Note: a E = embryo; O = ovule; O>E = ovule culture followed by embryo culture.
Source: Adapted from Ramsay et al. 1991.

obtained in this way — for example, between barley and secale (Fedak 1986), between *Lycopersicum* species and *Solanum* species (Sink and Reynolds 1986), among cucurbits (Jelaska 1986), and among legumes.

In vitro culture of anthers allows the regeneration of large numbers of haploid plants, from which, by chromosome doubling via colchicine treatment or via further in vitro culture, diploid plants can be obtained which are homozygous at all loci. Lines produced from anther culture of hybrids are obtained in less time and show greater variability than those obtained by selfing; several varieties produced using this method have already been released (Table 3). Direct culture of microspores has proved to be efficient in *Brassica* species (Tomes 1990). A high rate of occurrence of in vivo haploids is routinely obtained in barley through pollination with *Hordeum bulbosum*, and several varieties have been bred using this technique (Kasha and Reinbergs 1980).

Plants regenerated in vitro can show some modifications (somaclonal variation) as a result of the mutagenic effect of the culture or the chimeric nature of the cultured tissue. Useful traits have been obtained in variants derived from tissue or anther cultures (Table 4);

Table 3 Cultivars/germplasm, released or in field trials, originating from in vitro and in vivo induced haploids

Species/cultivar	Source[a]	Year	Characteristics
Brassica napus			
na[b]	M	1990	agronomic
Nicotiana tabacum			
CTDH 38	B	1988	verticillium wilt
CTDH 40	B	1988	verticillium wilt
HI, LI	A	1988	nicotine levels
Zea mays			
AC 4115	A	1987	agronomic
Hordeum vulgaris			
Mingo	B	1979	agronomic
Rodeo	B	1983	agronomic
Craig	B	1988	agronomic
Etienne	B	1989	agronomic
Winthrop	B	1987	agronomic
TBR 579-5	B	1990	agronomic
TBC 555-1	B	1990	agronomic
Gwylan	B	1989	agronomic
Oryza sativa			
Texmont	A	1990	earliness, yield
Triticum aestivum			
Florin	A	1985	agronomic

Note: a A = anther culture; M = microspore; B = maternal haploid.
b na = not available.
Source: Adapted from Tomes 1990.

Table 4 **Cultivars/germplasm, released or in field trials, originating from genetic variation in tissue culture**

Species/cultivar	Source[a]	Year	Characteristics
Brassica napus	DS	1988	chlorosulfuron, imidazilinone resistance
Nicotiana tabacum			
NCT G-51, NCT G-52	GV	1990	potato virus Y resistance
Capsicum annuum			
Bell sweet	GV	1987	low seeded
Lycopersicon esculentum			
DNAP-9	SV	1984	high solids in fruit
DNAP-17	SV	1987	fusarium race 2 resistance
Medicago sativa			
C2-M4	SV	1988	white flower, transposable element
Zea mays	DS	1990	imidazilinone resistance
Sorghum bicolor	SV	1989	fall armyworm resistance

Note: a DS = direct selection; GV = gametoclonal; SV = somaclonal.
Source: Adapted from Tomes 1990.

by adding metabolites and other substances to the media, it is possible to isolate resistant cells from which resistant plants can be obtained. Table 4 also lists two herbicide-resistant mutants of maize and *Brassica* obtained in this way and now in an advanced testing stage.

New plants can be obtained through somatic embryogenesis from pedicels, stems, leaves, roots, and other explants. This technique makes possible the production of "artificial seeds", provided that improved synchronization of the somatic embryo development, good propagule vigour, appropriate embryo desiccation, and encapsulation are achieved (Beversdorf 1990). The main problem with the technique is not somatic embryogenesis, now possible in several vegetable species (Table 5), but lack of plant regeneration from these embryos and poor viability after storage, handling, and transportation.

Table 5 **Somatic embryogenesis in vegetables**

Species	In vitro culture of
Apium graveolens	pedicel
Daucus carota	root
Foeniculum vulgare	stem
Petroselinum hortense	pedicel
Solanum melongena	hypocotyl, leaf
Brassica oleracea var *botrytis*	leaf
Cucurbita pepo	cotyledons, hypocotyl
Cucumis sativus	anthers (somatic tissue)

Crosses between species with different levels of ploidy can be accomplished by exploiting some meiotic mutants in which $2n$ gametes are formed; this makes it possible to cross $4x$ species with individuals belonging to $2x$ species characterized by this mutation. This phenotype can involve either or both the male and female gametes. To date, 35 plant genera or families have been reported to exhibit this phenomenon (Table 6), so it appears to be ubiquitous in the plant kingdom. Interspecific crosses based on this methodology have revolutionized the breeding of potato (Mendiburu and Peloquin 1977) and other crops, by the introgression of useful genes from wild species.

Table 6 Plant genera and families in which functional $2n$ gametes have been found

Male	Male and female	Female
Lotus	*Lactuca*	*Beta*
Cerasus	*Brassica*	*Parthenium*
Prunus	*Vaccinium*	*Citrus*
Pyrus	*Glycine*	*Dactylis*
Oryza	*Medicago*	*Hordeum*
	Fragaria	*Lolium*
	Malus	*Pennisetum*
	Rubus	*Secale*
	Capsicum	*Sorghum*
	Lycopersicon	*Triticale*
	Solanum	*Musa*
	Vitis	
	Manihot	
	Ananas	
	Panicum	
	Saccharum	
	Triticum	
	Zea	
	Orchidaceae	

Transformation can be accomplished through somatic hybridization or gene transfer. Several interspecific and intergeneric hybrids have been obtained through fusion protoplasts, from which some interesting genotypes have been selected. In *Brassica*, new germplasm derived from somatic hybridization (Barsby et al. 1987) and characterized by cytoplasmic male sterility and atrazine resistance has been released (Table 7).

About 20 crops, for which transformation and regeneration systems are available, can be genetically engineered at present (Table 8). All the transgenic plants that have been released following genetic engineering were obtained using *Agrobacterium tumefaciens* as the vector (Schell 1987). Transformed plants of tobacco, potato, tomato, soybean, and other species are now in field trials (Tomes 1990) to evaluate product quality and to test their tolerance to viruses, insects, and herbicides (Table 7).

Other delivery systems such as direct DNA uptake by electroporation or microprojectile bombardment by the so-called "gene gun" are being developed. The bottleneck in genetic

Table 7 Germplasm, released or in field trials, originating from gene transfer, using either somatic hybridization or genetic engineering

Species	Source[a]	Year	Characteristics
Brassica napus	S	1987	cytoplasmic male sterility, atrazine resistance
	G	1989	glyphosate resistance
Nicotiana tabacum	G	1988	alfalfa mosaic virus resistance
Solanum tuberosum	G	1988	virus resistance
	G	1988	phosphinothricin
Lycopersicon esculentum	G	1988	tobacco mosaic virus resistance
	G	1987	insect-*Bacillus thuringiensis*
	G	1988	fruit ripening
Linum sp.	G	1989	glyphosate resistance
Medicago sativa	G	1988	glyphosate resistance
	G	1990	alfalfa mosaic virus resistance
Glycine max	G	1989	glyphosate tolerance

Note: a S = somatic hybridization; G = gene transfer.
Source: Adapted from Tomes 1990.

Table 8 Crops for which transformation and regeneration systems are available

Group	Crop
Vegetables	carrot cauliflower celery cucumber potato tomato
Cereals	maize rye
Legumes	soybean alfalfa (lucerne) stylosanthes
Industrial	cotton flax rapeseed tobacco
Trees	poplar walnut

Table 9 The use of restriction fragment length polymorphisms (RFLPs) as an aid to plant breeding

Traits	Technique[a]	Reference
Simply inherited traits		
Resistance to TMV in tomato	RFLP markers linked to Tm-2 gene on chromosome 9	Tanksley et al. 1989
Resistance to dwarf mosaic virus in maize	1 major gene identified on chromosome 6	Barone et al. 1990
Resistance to *Heterodera schactii* in sugar beet	1 major gene identified by using additional lines	Barone et al. 1990
Resistance to *Globodera rostochiensis* in potato	1 major gene on chromosome 9	Barone et al. 1990
Resistance to small animal predators in *Trifolium repens*	analyses of Li locus controlling the presence of limarase, which functions as a defence mechanism	Hughes et al. 1990
Quantitative traits		
Resistance to insects due to 2-trideranone in tomato	RFLPs on three different linkage groups were correlated to this character	Nienhuis et al. 1987)
Soluble solids content in tomato	4 QTL increase the soluble solid concentration	Tanksley et al 1989
Fruit pH and fruit weight in tomato	6 QTL control fruit weight and 5 QTL control fruit pH	Tanksley et al 1989
Water use efficiency in tomato	3 RFLP markers were associated with this character	Martin et al. 1989

Note: a QTL = quantitative trai loci.

engineering is the limited number of genes that are available for application (at present, about 60) out of the approximately 1000 genes considered to be of interest for breeding.

The application of RFLPs has been useful in mapping beneficial genes and in the selection of qualitative and quantitative traits. This technique has been used to study disease and pest resistance and qualitative traits (Table 9).

The development of easily used techniques in immunology, based on monoclonal and polyclonal antibodies (Miller and Williams 1990), has allowed their practical application in diagnostic assays to detect pathogens and pests in plants, mycotoxins in grains, and pesticide residues in agricultural products. Nucleic acid probes are also used as diagnostic tools for pathogens, but in a limited way because of the current need for radioactive labels. The development of non-radioactively labelled probes of high sensitivity should be a major priority. Among the activities that would benefit from such improved diagnostic techniques in agriculture are: quarantine services; germplasm exchange; detection and monitoring of pathogens in crops, food, and grains; and control measures against diseases in the field.

Future Role of Biotechnology

In the near future, efforts should be devoted to obtaining transformed plants from apparently recalcitrant species. As noted above, there is a need to increase the number of isolated genes, particularly those conferring resistance to pests, diseases, and abiotic stresses, and for quality improvement. Attention should be given to genes coding for the reproductive process, such as genes for self-incompatibility and for male sterility, and to genes which influence the interaction between host and symbionts in nitrogen fixation. The development of an appropriate methodology for producing artificial seeds will open the way to considerable progress in agriculture through better establishment and uniformity of crops. Mapping the genomes of the most important crops should be one of the priorities of current and future genetic research. This will result in the production of new crop plants with desirable traits and in a better understanding of plant physiological processes.

One of the greatest concerns about biotechnology is patent legislation relating to genes and their use in plant breeding. With the high cost of biotechnological research, coupled with the large investments by private companies operating in this sector, patent legislation is necessary to protect new genetic products. However, there is a risk that while new biotechnology will certainly improve knowledge of crop genetics, patents could severely limit its application in crop improvement. This problem should not be under-estimated; a solution must be found that takes account of both private interests and agricultural progress.

References

Barone, A., E. Ritter, U. Schachtschabel, T. Debener, F. Salamini, and C. Gebhardt. 1990. Localization by RFLP mapping in potato of a major dominant gene conferring resistance to the potato cyst nematode *Globodera rostochensis*. *Molecular General Genetics* 224: 177-182.

Barsby, T.L., P.V. Chuong, S.A. Yarrow, S.C. Wu, M. Coumans, R.J. Kemble, A.D. Powell, W.D. Beversdorf, and K.P. Pauls. 1987. The combination of Polima cms and cytoplasmic triazine resistance in *Brassica napus*. *Theoretical and Applied Genetics* 73: 809-814.

Beversdorf, W.D. 1990. Micropropagation in crop species. Pages 3-12 in *Progress in Plant Cellular and Molecular Biology*. Kluwer Academic Publishers, Dordrecht, The Netherlands.

Fedak, G. 1986. Hordecale (*H. vulgare* x *Secale cereale* L.). Pages 371-386 in *Biotechnology in Agriculture and Forestry. 2. Crops I* edited by Y.P.S. Bajaj. Springer-Verlag, Berlin, Germany.

Hughes, M.A., A.L. Sharif, M.A. Dunn, E. Oxtoby, and A. Pancoro. 1990. Restriction fragment length polymorphism segregation analysis of the Li locus in *Trifolium repens* L. *Plant Molecular Biology* 14: 407-414.

Jelaska, S. 1986. Cucurbits. Pages 544-555 in *Biotechnology in Agriculture and Forestry. 2. Crops I* edited by Y.P.S. Bajaj. Springer-Verlag, Berlin, Germany.

Kasha, K.J., and E. Reinbergs. 1980. Achievements with haploids in barley research and breeding. Pages 215-230 in *The Plant Genome* edited by D.R. Davies and D.A. Hopwood. John Innes Charity, Norwich, UK.

Martin, B., J. Nienhuis, G. King, and A. Schaefer. 1989. Restriction fragment length polymorphism associated with water use efficiency in tomato. *Science* 243: 1725-1728.

Mendiburu, A.O., and S.J. Peloquin. 1977. The significance of 2*n* gametes in potato breeding. *Theoretical and Applied Genetics* 43: 53-61.

Miller, S.A., and R. Williams. 1990. Agricultural diagnostics. Pages 87-107 in *Agricultural Biotechnology: Opportunities for International Development* edited by G.J. Persley. CAB International, Wallingford, UK.

Nienhuis, J., T. Helentjaris, M. Slocum, B. Ruggero, and A. Schaefer. 1987. Restriction fragment length polymorphism analysis of loci associated with insect resistance in tomato. *Crop Science* 27: 797-803.

Ramsay, G., B. Pickersgill, and P.D.S. Caligari. 1991. The application of novel techniques in food, forage and pasture legumes to overcome genetic barriers which prevent the use of potential germplasm resources. Pages 421-460 in *Legume Genetic Resources for Semi-Arid Temperate Environments* edited by A. Smith and L. Robertson. Proceedings, International Workshop on Genetic Resources of Cool-Season Pasture, Forage, and Food Legumes for Semi-Arid Temperate Environments, 19-24 June 1987, Cairo, Egypt. ICARDA, Aleppo, Syria.

Schell, J.S. 1987. Transgenic plants as tools to study the molecular organization of plant genes. *Science* 237: 1176-1183.

Sink, K.C., and F.J. Reynolds. 1986. Tomato. Pages 319-344 in *Biotechnology in Agriculture and Forestry. 2. Crops I* edited by Y.P.S. Bajaj. Springer-Verlag, Berlin, Germany.

Tanksley, S.D., N.D. Young, A.H. Paterson, and M.W. Bonierbale. 1989. RFLP mapping in plant breeding: New tools for an old science. *Biotechnology* 7: 257-264.

Tomes, D.T. 1990. Current research in biotechnology with application to plant breeding. Pages 23-32 in *Progress in Plant Cellular and Molecular Biology* edited by H.J.J. Nijkamp, L.H.W. Van Der Plas, and J. Van Aartrijk. Kluwer Academic Publishers, Dordrecht, The Netherlands.

1.2

Beyond Mendel's garden: Biotechnology in agriculture

G.J. Persley

Agriculture and Rural Development Department, World Bank, 1818 H Street, Washington DC 20433, USA

Abstract

A study was commissioned by the World Bank, the Australian Government, and the International Service for National Agricultural Research (ISNAR), a member of the Consultative Group on International Agricultural Research (CGIAR), to examine a number of issues related to the use of biotechnology. These included: the likely impact of biotechnology on agricultural productivity and rural development; the socioeconomic and institutional issues which will affect the successful application of biotechnology; the changing roles of public and private sector research; intellectual property management; regulatory issues; and the role of international development agencies, including the World Bank. The study concluded that the applications of modern biotechnology were necessary in order to resolve constraints to the productivity for many tropical commodities and to minimize damage to the environment through, for example, reducing the use of pesticides. These commodities are "orphan commodities" in that, in terms of modern biotechnology, there has been little investment in them. Additional efforts focusing on these commodities are required by international development agencies.

This paper outlines the major findings of a study on agricultural biotechnology, sponsored by the World Bank, the International Service for National Agricultural Research (ISNAR), and the Australian Government, through the Australian International Development Assistance Bureau (AIDAB), and the Australian Centre for International Agricultural Research (ACIAR). These findings have been described by Persley (1990a) and in a recent technical report by the World Bank (1991). The edited text of the commissioned papers is available in Persley (1990b).

The study began in May 1988 and involved 66 collaborators from 18 countries. During 1988-89, 31 papers and 10 country studies were commissioned. Through these, the study addressed the following issues:

- The likely impact of biotechnology on agricultural production and rural development
- The socioeconomic and institutional issues which will affect its successful application
- The changing roles of public and private sector research
- Intellectual property management
- Regulatory issues
- The role of international development agencies

Definitions

Biotechnology is comprised of a continuum of technologies, ranging from traditional biotechnology to modern biotechnology (Figure 1). In this context, *biotechnology* is defined as "any technique that uses living organisms, or substances from those organisms, to make or modify a product, to improve plants or animals, or to develop microorganisms for specific uses" (OTA 1989).

Figure 1 Biotechnology continuum, from traditional biotechnologies to modern biotechnologies

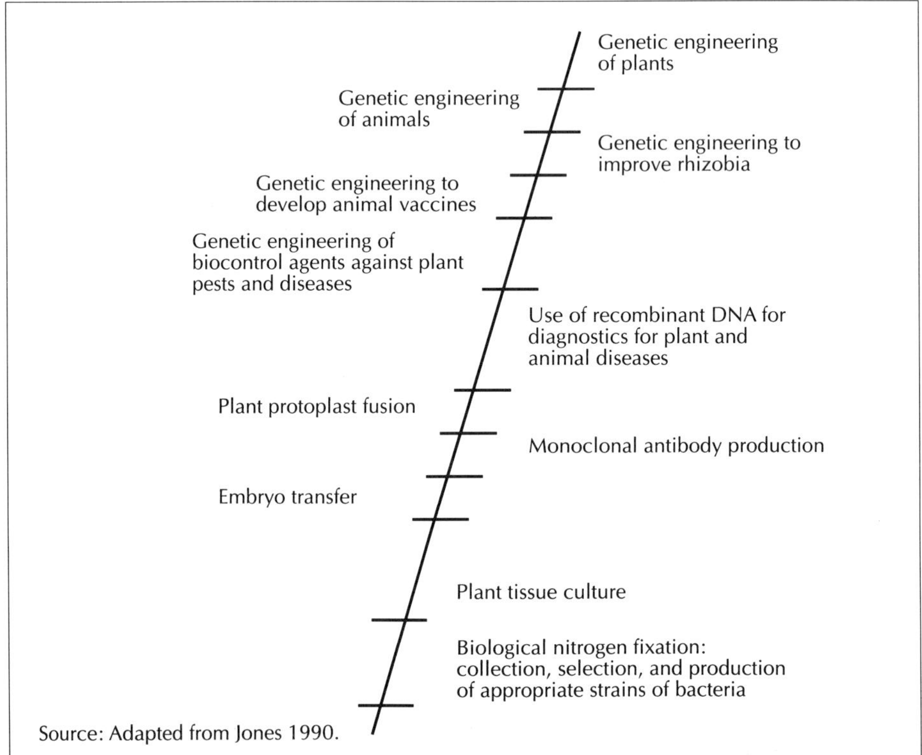

Source: Adapted from Jones 1990.

Traditional biotechnology covers well-established and widely used technologies based on the commercial use of living organisms. These include the biotechnologies currently used in brewing, food fermentation, conventional animal vaccine production, and many others. *Modern biotechnology* encompasses the uses of more recently developed technologies, particularly those based on the use of: recombinant DNA technology; monoclonal antibodies; and new cell and tissue culture techniques, including novel bioprocessing. *Recombinant DNA technology* is a series of enabling techniques for genetic engineering which allows the manipulation of the DNA, the essential genetic material in the cells. *Monoclonal antibodies* are specific diagnostic tools which allow the rapid detection of individual proteins produced by the cells. Recent advances in cell and tissue culture allow the rapid propagation of living cells.

The basic principle of genetic engineering is that the genetic material, the DNA, can be transferred from a cell of one species to another unrelated species, and it can be made to express itself in the recipient cells. The newly formed DNA in the recipient cells contains its own naturally occurring genes as well as the new gene. The new *recombinant DNA* is inserted into the chromosomes of the *transgenic* plant. The recipient cells are said to be *transformed* by the arrival of the new genetic information.

The current excitement about genetic engineering and its potential impact on agriculture stems from the ability of scientists to manipulate and control genes in new ways. These recent developments need to be seen from an historical perspective, as part of the evolution of the science of genetics — a science that originated in 1866 with Mendel's postulations on the inheritance of specific characteristics. The evolution of the science of genetics, leading to modern biotechnology, is shown in Table 1 (*overleaf*).

Technology Assessment

The first step in assessing the usefulness of new technology to agriculture is to identify the problems that have proved intractable using conventional approaches and that may benefit from its application. In analysing what needs to be done, it is important to combine the skills of agricultural scientists and economists, who understand the commodity and its needs, with the skills of molecular biologists, who see new ways to approach old problems. Cell and molecular biologists are the new partners of agricultural scientists, not their replacements.

Application of biotechnology to crop production

Biotechnology as applied to crop production is concerned with:

- Agricultural microbiology, to produce microorganisms beneficial to crop cultivation
- Cell and tissue culture, including the rapid propagation of useful microorganisms and plant species
- New techniques, based on the use of monoclonal antibodies and nucleic acid probes, for diagnosing plant pests and diseases anddetecting foreign chemicals in food
- Genetic engineering of plant species to introduce new traits
- New genetic mapping techniques, based on the use of restriction fragment length polymorphisms (RFLPs), as an aid to conventional plant breeding programmes

Table 1 The evolution of the science of genetics, leading to modern biotechnology

1866	Mendel postulates a set of rules to explain the inheritance of biological characteristics in living organisms.
1900	Mendelian law rediscovered after independent experimental evidence confirms Mendel's basic principles.
1903	Sutton postulates that genes are located on chromosomes.
1910	Morgan's experiments prove genes are located on chromosomes.
1911	Johannsen devises the term "gene", and draws a distinction between genotypes (determined by genetic composition) and phenotypes (influenced by environment).
1922	Morgan and colleagues develop gene mapping techniques and prepare a gene map of fruit fly chromosomes, ultimately containing over 2000 genes.
1944	Avery, MacLeod, and McCarty demonstrate that genes are composed of deoxyribonucleic acid (DNA) rather than protein.
1952	Hershey and Chase confirm the role of DNA as the basic genetic material.
1953	Watson and Crick discover the double-helix structure of DNA.
1960	Genetic code deciphered.
1971	Cohen and Boyer develop initial techniques for rDNA technology, to allow transfer of genetic material from one organism to another.
1973	First gene (for insulin production) cloned, using rDNA technology.
1974	First expression in bacteria of a gene cloned from a different species.
1976	First biotechnology firm established to exploit rDNA technology (Genentech, in USA).
1980	USA Supreme Court rules that microorganisms can be patented under existing law (Diamond v. Chakrabarty).
	Cohen/Boyer patent issued on the technique for the construction of rDNA.
1982	First rDNA animal vaccine approved for sale in Europe (colibacillosis).
	First rDNA pharmaceutical (insulin) approved for sale in USA and UK.
	First successful transfer of a gene from one animal species to another (a transgenic mouse carrying the gene for rat growth hormone).
	First transgenic plant produced, using an *Agrobacterium* transformation system.
1983	First successful transfer of a plant gene from one species to another.
1985	US Patent Office extends patent protection to genetically engineered plants.
1986	Transgenic pigs produced carrying the gene for human growth hormone.
1987	First field trials in USA of transgenic plants (tomatoes with a gene for insect resistance).
	First field trials in USA of genetically engineered microorganisms.
1988	US Patent Office extends patent protection to genetically engineered animals.
	First genetically modified microorganism approved for commercial sale as a biocontrol agent of a plant disease (crown gall of fruit trees in Australia).
1989	Human genome mapping project initiated.
1990	Plant genome mapping projects (for cereals and *Arabidopsis)* initiated.

Source: Adapted from OTA 1984 and Wyke 1988.

The likely early applications of new biotechnologies to crop production in the developing world are:

- New diagnostics, which could be made readily available and used in many countries.
- Genetic mapping of major tropical crops to benefit conventional breeding programmes.
- Plant disease resistance by genetic engineering of the host plant.
- New biocontrol agents for pest control, to reduce pesticide use.

Commodity analysis

Several commodities were assessed in terms of their current constraints and the likely availability of new technologies which overcome these constraints. Substantial progress may be expected in the short term (within 5 years) for potato, rapeseed, and rice; in the medium term (5-10 years) for banana/plantain, cassava, and coffee; and in the long term (10+ years) for coconut, oil palm, and wheat.

Some of these commodities have been termed "orphan commodities" — commodities in which there is little investment in modern biotechnology in industrialized countries, either because the commodity is not important in temperate areas or because there are no obvious profits for transnational companies. Yet they are important food or export commodities in the developing world (for example, cassava, banana/plantain). Special attention should be paid to these commodities in order to facilitate the application of biotechnology in an effort to resolve production constraints.

Socioeconomic impact

Significant economic impact of modern biotechnology on agricultural production is not expected before the year 2000. It will then become an increasingly important component of new technologies. This time frame is similar to that for conventional research, in that it takes about 10 years to develop a new plant variety, animal vaccine, or pharmaceutical.

Biotechnology is likely to change the comparative advantages between countries and between commodities, particularly export commodities. The application of new technologies to export commodities will improve their competitive position in the world market. The likely socioeconomic benefits of biotechnology in the developing world are an increase in the productivity of tropical commodities to meet future food needs, new opportunities for the use of marginal lands, and a reduction in the use of agrochemicals. However, there are also likely to be negative effects in that biotechnology offers the possibility of producing high-value products in tissue culture, displacing crops presently grown for export by the developing world. The potential for negative substitution effect should be monitored and remedial strategies developed when economically damaging effects are identified.

National strategies

The development of a national strategy that involves both the public and private sector is critical if effective use is to made of domestic and external resources. The development of a national biotechnology strategy is preferable to the development of a national institute of biotechnology. New initiatives are required, rather than new institutions. It may actually be a hindrance to the development and application of science in a country to establish exclusive biotechnology programmes. In a decade or less, today's modern biotechnology will be considered conventional technology. The aim of research managers should be to integrate new technologies into existing biological research programmes. As part of a national strategy, it may be more efficient to provide some common services from a central laboratory, especially where expensive equipment is required.

Training

Training is required at several levels:

- Bridging courses for research managers and mature scientists
- Post-doctoral fellowships
- Graduate training (MSc and PhD)
- Undergraduate courses

There is a particular need for bridging courses to introduce policymakers, research managers, and mature scientists to the new possibilities offered by modern biotechnology. This would allow more informed decisions to be taken on resource allocation. Post-doctoral fellowships in advanced laboratories are a key component of keeping abreast of the latest advances worldwide. Graduate and undergraduate training underpin the development of in-country capability in the basic biological sciences. Mobilizing the basic scientific skills usually found in universities to solve agricultural problems will require new policy and institutional arrangements and more financial resources.

Public/private sector collaboration

The major change in the funding of agricultural research in industrialized countries in the past decade has been the greatly increased role of the private sector, largely in modern biotechnology. It has been estimated that at least half the current funding of research and development (R&D) activities in agricultural biotechnology worldwide comes from the private sector. The major reason for this is that for many of the new technologies the process and/or the product is patentable. A company is able to appropriate many of the benefits of its research investments. Private companies are, therefore, much more likely to invest in modern biotechnology than in conventional agricultural research.

This trend towards greater private sector investment in agricultural research is well established in the major Organization for Economic Cooperation and Development (OECD) countries. It is an emerging trend in Latin America, and in parts of Asia and the Middle East, but there is little evidence of it as yet in Africa. In most developing countries, however, public sector investments are still the main source of finance for biotechnology.

There is a need for greater public/private sector collaboration in relation to biotechnology and its application to problems in the developing world. This will require: continued public sector investments from domestic and external resources; public/private sector partnerships; innovative funding mechanisms on the part of international development agencies; and involvement of both local private sector companies and transnational companies.

Intellectual property management

The major issue that will affect the application of biotechnology in agriculture is the protection of intellectual property. The lack of patent protection is a major disincentive for private sector (local and transnational) investment in biotechnology in the developing world. The advantages of the availability of intellectual property protection (by various mechanisms) are that it encourages the development of local research capability and greater in-country investments in biotechnology. The major disadvantage is that it involves giving proprietary protection to living organisms, which some consider to be part of the common heritage of mankind. Each country needs to weigh the benefits and costs of intellectual property rights in biotechnology and frame its policies accordingly. Model systems could be developed by countries that decide to take steps to protect their intellectual property.

Biosafety

Another important issue that will affect the application of biotechnology to agriculture concerns the regulations governing the release of model products. A safe and efficient regulatory environment is in itself a comparative advantage in biotechnology. Action has been taken in several countries to ensure environmental safety and public health in biotechnology applications in agriculture. Procedures have been developed to assess high, medium, and low levels of risk associated with the release of genetically engineered organisms, and guidelines formulated to govern national regulatory systems. These systems are based on a tiered structure of responsibility, involving: a national review body; institutional biosafety committees; and project biosafety reviews. Biosafety issues are discussed in more detail by Persley (1990a, 1990b) and in Paper 7.3 (pp 345-352 this volume).

World Bank investments

World Bank options for increased support for agricultural biotechnology include:

- Additional loans for specific biotechnology components of national agricultural research projects
- New joint ventures, involving the International Finance Corporation, to help local companies participate in the commercial development of biotechnology
- Additional targetted grant funds to the international agricultural research centres (IARCs) in order to increase their capability in biotechnology and their ability to assist national agricultural research systems in acquiring and adapting new technologies

International development agencies

There are three main policy options for international development agencies interested in facilitating the application of biotechnology to agriculture:

- Public sector support for biotechnology in association with other support for agricultural research in public sector institutions and universities. This includes the provision of training opportunities and research facilities, and would require little change from current policy and practice.

- Public and private sector support for biotechnology in public sector institutions in conjunction with collaborative research with an advanced laboratory in the public or private sector. Private companies may choose to participate in such collaborative activities either on a contract research basis or because participation in the project may confer commercial advantages that are not linked directly to the project itself.
- Joint commercial ventures, facilitating the establishment of viable joint ventures between public and private sector agencies in developing and industrialized countries. These partnerships need to include both R&D to produce new products and processes and the business systems necessary for the delivery of new products to farmers.

Agricultural biotechnology services

During the extensive consultations undertaken in several countries in 1989-91, it became clear that there are some common concerns among policymakers, research managers, and scientists in many countries. These include:

- Priority identification, to set achievable targets relevant to agricultural development
- Technology architecture, to design biotechnology appropriate to the solution of local problems
- Project design, to prepare projects aimed at delivering to farmers new products developed with the aid of biotechnology
- Intellectual property management, to facilitate the acquisition and adaptation of proprietary technologies
- Regulations, to enable the safe and efficient use of new biological products
- Information services, to facilitate access to the latest available information
- Training, at all levels
- International liaison, to encourage collaboration with various international activities, including the activities of IARCs and international development agencies.

These issues could be addressed by providing up-to-date advice and information to enable policymakers and research managers in individual countries to make informed decisions. These biotechnology advisory services need to be available before decisions are taken on whether to seek substantial external financing for national biotechnology programmes and, if so, what form this financial support should take. It is necessary to examine how such services could be made available at the national, regional, and international level. At the international level, this question is currently being examined by the Consultative Group on International Agricultural Research (CGIAR) task force on biotechnology (BIOTASK) and by a group of interested development agencies.

Conclusion

In an issue of *New Scientist* (17 November 1990), Professor Thomas Odhiambo, President of the African Academy of Science, stated that: "Africa needs science, not just technology.

The transfer of technology to Africa has demonstrably failed ... to solve Africa's problems in health and agriculture. Africa needs science transfer, so that the science can be tailored to solve the problems unique to the African environment." Molecular biology is a key component of this science. What is required in Africa in the future is much more first-class science, conducted by African scientists, to solve the problems of African agriculture.

References

Jones, K.A. 1990. Classifying biotechnologies. Pages 25-28 in *Agricultural Biotechnology: Opportunities for International Development* edited by G.J. Persley. CAB International, Wallingford, UK.

OTA. 1984. *Commercial Biotechnology: An International Analysis.* OTA-BA-218. US Government Printing Office, Washington DC, USA.

OTA. 1989. *New Developments in Biotechnology: Patenting Life.* Special Report OTA-BA-370. US Government Printing Office, Washington DC, USA.

Persley, G.J. 1990a. *Beyond Mendel's Garden: Biotechnology in the Service of World Agriculture.* CAB International, Wallingford, UK.

Persley, G.J. 1990b. *Agricultural Biotechnology: Opportunities for International Development* edited by G.J. Persley. CAB International, Wallingford, UK.

World Bank. 1991. *Agricultural Biotechnology Study: The Next Green Revolution?* Technical Paper No. 133. World Bank, Washington DC, USA.

Wyke, A. 1988. The genetic alternative: A summary of biotechnology. *The Economist* (30 April).

1.3

Status of research on constraints to cowpea production[1]

S.R. Singh, L.E.N. Jackai, G. Thottappilly, K.F. Cardwell and G.O. Myers

International Institute of Tropical Agriculture, PMB 5320, Ibadan, Nigeria

Abstract

This paper discusses the constraints to cowpea production in Africa (including insect pests, diseases, plant parasitic weeds, drought, and heat, and the possibilities of overcoming some of these constraints through the application of biotechnology.

Cowpea (*Vigna unguiculata* (L.) Walp) is an important food legume crop in the semi-arid regions of Africa. It is cultivated as a mixed crop, mainly with sorghum and millet, and forms a significant part of the diet of a large part of the African population. The major constraints to cowpea production are insect pests, diseases, plant parasitic weeds, drought, and heat.

Resistance Studies in Cowpea

Diseases

Emechebe and Shoyinka (1985) reviewed progress in the development of cowpea varieties with multiple resistance to fungal and bacterial diseases. The advanced cowpea lines TVx 3236 and TVx 82D-716, which have high yield potential, also have multiple resistance to

1 Contribution No. IITA/91/CP/16 from the International Institute of Tropical Agriculture, PMB 5320, Ibadan, Nigeria.

anthracnose, bacterial blight, brown blotch, web blight, and scab. It appears that resistance to fungal and bacterial diseases has been well established and that adequate sources of resistance are available within the cowpea germplasm. However, some fungal diseases are known to form new strains. Restriction fragment length polymorphism (RFLP) and polymerase chain reaction (PCR) technologies are potentially useful for determining differences between strains of *Colletotrichum*, for example. The use of RFLP or PCR technologies on the bacterial DNA could also provide a definitive answer to another unresolved question in phytopathological research on cowpea: whether the *Xanthomonas campestris* pathovars that cause distinct symptoms on cowpea — bacterial blight, bacterial pustule, and bacterial canker — are distinct species or strains.

Viruses

Singh and Ntare (1985) and Thottappilly and Rossel (1985) reviewed the development of cowpea lines with multiple resistance to different viruses infecting cowpeas. Several improved cowpea lines with resistance to as many as five viruses have been developed (Table 1).

Table 1 Improved cowpea lines with multiple virus resistance

	Disease[a]				
Variety	CYMV	CAbMV	CMeV	CuMV	SBMV
IT 81D-1137	R	R	MR	MR	MR
IT 82E-16	R	R	MR	R	R
IT 82D-889	R	R	MR	S	R
IT 83S-818	R	R	MR	R	R
IT 83D-442	R	R	R	R	R
IT 84D-449	R	R	S	R	R
IT 85F-867-5	R	R	R	R	R
IT 85F-2687	R	R	R	R	R

Note: a CYMV = cowpea yellow mosaic virus; CAbMV = cowpea aphid-borne mosaic virus; CMeV = cowpea mottle virus; CuMV = cucumber mosaic virus; SBMV = southern bean mosaic virus; R = resistant; MR = moderately resistant; S = susceptible.

Viruses of cowpea include cowpea aphid-borne mosaic, cowpea yellow mosaic, cowpea mottle virus, southern bean mosaic virus, cowpea mild mottle, cowpea golden mosaic, and cucumber mosaic viruses. Some of these, including cowpea aphid-borne mosaic and cowpea yellow mosaic viruses, exhibit pathogenic variation on the basis of differential infection of cowpea genotypes and of symptom type and severity. This variation complicates the strategy for resistance breeding, so it is essential that information on the distribution and economic significance of viruses and their strains be kept up-to-date.

Since its inception in 1977, the virology unit at the International Institute of Tropical Agriculture (IITA) has produced polyclonal antisera of high titre to a number of the viruses infecting African food crops. These antisera are being used for virus identification, rapid detection in plant extracts, and determination of relationships with other viruses. There are important viruses, such as cowpea golden mosaic, for which good antisera are not yet available. Some of these viruses are extremely difficult to purify in sufficient quantities for the conventional production of antisera in rabbits. A limitation in the use of polyclonal antisera is the inability to readily discriminate between strains of a virus. This limitation has been overcome by the development of monoclonal antibodies which, because of their nature and the method of selection, can be strain-specific.

The control of virus diseases is important for increasing food production in the tropics. In order to devise effective control methods, an understanding of the nature and distribution of virus strains is essential. Unfortunately, resistance is sometimes strain-specific, requiring several gene sources for the development of a suitable cultivar. Awareness of the existence of strains and knowledge of their distribution will facilitate development of breeding and screening strategies.

Insect pests

Resistance to insect pests in cowpea has been studied extensively. Most cowpea accessions have been screened for resistance to the major pests. High to moderate levels of resistance have been obtained to some of the pests, including the cowpea aphid (*Aphis craccivora*), flower bud thrips (*Megalurothrips sjostedti*), and cowpea bruchid (*Callosobruchus maculatus*). In the case of the legume pod borer (*Maruca testulalis*) and a pod-sucking bug (*Clavigralla tomentosicollis*), it has been difficult to identify useful sources of resistance.

The cowpea aphid is a seedling pest, but it also infests flowers and green pods on older plants. High levels of resistance to this pest have been identified. Several lines, including TVu 310, TVu $408P_2$, TVu 801, and TVu 3000, are highly resistant to aphid infestation (Singh 1977; Singh and Jackai 1985). Resistance to the cowpea aphid has been successfully incorporated into several breeding lines, eliminating the need for other control measures.

Flower bud thrips is a major pest of cowpea in Africa. It infests flower buds and often causes total crop loss (Singh 1987). Only limited success has been attained in identifying a high level of resistance. One line (TVu 1509) has been identified as moderately resistant (Singh 1977) but the level of resistance on its own is not sufficient to protect the crop from damage by thrips. However, relatively little insecticide is required to control flower bud thrips on TVu 1509 than on other varieties. This resistance has been incorporated into some advanced breeding lines, including TVx 3236 and IT 4S-2246.

The legume pod borer infests stems, buds, flowers, and green pods. The greatest damage is caused to flowers, while the most obvious damage appears on the pods. Several varieties with better levels of resistance than commercial varieties have been identified (for example, MRx 10-85S, IT 83D-378, MRx 17-85S, MRx 55-85S, Zonkwa-1, and Zonkwa-2) but many of these have generally unacceptable seed characteristics and agronomic traits. Resistance is achieved mainly through seed size (many are small), accelerated maturity, pod angle, and canopy structure. Many resistant genotypes have purple pods and it is possible that plant pigmentation plays a role or that the purple pods are the result of a linked gene.

Pod-sucking bugs constitute a complex of up to nine species, *Clavigralla tomentosicollis* being the most dominant and damaging in much of Africa. Young developing pods are particularly susceptible; infestation causes the pods to abort or dry up and drop from the plant. Extensive entomological work has identified a number of genotypes with moderate levels of resistance (for example, TVu 1 and TVu 1890). These genotypes and several others with much reduced damage to seeds are generally small-seeded, with the seeds purple or black and possessing a smooth coat. These characteristics are not acceptable to consumers in West Africa, where most cowpea is produced and consumed. Cowpea resistance to pod-sucking bugs also seems to be associated with seed number, pod-wall toughness (Chiang and Jackai 1988), and, occasionally, pubescence (Jackai and Oghiakhe 1989). These criteria are used in current screening techniques (Jackai 1990).

The cowpea bruchid is a serious pest of cowpea in storage. At the farm level, complete infestation often occurs within 6 months of storage. Resistance to cowpea bruchid has been studied extensively at IITA. A moderate level of resistance has been identified in TVu 2027 (Singh 1977). After 3 months' storage, TVu 2027 showed about 20% infestation while other cowpea varieties showed 66-96% infestation. Similarly, during 6 months' storage, TVu 2027 showed about 71% infestation while the other varieties were already completely destroyed, indicating the need for better protection over a longer period. This resistance has been incorporated into most advanced breeding lines.

In order to identify higher level of resistance, many wild *Vigna* species have been screened for resistance to the major insect pests (Table 2).

Table 2 Resistance in wild *Vigna* species to major pests of cowpea

Vigna species	Cowpea aphid	Flower bud thrips	Legume pod borer	Pod-sucking bugs	Cowpea bruchid
V. luteola	S[a]	R/S	R/S	R/S	R
V. vexillata	R/S	R/S	R	R	R
V. oblongifolia	S	R/S	R	R/S	R/S
V. kirkii	R	R/S	nt	nt	S
V. unguiculata ssp. *dekindtiana*	S	R/S	R/S	R/S	S
V. racemosa	R	R/S	nt	nt	R/S
V. ambacensis	S	R/S	R	R	S
V. reticulata	R	R/S	R	S	R/S

Note: a R = resistant; S = susceptible; R/S = having both resistant and susceptible accessions; nt = not tested.

Parasitic weeds

Striga gesnerioides and *Alectra vogelii* are two major parasitic weeds infesting cowpea. Striga is more common in West Africa and severe yield losses have often been reported (Aggarwal and Ouedraogo 1989). *A. vogelii* has been reported from West and southern Africa. It causes damage similar to that of striga, with infested cowpea plants suffering serious yield losses.

At least four cowpea lines resistant to striga have been identified (Aggarwal et al. 1984; Aggarwal et al. 1986; Aggarwal and Ouedraogo 1989; Parker and Polniaszek 1990) and the presence of different physiological strains of striga in different regions has been reported. Resistant cowpea cultivars exhibit varying reactions to these strains; however, B301, a cowpea line from Botswana, has shown resistance to a number of striga populations in laboratory tests (Parker and Polniaszek 1990).

The presence of various physiological strains of striga creates an enormous task in breeding for resistance. Preliminary observations by cowpea scientists in West Africa indicate the likelihood of finding further physiological strains of striga, even if the frequency of occurrence of strains has not been fully studied. The use of isozymes and RFLPs is helping with strain identification. RFLPs could also be used to develop an understanding of population genetics in striga and to follow the genetic basis of epidemics.

Stress Physiology

Drought and heat are major constraints to cowpea production. Most screening methodologies are field based. However, methods for the identification of genes for resistance to drought and heat, using RFLPs, have a potential in plant breeding.

Cowpea Germplasm and Seed Health

There are about 15 000 accessions of cowpea in the germplasm unit at IITA, although some could be duplicates. Because RFLP analysis is able to characterize the genotype of individuals with high precision, it may be an effective tool for classifying this material.

Another important role for biotechnology is in seed health management. Several fungal and bacterial pathogens are known to be seed-borne. IITA, as well as other members of the Consultative Group on International Agricultural Research (CGIAR), has as part of its responsibility the movement of both breeders' germplasm and germplasm bank material. There are two ways in which biotechnology might help in this respect. First, the detection of various kinds of pathogens within a seedlot might be faster and more accurate with biotechnological methods. Second, methods to distinguish between live and dead organisms on a seed surface would allow determination of seed treatment efficacy, development of seed treatment protocols, and effective application of phytosanitary certification.

An RFLP map of the cowpea genome is expected to help identify specific markers for traits relating to pest resistance and, ultimately, to help in the selection of desirable genotypes. It is possible that some of the constraints discussed here could be overcome with the help of the molecular techniques, including RFLPs and DNA fingerprinting, described elsewhere in this volume.

Discussion

PERSLEY: Of the various pests and diseases you have listed, which would you nominate as the most important?

MYERS: If I could list three, two pathological and one entomological: bacterial blight (bacterial/fungal disease); maruca pod borer (insect); and striga (parasitic weed).

LADEINDE: How important is the leaf miner in cowpea production and what are you doing presently to control it?

MYERS: It is not a major problem in West and Central Africa, but is more important in East Africa. At present, no concerted effort is being made on breeding resistance to the leaf miner.

References

Aggarwal, V.D., N. Muleba, I. Drabo, J. Souma, and M. Mbewe. 1984. Inheritance of *Striga gesneriodes* resistance in cowpea. Pages 143-147 in *Proceedings, Third International Symposium on Parasitic Weeds.* ICARDA, Aleppo, Syria.

Aggarwal, V.D., S.D. Haley, and F.E. Brockman. 1986. Present status of breeding cowpea for resistance to *Striga* at IITA. Pages 176-180 in *Proceedings, Workshop on the Biology and Control of Orobanche* edited by S.J. ter Borg. LM/VPO Wageningen, The Netherlands.

Aggarwal, V.D., and J.T. Ouedraogo. 1989. Estimation of cowpea yield loss from *Striga* infestation. *Tropical Agriculture (Trinidad)* 66: 91-92.

Chiang, H.S., and L.E.N. Jackai. 1988. Tough pod wall: A factor involved in cowpea resistance in pod sucking bugs. *Insect Science and Its Application* 9: 389-393.

Emechebe, A.M., and S.A. Shoyinka. 1985. Fungal and bacterial diseases of cowpeas in Africa. Pages 173-192 in *Cowpea Research, Production and Utilization* edited by S.R. Singh and K.O. Rachie. John Wiley, Chichester, UK.

Jackai, L.E.N. 1990. Screening of cowpeas for resistance to *Clavigralla tomentosicollis* Stal (Hemiptera: Coreidae). *Journal of Economic Entomology* 83: 300-305.

Jackai, L.E.N., and S. Oghiakhe. 1989. Pod wall trichomes and resistance of two wild cowpea, *Vigna vexillata*, accessions to *Maruca testulalis* (Geyer) (Lepidoptera: Pyralidae) and *Clavigralla tomentosicollis* Stal (Hemiptera: Coreidae). *Bulletin of Entomological Research* 79: 595-605.

Parker, C., and T.I. Polniaszek. 1990. Parasitism of cowpea by *Striga gesnerioides:* Variation in virulence and discovery of a new source of host resistance. *Annals of Applied Biology* 116: 305-311.

Singh, S.R. 1977. Cowpea cultivars resistant to insect pests in world germplasm collection. *Tropical Grain Legume Bulletin* 9: 3-7.

Singh, S.R. 1987. Host plant resistance for cowpea insect pest management. *Insect Science and its Application* 8: 765-769.

Singh, S.R., and L.E.N. Jackai. 1985. Insect pests of cowpeas in Africa: Their life cycle, economic importance and potential for control. Pages 217-231 in *Cowpea Research, Production and Utilization* edited by S.R. Singh and K.O. Rachie. John Wiley, Chichester, UK.

Singh, B.B., and B.R. Ntare. 1985. Development of improved cowpea varieties in Africa. Pages 105-115 in *Cowpea Research, Production and Utilization* edited by S.R. Singh and K.O. Rachie. John Wiley, Chichester, UK.

Thottappilly, G., and H.W. Rossel. 1985. World-wide occurrence and distribution of virus diseases. Pages 155-171 in *Cowpea Research, Production and Utilization* edited by S.R. Singh and K.O. Rachie. John Wiley, Chichester, UK.

1.4

Analysis of the need for biotechnology research on cassava, yam, and plantain[1]

R. Asiedu, S.Y.C. Ng, D. Vuylsteke, R. Terauchi and S.K. Hahn

International Institute of Tropical Agriculture, PMB 5320, Ibadan, Nigeria

Abstract

Cassava, yam, and plantain are staple food crops essential to food security in the humid and subhumid tropics of Africa. However, these crops have long growth cycles, high perishability and slow multiplication rates of propagation materials, and they are subject to a number of biological stresses. These stresses include diseases, insects, mites, nematodes, and weeds. There are gaps in our knowledge, especially in the area of genetics, which influence the pace of breeding. Intensification of research is needed in cytogenetics, molecular genetics, in vitro culture, cryopreservation, and disease diagnosis.

Cassava (*Manihot esculenta* Crantz), yam (*Dioscorea* spp.), and plantain (*Musa* spp.) are major food crops in much of humid and subhumid Africa and a major source of energy for millions of people in these regions. Their adaptation to the food and farming systems and their multiplicity of uses make them indispensable to food security. They are grown mainly by subsistence farmers, but commercialization is increasing in some countries.

The planting materials of the three crops have slow multiplication rates and, like the produce, are bulky to transport. All three crops are highly perishable and thus suffer severe post-harvest losses; also, their long growth cycles expose them to various biotic and abiotic stresses, some of which could be avoided by short-duration crops. The major biological constraints to increased and sustainable production are diseases, insects, mites, nematodes, and weeds. These constraints, in conjunction with the limited use of inputs in traditional cultivation systems, have kept the yield levels of most cultivars far below their potential.

---

1 Contribution No. IITA/91/CP/17 from the International Institute of Tropical Agriculture, PMB 5320, Ibadan, Nigeria.

Constraints to Production and Research

Cassava

African cassava mosaic virus (ACMV) is widespread and is the most important insect-borne disease of cassava in Africa. Caused by a gemini virus (Bock and Woods 1983) and transmitted by the white fly *Bemisia tabaci*, it can cause yield losses of up to 95% in Africa. The International Institute of Tropical Agriculture (IITA) has successfully incorporated resistance to ACMV into high-yielding cultivars. More research is required to identify any existing pathogenic variations and to explain the mechanism(s) of resistance, as well as the influence of environmental factors, such as ambient temperature, on symptom expression.

Cassava bacterial blight disease (CBB) occurs in many countries in South America and Africa, often resulting in complete crop failure in susceptible cultivars. It is caused by *Xanthomonas campestris* pv *manihotis* and leads to leaf spot, complete wilt, and tip dieback. Some clones with resistance to CBB have been developed. Cassava anthracnose and cassava root rot diseases are also widespread and can cause significant reductions in yield.

Cassava mealybug (*Phenacoccus manihoti*) and cassava green mite (*Mononychellus* spp.) are important dry-season pests of cassava. Cassava mealybug (CM) can reduce tuber root yield severely, especially in late-planted crops. Cassava green mite (CGM) can cause losses of over 40% (Nyiira 1975). Both pests are being tackled through host plant resistance breeding and biological control. Cassava varieties with moderate resistance have been developed or identified. Resistance has been shown to be associated mainly with pubescence of young leaves (Hahn et al. 1989; Kanno et al., 1991), but there are indications that antibiosis, preference, and tolerance are also involved. Additional sources of resistance are being sought from new introductions of cassava germplasm and wild *Manihot* species from Latin America. Hybrids produced recently at IITA between cassava and *Manihot tristis* (currently undergoing second-season evaluation) are fairly promising in terms of resistance to CGM. Biological control of the mealybug has been successful.

Cassava contains cyanogenic glucosides which release hydrocyanic acid (HCN) during hydrolysis, following tissue damage. Although varietal differences in levels of total cyanide have been established, no acyanogenic cultivar has been reported. The presumed relationship between HCN content and taste of the tuberous root (sweet or bitter) has led to confusion in the classification of cassava into low or high cyanide types. Research is required to clarify the situation and to positively identify what compounds account for bitterness in cassava. Moreover, the methodologies used for determining total HCN in the crop have produced variable results. Even with the best of techniques, the high variability resulting from environmental influences on field-grown materials renders most of the existing procedures in breeding programmes inefficient. There is an urgent need for more efficient screening procedures for identifying low HCN or acyanogenic clones. It is necessary to clarify the role the cyanogenic glucosides and/or their breakdown products play in the plant. Meanwhile, we can intensify molecular and conventional approaches to modify the genetic capacity of the plant to produce the glucosides. Another approach would be to increase the activity of endogenous degradative enzymes, such as linamarase, in the plant so that the glucosides can be rapidly and completely degraded for the release of the volatile toxin, hydrogen cyanide, during processing (M. Bokanga, IITA, 1990, pers. comm.). Ultimately, the best course may be to develop varieties that are low in cyanogenic glucosides but high in the appropriate degradative enzymes.

High dry-matter content of the tuberous root is desirable, but it seems to be negatively correlated with fresh tuberous root yield (Whyte 1987). Also, high dry-matter content has been associated with susceptibility to post-harvest physiological root deterioration (Jennings and Hershey 1985). Further resarch is needed on the physiological, biochemical, and genetic bases to determine the feasibility of combining high tuberous root yield, high root dry-matter content, and slow post-harvest deterioration in one genotype.

For millions of Africans, the preferred way of using cassava is direct consumption after roasting or boiling the fresh tuberous root (Hahn 1989). In some areas, the boiled root is pounded into a thick paste, often in combination with boiled plantain or cocoyam. Mealiness of the boiled tuberous root, as well as low cyanide content, is essential for these uses. Very little is known, however, about the biochemical/biophysical basis of mealiness. There is variation between cassava clones and there may be interactions between clone, age, and climate in their influence on the trait. It is important to gain a better understanding of this valuable characteristic in order to develop the ability to control it or breed more efficiently for it. Selection is hampered by the poor correspondence between young and adult plant expressions of the characteristic and the influence of climatic changes.

Cassava tuberous roots are low in protein. Attempts to breed for a higher level of protein have included the use of closely related species such *Manihot melanobasis* (Bolhuis 1953). The debate over whether to continue this effort using molecular approaches or to accept the main physiological function of cassava roots as storage organs for starch is unresolved. It has been argued that cassava is normally consumed with a protein source, such as fish or legumes; however, these accompaniments are often in short supply in communities with high dependence on cassava. On the other hand, the possibility of a trade-off between root yield and protein content should be noted. For the time being, it might be a good idea to refine fermentation processes for enhancement of the protein content of cassava products.

As cassava is cultivated under increasingly marginal conditions, its association with mycorrhizae, reported to boost its performance, ought to be investigated further.

Yam

The major yam species grown in Africa are: white yam (*Dioscorea rotundata*), yellow yam (*D. cayenensis*), water yam (*D. alata*), and trifoliate yam (*D. dumetorum*). The general preference for large tubers necessitates large planting setts, large mounds, and staking. The availability of enough good quality planting setts has often been the most serious constraint to production. The minisett technique (Otoo et al. 1985) has gone a long way towards solving the problem in the yam belt of West Africa. Sprouting leads to loss in tuber weight and quality. Some extension of tuber dormancy has been achieved through the application of gibberellic acid (GA3) (Wickham et al. 1984).

Yam mosaic virus and water yam chlorosis cause yield loss through their effects on the shoot; nematodes and yam beetles affect the tubers directly. It is likely that yam beetles would be vulnerable to alpha-amylase inhibitors (present in other foods, such as common bean), as well as to digestive proteinase inhibitors (L.L. Murdock, Purdue University, 1990, pers. comm.). Studies on the impact of such inhibitors on the growth, development, and survival of the insect when it feeds on foods containing them could lay the foundation for introducing insect resistance into yam by genetic engineering. Root-knot nematodes affect marketability and storability of tubers through discoloration and galling of the surface.

Poor flowering and asynchrony of flowering has been a major obstacle to hybridization in yams. Many cultivars do not flower at all. Flowering is erratic in some cultivars which appear to be sensitive to climatic changes in a manner not well understood. Even more frustrating are the situations where flowers of only one sex are available. A lot more work needs to be done to gain an understanding of the physiology of flowering. The role of premature necrosis of the shoot, which prevents the plants from reaching the basal vegetative phase desirable for flowering, needs further clarification.

Plantain

Black sigatoka is the most important disease of plantain at present. This virulent fungal disease, caused by *Mycosphaerella fijiensis*, is threatening plantain and cooking banana production in many African countries. IITA has identified resistant diploid wild bananas and produced resistant hybrids from crosses between susceptible plantains and a resistant wild diploid banana. The relatively rapid decline in plantain productivity in Africa requires further research. Current thinking is that it is influenced by rising mats, attacks by the banana weevil and nematodes, and declining soil fertility. Lodging can lead to serious losses. Poor seed set and viability impose restrictions on genetic improvement through hybridization.

Biotechnology Research

The need to intensify biotechnological research and to initiate new research on cassava, yam, and plantain cannot be overemphasized. In vitro culture is critical for rapid multiplication, virus elimination, germplasm exchange, and germplasm conservation. Somaclonal variation can be useful as an additional source of genetic variability, but it can be a nuisance in germplasm conservation. More research is needed on the causes and frequency of the phenomenon in plantains. Meanwhile, molecular markers could play a useful role in rapid identification of off-types. Embryo culture is a routine procedure used for many crops. In cassava and plantain, it has been very successful using fairly mature seeds, whereby the embryos were helped to germinate more readily, free from seed coat obstructions (Ng, Paper 3.4, pp 135-141 this volume; Vuylsteke, Paper 3.5, pp 143-150 this volume). Similar procedures could be used for yams. It would be even more useful to develop appropriate media that could support very young embryos of the three crops, so that these could be rescued from young fruits in order to accelerate hybridization and prevent losses through premature embryo abortion, especially in interspecific crosses, or damage to the plant.

Protoplast culture needs to be perfected to facilitate genetic transformation. Regeneration remains a bottleneck, depriving these crops of some of the new and exciting avenues for gene transfer. The promising results from somatic embryogenesis could be extended to serve as an additional avenue. Also, promising physical methods, such as using the particle accelerator to transfer DNA into meristems, ought to be pursued vigorously.

It would be useful to extend the successful cryopreservation of plantain to yams and cassava for long-term preservation without loss of totipotency or genetic integrity.

Because they are vegetatively propagated crops, the dangers of disease transmission are rather acute. Modern and efficient methods of diagnosing disease agents, including latent viruses, are critical to the safe exchange of germplasm.

Despite years of research on these crops, their reproductive biology has not been clarified. Many questions remain concerning control of flowering, compatibility of parents, and fruit development. There is poor understanding of the genomes present in the cultigens and their related wild species. The differences in ploidy, which have important influences on cross-compatibility and success of backcrossing, have not been studied systematically. The need for these basic cytogenetic studies can no longer be ignored. They would complement and accelerate studies on molecular genetics towards the ultimate goal of understanding these crops well enough to enable planned and precise genetic manipulation.

Genetic analysis is at an early stage for all three crops. The genetic control of only a few traits is known (Hahn et al. 1989) and few morphological markers have been established. This situation may be explained partly by the long growth cycles of these crops, as well as the complexity of their genetic analyses. These factors, coupled with the complexities of screening for economically important traits, call for the urgent establishment of genetic markers to help breeders make more rapid progress. For instance, linkage of biochemical/ molecular markers to agronomic traits and applying them to plants at the seedling or early growth stages would considerably improve the efficiency of selection. The production of haploids through a culture of anthers/microspores or unpollinated ovaries could open up another avenue for simpler genetic analyses. Phylogenetic studies on cassava, yam, and plantain need to be accelerated using the new techniques of molecular biology.

As the cultivation of cassava, yam, and plantain shifts to marginal lands, biotechnology has an increasingly important role to play in the development of these essential crops. They are not just sources of food for the people of Africa; they constitute an integral part of African culture.

Discussion

ENE-OBONG: In cassava, can the problem of determining the nature of inheritance of resistance to mosaic virus be solved without a good knowledge of the plant's chromosomes? Can't the problem be associated with the ploidy nature, as we are dealing with polyploids?

ASIEDU: The problem is not one of determining the inheritance of resistance but of understanding the mechanism by which the resistance is expressed. In terms of inheritance, we already know that it is controlled by recessive genes.

FAUQUET: There is not one resistance to African cassava mosaic virus (ACMV) but several. A viral infection is the result of a series of events, beginning with the virus inoculation and ending with the spread of the items to the next generation. Several genes are controlling these different steps of "resistance" and their expressions can change according to the growth of the plant and the environmental conditions.

ROBERTSON: There is a simple answer as to why a virus may cause symptoms which then disappear. We have found, in both the laboratory and the field, that expression of a symptom can be related to temperature. When tissues are colder (about 25°C) virus symptoms appear, while at 28-30°C they can disappear. In a sense, the plant wins the battle against the virus when it is at its most compatible temperature.

MURDOCK: You mentioned the yam tuber beetle. This insect apparently thrives on a diet rich in carbohydrate (starch) and poor in protein. It is probably vulnerable to alpha-amylase inhibitors present in other foods, such as common bean, as well as to digestive proteinase inhibitors. Studies of the impact of such inhibitors on the growth, development, and survival of the insect when it feeds on diets containing them could lay the foundation for introducing insect resistance into yam by genetic engineering. To initiate such studies, some kind of dietary bioassary would be required. Can this insect be reared in the laboratory, and, in particular, is there a diet available on which it can be reared?

MBANASO: My comment is on increasing the protein in cassava as an area for biotechnology research for cassava improvement. I suggest this area be given a low rating in priority since cassava-based products are usually eaten with protein-enriched sauces.

ASIEDU: We ought to separate the two issues of gene splicing or transformation techinques and techniques such as the application of moleculer markers which assist breeders in doing what they are already doing by increasing their efficiency.

ROBERTSON: In the debate on whether we need a strong breeding programme before we go to genetic manipulation, it is natural that plant breeders fear that when genetic engineers make their "wild" claims they may be seen to be out of a job or, worse, restricted in their funding. Nevertheless, especially where countries have weak breeding programmes, it is possible in some cases (for example, virus resistance, insect resistance, altering protein levels and quality) to take local cultivars and improve them by the "quick fix" of single gene addition. Plant breeders need not fear, they will be needed forever. Yet simple additions can be very useful and do not, in fact, need already established breeding programmes to be in place before the genetic engineers can begin their work.

References

Bock, K.R., and R.D. Woods. 1983. The etiology of African cassava mosaic virus disease. *Plant Disease* 67: 994-995.

Bolhuis, G.G. 1953. A survey of attempts to breed cassava varieties with a high content of proteins in the roots. *Euphytica* 2: 107-112.

Hahn, S.K. 1989. An overview of African traditional cassava processing and utilization. *Outlook on Agriculture* 18: 110-118.

Hahn, S.K., J.C.G. Isoba, and T. Ikotun. 1989. Resistance breeding in root and tuber crops at the International Institute of Tropical Agriculture (IITA), Nigeria. *Crop Protection* 8: 147-168.

Jennings, D.L., and C.H. Hershey. 1985. Cassava breeding: A decade of progress from international programmes. Pages 89-116 in *Progress in Plant Breeding* (Vol. 1) edited by G.E. Russell. Butterworths, London, UK.

Kanno, H., A.G.O. Dixon, R. Asiedu, and S.K. Hahn. 1991. Host plant resistance of cassava green spider mite (*Mononychellus tanajoa*). Pages 103-106 in *The Role of Root Crops in Regional Food Security and Sustainable Agriculture* edited by M.N. Alvarez and R Asiedu. Proceedings of a Southern Africa Regional Workshop, Mansa, Zambia, 1990. IITA, Ibadan, Nigeria.

Nyiira, Z.M. 1975. Advances in research on the economic significance of cassava green mite (*Mononychellus tanajoa*) in Uganda. Pages 27-29 in *Proceedings, Workshop on International Exchange and Testing of Cassava Germplasm in Africa, IITA, Ibadan, Nigeria*. IDRC-063e.

Otoo, J.A., D.S.O. Osiru, S.Y.C. Ng, and S.K. Hahn. 1985. *Improved Technique for Seed Yam Production*. IITA, Ibadan, Nigeria.

Wickham, L.D., Passam, H.C., and Wilson, L.A. 1984. Dormancy responses to post-harvest application of growth regulators in *Dioscorea* species. 1. Responses of bulbils, tubers and tuber pieces of three *Dioscorea* species. *Journal of Agricultural Science (Camb.)* 102: 427-432.

Whyte, J.B.A. 1987. Breeding cassava for adaptation to environmental stress. Pages 147-176 in *Cassava Breeding: A Multidisciplinary Review* edited by C.H. Hershey. Proceedings of a Workshop held in the Philippines, 4-7 March 1985. CIAT, Cali, Colombia.

1.5

Constraints in food and nutrition research[1]

M. Bokanga

International Institute of Tropical Agriculture, PMB 5320, Ibadan, Nigeria

Abstract

For several millennia, biotechnology has been at the service of mankind, particularly in the production of fermented foods and beverages. This paper highlights three areas of food and nutrition research that will benefit from advances in biotechnology in Africa. First, the development of starter cultures for African fermented foods and beverages is the key to their expansion. Starter cultures reduce fermentation time and improve and stabilize the nutritional and organoleptic quality of fermented products. Second, cassava, a major food staple in Africa, contains cyanogenic glucosides that can produce the toxic hydrogen cyanide. Fermentation is one of the best methods of detoxifying cassava. The production of cassava varieties with low amounts of cyanogenic glucosides will be facilitated by a better understanding of the molecular mechanisms involved in their biosynthesis. Third, cassava, yam, sweet potato, plantain, and banana, the most important crops in tropical Africa, are highly susceptible to post-harvest deterioration. The texture of foods derived from these crops is the major organoleptic attribute assessed by consumers. Discoloration in cassava and yam and ripening of plantain and banana are enzyme-mediated processes. Genetic manipulations of key enzyme systems can lead to varieties with improved post-harvest characteristics.

Recent advances in molecular biology have provided the means to alter the genetic make-up of living organisms in order to enhance desirable traits and/or suppress unwanted defects. Other papers in this volume highlight the many techniques of genetic engineering available to plant breeders for introducing new genes into various crops. These techniques may also find applications in food and nutrition research. This paper illustrates the need for biotechnology research on traditional food fermentations, on cyanogenesis in cassava, and on the post-harvest characteristics of tropical food crops.

1 Contribution No. IITA/91/CP/18 from the International Institute of Tropical Agriculture, PMB 5320, Ibadan, Nigeria.

Biotechnology and Food

Biotechnology in its most fundamental sense — the use of biological systems to obtain useful products — has been practised by mankind for the production of food from time immemorial. There is evidence dating back to 6000 BC that the use of yeasts in baking and brewing was already well established then, and that fermented foods and beverages were an essential part of the diet of people in ancient civilizations (Miller 1982).

This should not come as a surprise; microorganisms evolved millions of years before man, and all the chemical reactions needed for fermentation were already established when man appeared. In a way, he had to compete with microbes for his food supply. Whenever they beat him to it, they sometimes produced unpleasant aromas, bad flavours or toxins causing illness or even death; the food was considered bad and man learned to avoid it. In other instances, however, microbes produced attractive aromas, flavours, and textures; man learned to appreciate and desire such foods, and to domesticate the responsible microbes in order to produce these foods at will. Over the years, man has relied on food fermentation not only to please his palate, but also to preserve food, to improve its digestibility, and to enrich it with vitamins, proteins, and amino acids (Steinkraus 1989). Several of these fermented foods, such as bread, wine, beer, cheese, yoghurt and soy sauce, have today given rise to multimillion-dollar businesses.

Traditional African Food and Beverage Fermentations

A host of fermented foods and beverages are found in African villages. These products have not, however, enjoyed an expansion similar to that of the fermented products of Europe or Japan, for example. Nearly all African food fermentations rely on the fortuitous presence of microbes on the food and on prevailing favourable conditions for the production of the desired product. In some instances, a small amount of a previous batch is kept and used to inoculate the next, but the fermentation is allowed to follow its natural course, with little or no attempt to control it. As a result, the flavour, aroma, and texture of the fermented product varies with the season, location, and producer.

Fermentation can considerably extend the shelf-life of a food item, induce desirable flavours, and increase the nutritive value. The preservative effect results mainly from a reduction of pH following the production of organic acids (mainly lactic and acetic),from the production of ethanol, and possibly from the formation of bacteriocins by fermentation organisms to inhibit other microbial species. Thus, the key to successful fermentation is to increase the number and/or activity of desirable microorganisms and to suppress the growth of pathogenic and spoilage organisms.

In recent years, a number of studies have identified the microorganisms associated with some traditional fermentations. In the natural ecology of the fermentation, there are many microbial species that probably do not contribute to the quality of the fermented product; some may even reduce the quality of the food, while others may represent a health hazard. For instance, in the submerged fermentation of cassava commonly practised in Zaire for the production of *kwanga* (or *chikwanghe*), 15 microbial species belonging to 10 genera have been isolated (Bokanga 1989). Three of these species are known pathogens for humans. Two other species were found to be essential to the fermentation process. When a mixture of pure cultures of those two species was used to inoculate cassava aseptically, a product similar to the traditional *kwanga* was obtained. The detoxification of cassava, a major benefit from the fermentation of this food, was achieved in both the traditional and the pure culture process, although the fermentation time was reduced by half when pure cultures were used.

In 1977, severe staphylococcal food poisoning outbreaks associated with the consumption of fermented meat products were recorded in the USA. The use of starter cultures containing known microbial cultures was recommended. The benefits of this practice were an end to the staphylococcal poisoning, a reduction of processing time, and a great improvement in product quality and consistency (Bacus 1984).

With the use of starter cultures, the quality of fermented foods can be kept constant and other fermentation characteristics can be better controlled. The production of these starters is not necessarily a complex and sophisticated operation. For centuries, starter cultures known as *koji* and *ragi* have been made by housewives for the production of the fermented soybean foods *miso* in Japan and *tempe* in Indonesia. The production of starter cultures can give rise to very profitable businesses. *Koji* production, for instance, has been industrialized in Japan and, in 1985, industrial *koji* was being used in 1700 plants to produce over half a million tonnes of *miso* (Steinkraus 1989).

Once the use of starter cultures is adopted, a strain improvement research programme needs to be developed. This programme should not only use the conventional means of natural and induced mutations followed by selection and enrichment, but it should also take advantage of gene manipulation techniques. Transformation systems for bacteria and yeasts are well established and afford higher success rates than plant transformation systems. However, the question of using genetically engineered organisms directly in the food supply will have to be addressed by appropriate safety authorities. The study of fermentation microorganisms can also lead to the identification of microbial strains particularly suited for the over-secretion of valuable biological products such as enzymes, amino acids, vitamins, antibiotics, and flavour enhancers (Sasson 1986). These microorganisms could be the basis of very profitable businesses.

Cassava and Cyanide Toxicity

Cassava is of paramount importance for people in sub-Saharan Africa, for whom it may provide up to 60% of the dietary caloric intake. The major drawback in the use of cassava is that it contains the cyanogenic glucosides linamarin and, to a lesser extent, lotaustralin, which upon hydrolysis release hydrocyanic acid (HCN).

The submerged fermentation of cassava has been shown to be efficient in removing linamarin from cassava (Bokanga et al. 1990). Other processes that include fine grinding of fresh plant tissues (for instance, pounding cassava leaves or grating cassava roots) are also efficient in detoxifying cassava. This is facilitated by the fact that all plant tissues containing linamarin also possess the enzymes capable of degrading linamarin. The degradation of linamarin occurs in two steps: first, linamarin is hydrolyzed by the enzyme linamarase, giving rise to glucose and acetone cyanohydrin; under slightly acidic, neutral, or alkaline conditions, the latter compound spontaneously decomposes to acetone and HCN, which is rapidly lost by volatilization (HCN boiling temperature is 25.7°C). A second enzyme, hydroxynitrile lyase, has been shown to also catalyze the decomposition of cyanohydrins (Kojima et al. 1979; Thayer and Conn 1981).

For food processing methods which involve crushing cassava tissues, it may be more important to develop varieties containing high amounts of linamarase and hydroxynitrile lyase or high levels of enzyme activity rather than low levels of linamarin. For other processes, such as boiling or drying of intact or large pieces of cassava, it is necessary to develop cassava varieties containing low levels of linamarin. Conventional breeding programmes have been addressing this issue. Screening methodologies have relied on the determination of linamarin content at harvest. A great difficulty has been the uneven distribution of cyanide in plants of the same clone and in organs of the same plant. Even within an organ, such as a tuberous root, linamarin is so irregularly distributed that obtaining a representative sample is a very delicate operation. Furthermore, the effect of the environment, season, age of the plant, and cultural conditions on the accumulation and distribution of linamarin in cassava plant tissues is not clearly understood.

A new and better way of screening cassava for cultivars containing low levels of linamarin needs to be developed. Such an approach could be based on specific genes or their products, on the analysis of isozyme systems, or on the use of DNA probes and restriction fragment length polymorphism (RFLP) markers. It is necessary to understand the metabolism of cyanogenic glucosides in the plant and the mechanism involved in their translocation to, and accumulation in, specific plant organs. What is currently known has been derived mainly from studies conducted on plant species belonging to genera other than *Manihot*, namely *Sorghum*, *Linum*, and *Trifolium*, and it needs to be validated in *Manihot*.

The purification of linamarin biosynthetic enzymes represents a formidable challenge. Four of the eight enzymes in the sequence are membrane-bound, and any attempt to separate them from the membrane results in a loss of activity, making their purification difficult (McFarlane et al. 1975; Hösel and Nahrstedt 1980; Cutler and Conn 1981; Halkier and Møller 1990; Halkier et al. 1991).

Linamarin has been located in the vacuoles of sorghum plant cells (Saunders et al. 1977; Saunders and Conn 1978). Its location in cassava plant cells is yet to be confirmed. The degradative enzyme linamarase has been located in cassava in the plant cell wall (Mkpong et al. 1990). Assuming that the linamarin is intracellularly located, it would still be necessary to know how the linamarin would exit from the leaf cell where it is believed to be synthesized without being hydrolyzed by the linamarase in the cell wall as a prelude to its transport to other parts of the plant.

Another interesting question arises from the observation that in low-cyanide varieties, so called because of the low linamarin content in the central pith of the tuberous roots, the linamarin content in the peel can be as high as, or even higher than, the linamarin content in the peel of high-cyanide varieties. Some mechanism must be preventing the accumulation of linamarin in the central pith of the tuberous roots of low-cyanide varieties. An understanding of this mechanism will help in the development of new breeding strategies for low linamarin content in cassava.

Improvement of Post-Harvest Characteristics

Cassava, yam, sweet potato, plantain, and banana constitute the major food staples in tropical Africa. The high moisture content of these commodities at harvest makes them very susceptible to deterioration during storage. In particular, discoloration in cassava starts within 48 hours after harvest, and the market value of the boil-and-eat type decreases sharply with the appearance of discoloration. Deletion of the gene(s) for key enzyme(s) responsible for the discoloration could lead to the development of cultivars with an extended shelf-life. This approach could also be applied to yam cultivars which undergo an enzymatic browning reaction during processing (Onayemi 1986).

Plantain and banana, when used as a staple, are often preferred green, and they lose their market value as they ripen. Cultivars with delayed ripening would have a longer shelf-life and provide an increased return to the farmer. An understanding of the regulation of ripening enzymes at the genetic level will be necessary for the development of such cultivars.

An important organoleptic characteristic common to African foods made from cassava, yam, plantain, and banana is texture. In the case of cassava and yam, for example, the ability of boiled tuberous roots to be pounded into a smooth and more or less sticky mass is of paramount importance to consumers in West Africa. Certain varieties of cassava lack this ability completely, while others lose it progressively, especially after the onset of the rainy season. Quite often, such varieties, although they may be high yielding, are not accepted by farmers who prefer pounded cassava. The factor responsible for this characteristic is not yet known and needs to be researched. The control of this trait will ensure that a greater number of farmers are attracted by improved varieties developed with other desirable traits, such as high yield and disease and pest resistance.

Conclusion

Biotechnology research will be useful in the improvement of microbial strains found in African food fermentations when the use of starter cultures becomes an established procedure. For the moment, techniques of molecular biology are needed to support the development of new breeding strategies for the improvement of tropical crops such as cassava, yam, sweet potato, banana, and plantain. These strategies should give priority not only to resistance to pests and diseases but also to the food quality and post-harvest handling of these crops. Biotechnologically developed crops should conserve or improve those characteristics preferred by consumers. After all, an increase in crop yield is meaningless if the commodity produced is not consumed.

Discussion

ENE-OBONG: As a general comment on the reduction of linamarin in cassava, consideration should be given in the long term to incorporating into cassava the genes for linamarase and rhodanase from insects such as the variegated grasshopper, *Zonocerus variegatus*.

References

Bacus, J. 1984. Update: Meat fermentation 1984. *Food Technology* 38(6): 59.

Bokanga, M. 1989. Microbiology and biochemistry of cassava fermentation. PhD dissertation, Cornell University, Ithaca, New York, USA.

Bokanga, M., S.K. O'Hair, K.R. Narayanan, and K.H. Steinkraus. 1990. Cyanide detoxification and nutritional changes during cassava (*Manihot esculenta* Crantz) fermentation. Pages 385-392 in *Proceedings, 8th Symposium of the International Society for Tropical Root Crops, 30 October-5 November, 1988, Bangkok, Thailand* edited by R.H. Howeler.

Cutler, A.J. and E.E. Conn. 1981. The biosynthesis of cyanogenic glucosides in *Linum usitatissimum* (linen flax) in vitro. *Archives of Biochemistry and Biophysics* 212(2): 468-474.

Halkier, B.A., and B.L. Møller. 1990. The biosynthesis of cyanogenic glucosides in higher plants. Identification of three hydroxylation steps in the biosynthesis of dhurrin in *Sorghum bicolor* (L.) Moench and the involvement of aci 1--nitro-2-(p-hydroxyphenyl)ethane as an intermediate. *Journal of Biological Chemistry* 265: 21114-21121.

Halkier, B.A., J. Lykkesfeldt, and B.L. Møller. 1991. 2-nitro-3-(p-hydroxyphenyl)propionate and aci-1-nitro-2-(p-hydroxyphenyl)ethane, two novel intermediates in the biosynthesis of the cyanogenic glucoside dhurrin in *Sorghum bicolor* (L.) Moench. *Proceedings of the National Academy of Science, USA* 88: 487-491.

Hösel, W., and A. Nahrstedt. 1980. In vitro biosynthesis of the cyanogenic glucoside taxiphyllin in *Triglochin maritima. Archives of Biochemistry and Biophysics* 203(2): 753-757.

Kojima, M., J.E. Poulton, S.S. Thayer, and E.E. Conn. 1979. Tissue distributions of dhurrin and of enzymes involved in its metabolism in leaves of *Sorghum bicolor. Plant Physiology* 63: 1022-1028.

McFarlane, I.J., E.M. Lees, and E.E. Conn. 1975. The in vitro biosynthesis of dhurrin, the cyanogenic glucoside of *Sorghum bicolor. Journal of Biological Chemistry* 250(12): 4708-4713.

Miller, M.W. 1982. Yeasts. In *Prescott and Dunn's Industrial Microbiology* edited by G. Reid (4th edn). AVI Publishing Co., Westport, Connecticut, USA.

Mkpong, O.E., H. Yan, G. Chism, and R.T. Sayre. 1990. Purification, characterization, and localization of linamarase in cassava. *Plant Physiology* 93: 176-181.

Onayemi, O. 1986. Some factors affecting the quality of processed yam. *Journal of Food Science* 51: 161-164.

Sasson, A. 1986. *Quelles Biotechnologies pour les Pays en Dévelopment?* Unesco, Paris, France.

Saunders, J.A. and E.E. Conn. 1978. Presence of the cyanogenic glucoside dhurrin in isolated vacuoles from sorghum. *Plant Physiology* 61: 154-157.

Saunders, J.A., E.E. Conn, C.H. Lin, and C.R. Stocking. 1977. Subcellular localization of the cyanogenic glucoside of sorghum by autoradiography. *Plant Physiology* 59: 647-652.

Steinkraus, K.H. 1989. *The Industrialization of African Fermented Foods.* Marcel Dekker, New York, USA.

Thayer, S.S., and E.E. Conn. 1981. Subcellular localization of dhurrin ß-glucosidase and hydroxynitrile lyase in the mesophyll cells of sorghum leaf blades. *Plant Physiology* 67: 617-622.

1.6

The need for a biotechnological approach in plantation crop research

A.I. Robertson

Department of Crop Science, University of Zimbabwe, PO Box 167, Mount Pleasant, Harare, Zimbabwe

Abstract

The gaps in available techniques to rapidly propagate plantation crops, especially in relation to the needs of developing countries, are discussed in this paper. The author suggests that each country would need to gradually build up its own capability, learning from the problems encountered in its indigenous crops.

Plantation crops tend to be those crops which are grown by large commercial concerns, using modern methods of mechanization, chemical applications, and irrigation if needed. These crops include cocoa, tea, coffee, rubber, papaya, and the monocots banana, sugar cane, date, oil palm, and coconut. Apart from dates, these crops grow mainly in the tropics and thus irrigation is seldom needed. However, one of the challenges that biotechnology faces is that of extending the range of these plantation crops so that they may be grown farther from their original climatic range; frost-tolerant coffee is an example.

Multiplication of Elite Selections to Meet Demand for Planting Material

Tea can be propagated successfully and easily by cuttings and thus there is no need for a special approach for multiplying elite clones. Similarly, for sugar cane perfectly adequate methods of vegetative propagation are now available. Clonal tea was a major advance in its time but normal vegetative propagation is now routine. Nevertheless, researchers in Sri Lanka have developed micropropagation methods, and thus the techniques are available if needed.

Papaya has recently become popular mainly because of the development of the attractive "Solo" cultivar. There has been a demand for rapid propagation because half of the seedlings that germinate are male only and therefore non-productive beyond providing pollen. This can be wasteful in terms of land, time, and effort. Litz (1984) documented the method for producing large numbers of female plantlets. This is now common practice in India and elsewhere. Should someone select or engineer another attractive cultivar, there will again be a scramble to produce planting material quickly. The methods are available unless cultivar requirements change with genotype improvement. In our laboratory, we found initial surface sterilization to be a real problem and the published methodology not easy to replicate. If there is an urgent need to replicate this methodology, it may be advisable to go to one of the more advanced laboratories and learn at the bench.

As yet, cocoa and rubber are not easy to multiply in tissue culture, certainly not in large numbers with routine techniques.

Surprisingly perhaps, with some monocots — banana, oil palm, and date — the methods for rapid and effective propagation have been established. For coconut, propagation is only now beginning to be successful. In Morocco and South Africa, commercial companies are beginning to offer test-tube or hardened-off in vitro plantlets of banana, oil palm, and date. For plantation crops, the technology is available for massive increases in numbers. However, here we should take heed of Unilever's experience with oil palm. After pioneering work (considered worthy of the OBE received by the company board member responsible for this work) to propagate elite oil palm in the test tube, large numbers of clones derived from a single plant were planted out in Malaysia. It was only some years later, as the clones approached maturity, that a "fickle" gene, one that expresses itself only under certain circumstances, devastated the programme by manifesting itself as a physiological defect, reducing yield after a period of partial drought earlier in the growth phase. This was a hard lesson, but a useful one. The answer, presumably, is to micropropagate from several original plants, which are chosen from the genepool to be as different as possible, while retaining the agronomic qualities desired. Such precautions should be taken for each and every clonally propagated crop.

Coffee is an interesting crop in that it is the source of more export earning power for the developing world than any other commodity, with the exception of oil. As such, it is vital to many economies. More than half the foreign currency earning power of Ethiopia, Uganda, Burundi, and Rwanda, for example, comes from coffee. In Zimbabwe, it is a valuable plantation crop, grown on large estates; in many countries, however, it is a small-scale crop, grown by thousands of farm families on plots of less than a hectare. Until recently, coffee was propagated by seed. Now, Brazil's national agricultural research programme, Empresa Brasileira de Pesquisa Agropecuária (EMBRAPA), uses embryo culture to produce clonal plants. Singapore is offering clonal micropropagation through Plantek International, and a number of laboratories are working on genetic improvement through added genes.

The Production of New Cultivars: Somaclonal Variation

For the reasons given above, coffee is perhaps a good vehicle for looking more generally at how biotechnology can help improve plantation crops. Coffee is plagued by a varied spectrum of pests and diseases; among the most important in Africa is coffee berry disease, which is caused by a fungus. In Zimbabwe, we have recently been invaded by this scourge. We are not even sure of the vector (air-borne, or insect-carried, or spread by mammals?), and would welcome germplasm that conferred resistance. As far as we know, there is no such germplasm as yet.

Most of the organizations that aim to commercialize molecular biology are hard at work on the many problems of fungal diseases. One low-tech approach is to raise calluses in vitro, expose them to the fungus or to the fungal toxin if it can be purified, and see if any callus cells survive and multiply. If so, this somaclonal variation may be used to generate resistant plants. The successful use of this general method has been reported for potato and recently for rice but it appears that no commercial cultivars have been released after this kind of screening for improved germplasm. In coffee, at least, there does not appear to be a laboratory that can regenerate plantlets direct from callus, although somatic embryogenesis does come through callus. However, it comes from callus originating in organized tissues, so that the callus cannot be used to look for appropriate somaclonal variation.

Thus, there are major technical barriers yet to be overcome. This is true of all plantation crops to varying degrees. It seems clear that a major effort has to be initiated before practical returns of agronomic significance, as has occurred in other crops, can be expected in plantation crops.

Fungal problems are ubiquitous in the plantation crops mentioned. If somaclonal variation could be employed and if regeneration systems were available for these crops, these techniques would provide a feasible approach that biotechnology, in the form of tissue culture, could offer even the less sophisticated laboratories of Africa. After all, we do have a ready supply of the crops' meristematic tissues and of the fungal pathogens.

Protoplast Fusion

For coffee, there are reports that hybrids ("arabusta") have been produced by fusing the protoplasts of *Coffea canephora* ("robusta") and *C. arabica*, resulting in a plant that is agronomically good and has rust resistance brought in by *C. canephora*. This is a very practical result from protoplast fusion. However, it is still difficult to get details of commercial plantings of such advances.

This fact illustrates one of the practical difficulties in southern African with regard to applying biotechnology to their particular needs. We find that as new knowledge approaches the commercial hurdle, the information sometimes dries up, and we do not know whether to interpret silence as success that has gone commercial and secretive or an indication that problems are still blocking progress. It is particularly difficult for a nation such as Zimbabwe, with limited funds, to know where to invest scarce resources. Should we try to defend our market share, small though it be, in a crop such as coffee, by research in micropropagation, somaclonal variation or genetic manipulation, or as much of all three as we can afford?

The doubt is raised because we hear reports or rumours that some group or other has cracked it, and so there is no point going on ourselves. For example, in the USA trials have been approved for coffee using the antisense technique. This method introduces a gene that programmes for a mRNA that is complementary to a normal mRNA and so it sticks to it and stops it from producing the normal protein. Using this route, decaffeinated coffee might be able to be grown to replace processed coffee by blocking one of the essential steps in the biochemistry of caffeine synthesis in the plant or bean. The nation that produces (and patents?) that coffee cultivar first will have a considerable advantage in the market place.

Here I digress briefly into politics to suggest that this is where a high critical mass, non-profit, molecular biological approach is urgently needed, one that is explicitly committed to helping the orphan economies, not just the orphan crops. Coffee is no orphan — it involves big money — but, at least at the suppliers' end of the industry, it also involves poor countries. If it is left to commercial interests to develop the next generation of coffee cultivars, a number of the indebted nations of the world could go bankrupt. Again, most of the less-developed laboratories of Africa can cope with *Agrobacterium* transformations. Coffee could be transformed by *Agrobacterium* (V. Masona, unpubl.) and kanamycin-resistant coffee plantlets produced. Non-profit organizations could act as brokers for genes that are available in the public domain. In the past, materials arising from discoveries made in the universities of the world were passed to whoever asked for them. Now we are frequently referred to lawyers who are looking for profit for their funders. In the developing world we do not have the time, money or skills to deal with such demands, and thus, in Zimbabwe at least, we need to set up a non-profit organization. There could be a role for the international agricultural research centres (IARCs) or perhaps the Food and Agriculture Organization of the United Nations (FAO) in this field, as brokers of good germplasm in plasmids.

Genetic Improvement

The most sophisticated agricultural biotechnology — adding genes to crops and switching them on in a useful manner, at a useful time — is also the most expensive, and it requires a large group of scientists to make it work. This biotechnology has already put about 100 new cultivars into field trials in the USA, and possibly about the same number in Europe. There is no question that this kind of molecular biology will soon help protect crops, particularly tropical plantation crops, from a wide range of pests and diseases. It is my hope that Africa will build the capacity to be involved in this activity, not just as a spectator but as a participant. In Zimbabwe there is now a 2-year MSc course in biotechnology, initiated precisely to begin to meet this need. As our graduates become more capable, we will establish a non-profit institute that will serve the needs of Africa for sophisticated biotechnology in all its varied applications. I hope that some outside Zimbabwe will find time to participate too.

Conclusion

I realize that I have assumed that the biotechnology we are talking about has to be indigenous. This is a matter of dignity and politics. Given this assumption, my impression is that each country and organization, whether national, university or commercial, should build as usual from the bottom up. Initially, this needs a low-tech tissue culture approach (there will always be a need for the propagation of new cultivars), followed fairly quickly by the addition of increasingly sophisticated (and expensive) DNA manipulatory techniques. The low-tech stage should justify and finance the next stage — it is a good test of competence. Most nations have particular crops that need improvement, but no one else is going to improve those crops for them, at least not without requiring payment. So indigenous scientists can learn from work on their own crops, then help their neighbours and utlimately compete in the real world. In this way, the necessary skills and commitment can be built up to provide economic independence.

Discussion

ENONUYA: The presentations by Drs Bokanga and Robertson highlighted the problem of crop research data in Africa. There is hardly any facility available where scientists working on a crop within the region could have access to all the scientific developments, prior to acquiring or properly employing biotechnology, whether high or low. This problem needs to be tackled.

References

Litz, R.E. 1984. Papaya. Pages 349-368 in *Handbook of Plant Cell Culture* edited by W.R. Sharp, D.A. Evans, P.V. Ammirato, and Y. Yamada. Macmillan, New York, USA.

1.7

Constraints in the accessibility and use of germplasm collections[1]

N.Q. Ng and S. Padulosi

International Institute of Tropical Agriculture, PMB 5320, Ibadan, Nigeria

Abstract

This paper discusses the problems associated with the lack of precise techniques for measuring the diversity of a crop species, selecting materials for accession, and determining the appropriate size for a core collection, as well as the boundaries of a genepool. The authors list the most important constraints to accessibility and use of germplasm collections, and indicate areas where biotechnology techniques can offer help. Among the research areas considered important for better utilization of germplasm are taxonomy of wild species, techniques for determination of a representative collection, reproductive biology and germplasm enhancement, conservation techniques and genetic stability under storage, evaluation for specific agronomic traits (particularly for stress resistance), and non-destructive pathogen testing techniques.

The genetic resource of a crop consists of a collection of seed samples or vegetative clonal propagules of current improved cultivars, landraces, elite breeding lines or genetic stocks, and related wild and weedy species. This represents an important heritage of mankind and the most fundamental natural source of materials for breeding crop varieties and for biological research. There has been a global effort over the past decade to collect and preserve the germplasm of many crop species in response to the fear that much of their genetic diversity could disappear because of pressures exerted by mankind.

The number of genebanks and the size of collections have increased rapidly during the past two decades (Hanson et al. 1984; Plucknett et al. 1987). Recent emphasis on the

1 Contribution No. IITA/91/CP/19 from the International Institute of Tropical Agriculture, PMB 5320, Ibadan, Nigeria.

collection of wild and weedy relatives of crops will certainly increase the complexity and size of the collections. This no doubt poses problems to some genebank managers, who are charged with the responsibility of multiplying, maintaining, conserving, documenting, and distributing the germplasm. Consequently, a large proportion of the germplasm has not yet been evaluated or utilized (Perrino and Monti 1991). Some policy makers consider that genebanks are expensive to maintain and, in many developing countries (particularly in Africa), the conservation of plant genetic resources is not a high priority (Ng 1982).

Diversity and Size of Germplasm Collections

A simple and precise technique for measuring the overall genetic diversity of a crop is not yet available, and no single approach is yet commonly considered best for measuring diversity. Thus it is difficult to determine the size of a collection necessary to represent the diversity existing within a crop. At present, most germplasm conservationists and breeders believe that the larger the collection (excluding duplication), the greater the chances of finding desirable characters for use in crop improvement.

Taking cowpea as an example, and applying Harlan and de Wet's (1971) definition, all cultivars of *Vigna unguiculata*, as well as its wild forms belonging to the subspecies *dekindtiana*, *stenophylla*, and *tenuis*, should constitute the primary genepool (Ng 1990). Successful crosses between cowpea and other species within the genus have not been reported, despite numerous attempts, and thus no other species can at present be considered to belong to the secondary or tertiary genepool of cowpea. With more basic research, however, some wild *Vigna* species may eventually be crossed successfully with cowpea, broadening the genetic base available for cowpea improvement. In addition, with advances in biotechnology, it is now possible to utilize genes isolated from any source, regardless of whether they belong to the same species, genus or family.

However, conventional hybridization techniques continue to play the most important role in plant breeding, and it is still necessary to distinguish a crop's primary, secondary, and tertiary genepool. Once this is done, the relative sizes of these genepools in a collection will need to be determined for each crop, taking into account its diversity and ecogeographical spread. Frankel and Brown (1984) suggested that "core collections" should represent subsets of a whole collection, but the selection of accessions for a "core collection" remains difficult. Molecular techniques may help simplify these tasks.

Constraints to Accessibility and Use

There are many practical, technical, social, and economic constraints to the accessibility and use of germplasm collections. In this paper we discuss some constraints we consider important, affecting the accessibility and use of germplasm collections, and the strategic research needed to overcome these constraints.

Economic constraints

Economic constraints are often the most serious in genebank operations, especially in developing countries. They affect the most basic operations, such as seed multiplication, processing, storing, and distribution. Money is essential to pay electricity bills and for repairs of equipment. Even if the purchase and installation of equipment for genebanks is funded by donor(s), in many cases national genebanks cannot afford the running costs.

Genetic resources have always been treated as public property and as such public funds have been required to collect and manage them. Unfortunately, only a few countries have been willing or able to invest in such public assets.

National plant quarantine regulations

National plant quarantine regulations are a necessary precaution against the accidental spread of pests and diseases. However, they can also hinder the flow of germplasm between genebanks, from genebanks to breeders and other users, and from plant explorers to genebanks. There have been numerous reports of losses of material because of delays or errors in quarantine.

Legislation for plant variety rights

Plant variety rights (PVR) have been evolved to recognize the contribution of a breeder in developing a variety and to protect the interests of such an investment for some years. Some commercial companies and researchers have isolated genes controlling traits of special interest, such as herbicide resistance or trypsin inhibition (found in cowpea), and then patented them (that is, claimed the intellectual property rights on them) in order to develop crop varieties for commercial purposes. Such restrictions encourage private enterprises to carry out research on exploiting genetic resources for crop improvement. However, it has been claimed that intellectual property rights legislation can be a major factor in restricting the free exchange of germplasm and some countries have reacted by enacting legislation prohibiting the export of germplasm (Mooney 1979).

Lack of documentation

Documentation and description of germplasm material held in genebanks enables curators or users to know what material is available for use and the variability that exists within it. Lack of documentation for accessions held in a genebank is one of the most important constraints limiting the use of germplasm collections (Perrino and Monti 1991). A computerized documentation system that provides efficient storage and retrieval of genetic resources data will ehance the accessibility of the information for users and curators.

Lack of evaluation and characterization

Characterization consists of recording those characteristics that are highly heritable, can be easily seen by the eye, and are expressed in all environments (IBPGR 1981). These are usually represented by the plant's morphological traits. Characterization can also include more sophisticated attributes generated by chemical, isozyme, restriction fragment length polymorphism (RFLP), and other molecular level analyses (Simpson and Withers 1986; Flavell 1991). These analyses are being used by some breeders, as well as by taxonomists, to study genetic diversity in depth with the aim of identifying differences and interrelationships among living organisms. This type of characterization has generated much interest, and its potential application, particularly in drawing linkage maps and identifying gene markers for selection, is considerable. However, such data are not available in most genebanks; in many of them even plant morphological data are inadequate.

The evaluation of germplasm frequently includes recording traits of agronomic interest, such as resistance to pests and diseases, and tolerance to physiological stresses that are influenced by the environment. Frankel (1989) also included cytogenetics and evolution studies in the genetic resources evaluation. Evaluation data are most sought after by plant breeders (Peters and Williams 1984). However, a multidisciplinary approach involving taxonomists, geneticists, physiologists, pathologists, entomologists, and other experts is essential (Perrino and Monti 1991). Genebank curators could play a major role in coordinating these activities by documenting information received from different evaluators and making it available to germplasm users. The evaluation of the diversity existing in germplasm is essential if the full potential value of genebank collections is to be revealed.

Unfamiliarity with collections

In order to exploit existing germplasm collections, familiarity with what exists on the part of both users and curators is essential. Taxonomic classification of the material must be correct. Strategies for collection, multiplication, and evaluation of germplasm vary with the breeding system of a particular genepool. Some species are inbreeders, other are outbreeders, but some can be both. The choice of a correct strategy is more problematic for those species that are assumed to be inbreeders because they actually undergo a significant degree of outbreeding.

Germplasm Strategic Research

The elements of research important for germplasm conservation and use are:

- Taxonomy, including various techniques for determining the representativeness of a collection
- Reproductive biology in germplasm enhancement
- Conservation techniques and genetic stability under genebank storage conditions
- Evaluation of specific agronomic traits, particularly of stress resistance
- Non-destructive pathogen testing

The classification of wild relatives of crop plants and the determination of their interrelationships require further study in many cases, as a large number of variants of species not described previously are being collected and acquired by genebanks. In addition to conventional methods used in classification, molecular techniques, such as DNA hybridization, restriction fragment length polymorphism (RFLP), and polymerase chain reaction could help in classifying and detecting variability in germplasm. All available techniques should be utilized to suit individual species.

Some material that has been collected from its place of origin cannot be reproduced in the environment in which the genebank is located. This problem is usually associated with plant physiology and reproductive biology. Research on reproductive biology will help scientists to choose the most appropriate method for reproducing or multiplying the germplasm and for devising techniques for crossing or overcoming crossing incompatibility that may be encountered in interspecific crosses. At present, breeders are reluctant to use wild species or even landraces in their breeding programmes because of the difficulties they encounter in transferring desired genes to the background of their improved materials.

Research on more efficient conservation techniques for some crop germplasm is required, particularly for vegetatively propagated crops and those whose seeds are recalcitrant. Improvements in conservation techniques will provide greater security for germplasm collections and reduce the cost of their management. We should also be concerned about the genetic stability of germplasm under storage; ways to prevent changes in the genetic constitution of original samples through mutation and selection and sensitive methods for detecting such variation are important subjects for future studies. In addition to genetic instability, germplasm under storage conditions can undergo disruptive biochemical changes from the action of free radicals; this aspect and its important implications for genebank management need further study (Benson 1990).

The problems associated with the evaluation of germplasm for effective utilization and with its documentation have been discussed by Williams (1989) and Perrino and Monti (1991). We wish to highlight the lack of information on alternative sources of genes that control traits of particular interest to breeders. Plant breeders, entomologists, and pathologists are usually reluctant to spend the time and effort to further evaluate more germplasm accessions once a promising source of germplasm with a particular desirable trait has become available. Yet, in order to broaden the genetic base of the species, it is important to further evaluate the germplasm for the identification of additional genes or alternative sources that control a particular trait for use in crop improvement.

Sensitive and non-destructive pathogen testing techniques are needed to detect a range of pathogens in germplasm, to prevent their spread through germplasm exchange and, at the same time, to accelerate the process of testing. When these techniques become available, they will alleviate one of the most important constraints to the availability of germplasm.

References

Benson, E. 1990. *Free Radical Damage in Stored Plants Germplasm.* IBPGR, Rome, Italy.

Flavell, R.B. 1991. Molecular biology and genetic conservation programmes. *Biological Journal of the Linnean Society* 43: 73-80.

Frankel, O.H. 1989. Principles and strategies of evaluation. Pages 245-260 in *The Use of Plant Genetic Resources* edited by A.H.D. Brown, D.R. Marshall, O.H. Frankel, and J.T. Williams. Cambridge University Press, Cambridge, UK.

Frankel, O.H. and A.H.D. Brown. 1984. Plant genetic resources today: A critical appraisal. Pages 249-257 in *Crop Genetic Resources: Conservation and Evaluation* edited by H.H.W. Holden and J.T. Williams. Allen and Unwin, London, UK.

Hanson, J., J.T. Williams, and R. Freund. 1984. *Institutes Conserving Crop Germplasm: The IBPGR Global Network of Genebanks*. IBPGR, Rome, Italy.

Harlan, J.R., and J.M.J. de Wet. 1971. Towards a rational classification of cultivated plants. *Taxon* 20: 509-517.

IBPGR. 1981. *Groundnut Descriptors*. IBPGR/ICRISAT, Rome, Italy.

Mooney, P.R. 1979. *Seeds of the Earth — A Private or Public Resource*. Mutual Press, Ottawa, Canada.

Ng, N.Q. 1982. *Survey of African Plant Genetic Resources*. IITA, Ibadan, Nigeria.

Ng, N.Q. 1990. Recent developments in cowpea germplasm collection, conservation, evaluation and research at the Genetic Resources Unit, IITA. Pages 13-28 in *Cowpea Genetic Resources* edited by N.Q. Ng and L.M. Monti. IITA, Ibadan, Nigeria.

Perrino, P., and L.M. Monti. 1991. Characterization and evaluation of plant germplasm — A problem of organisation and collaboration. Page 71-83 in *Crop Genetic Resources of Africa* (Vol. II) edited by N.Q. Ng, P. Perrino, F. Attere and H. Zedan. IITA, Ibadan, Nigeria.

Peters, J.P., and J.T. Williams. 1984. Towards better use of genebanks with special reference to information. *FAO/IBPGR Plant Genetic Resources Newsletter* 60: 22-31.

Plucknett, D.L., N.J.H. Smith, J.T. Williams, and N.M. Anishetty. 1987. *Genebanks and the World's Food*. Princeton University Press, New Jersey, USA.

Simpson, M.J.A., and L.A. Withers. 1986. *Characterization using Isozyme Electrophoresis. A Guide to the Literature*. IBPGR Technical Report. IBPGR, Rome, Italy.

Williams, J.T. 1989. Principles and strategies of evaluation. Pages 235-244 in *The Use of Plant Genetic Resources* edited by A.H.D. Brown, D.R. Marshall, O.H. Frankel, and J.T. Williams. Cambridge University Press, Cambridge, UK.

1.8

Biotechnology for kola improvement

A.C. Adebona

Department of Botany, Obafemi Awolowo University, Ile-Ife, Nigeria

Abstract

The production of kola nut (*Cola* spp.) is hampered by low seed germination resulting from poor development of fruit/seed/cotyledon, arising from crossing incompatibility. To explore ways of overcoming this problem, the author suggests the use of somatic embryogenesis and plant somatic cell hybridization.

The genus *Cola* Schott and Endl. is indigenous to West Africa (Chevalier and Perrott 1911; Bodard 1955). It was first described by Schott and Endlicher in 1832, who separated it from the genus *Sterculia* Linn. (Eijnatten 1964). Keay (1958) described 42 species of *Cola*; Bodard (1962) estimated that there were about 90 *Cola* species, distributed between about 12° north and south of the equator in Africa. At present, about 50 *Cola* species have been recorded in West Africa (Eijnatten 1969) but only two are of economic and cultural importance in Nigeria: *C. nitida* (known in Yoruba as *obi gbanja*) and *C. acuminata* (*obi abata*). Nigeria is the world's largest producer of kola nuts, with about 88% coming from the western part of the country (Quarcoo 1969; Eijnatten 1969).

The kola tree is cultivated mainly for its nuts, which contain alkaloids (caffein, theobromine, and kolanin). Because these alkaloids dispel sleep, thirst, and hunger, the nuts are used widely as a masticatory for their stimulatory effects. Although *C. acuminata* nuts are second to *C. nitida* in terms of consumption, *C. acuminata* nuts have a high social and ceremonial value among the Yoruba people (Russell 1955) and in the southern states of Nigeria.

Production and Research Constraints

The seeds of *C. acuminata* usually have three to five (rarely two or six) cotyledons, while those of *C. nitida* have two. The main flowering season is August-September, which may extend into October or later; the minor flowering season lasts from January to March. There is no report of any barrier, physiological or structural, to interspecific pollination and fertilization between the two species. However, F_1 hybrids, which are abundant in nature, are sterile (Okoloko and Jacob 1971). The low yield of *Cola* trees on some farms may be due to the poor performance of these hybrids, although other factors such as incompability, inadequate pollination, and the occurrence of relatively few hermaphroditic (functionally female) flowers could also reduce the yield. It is known that while *C. nitida* takes 5-7 years to fruit (Russell 1955), hybrids flower within 3-5 years of planting in the field, although the flowers have been reported to be sterile (Jacob and Opeke 1969). *C. acuminata* takes an even longer period to fruit.

A major problem confronting research on kola is the slow and uneven rate of seed germination (Clay 1964; Dublin 1965; Eijnatten 1964; Eijnatten and Odegbaro 1966). It is possible to eliminate the problem of late fruiting and poor germination through propagation by means of cuttings (Eijnatten 1964).

Pollen grains from hermaphroditic flowers are not functional, although they are viable; such flowers are therefore functionally female (Eijnatten 1964, 1969). Jacob and Scott-Emuakpor (1975) induced pollen fertility in the sterile interspecific hybrid of *C. acuminata* and *C. nitida* by using UV-irradiated pollen of the hybrid on the female flowers of *C. nitida*. The seeds (nuts) which were obtained from this backcross were di- and tricotyledonous. When UV-irradiated pollen from *C. nitida* was used to pollinate the hybrid, no fruit was set. The authors did not report having used UV-irradiated pollen for selfing in either the hybrid or in *C. nitida*. Where the pollen grains and embryo sacs are fertile, sterility of the hybrid could be attributable to the incompatible interaction of genes or gene complexes (Stebbins 1971).

Two general types of incompatibility are expressed in higher plants: the gametophytic or haplo-diplo type, in which the incompatibility depends upon the genotype of the gametophyte; and sporophytic or diplo-diplo type, in which the incompatibility is impressed upon the gametophyte by its sporophytic parent.

In *Theobroma cacao* (Sterculiaceae), the incompatibility reaction has been shown to be sporophytically determined, the reaction being between the haploid pollen grains and the diploid stylar tissue. The site of the incompatibility reaction in *T. cacao* lies not in the style but in the embryo sac (Knight and Rogers 1955). This was the first example reported among flowering plants of an incompatibility mechanism based on genetic control of success or failure of syngamy.

It is known that induced polyploidy may circumvent the problem of sterility in interspecific hybrids, especially when sterility is traceable to structural differences between the constituent genomes of the hybrid (Stebbins 1971; Olorode and Olorunfemi 1973).

In reciprocal *C. nitida* x *C. milleni* crosses, Morakinyo (1978) reported 80% fruit set on *C. milleni* and 10% on *C. nitida*. None of the seeds from either of the female parents germinated. Longitudinal and transverse sections of *C. nitida* (male) x *C. milleni* (female) seeds showed tiny and fragile cotyledons in the hybrid, a layer of tissue around the cotyledons that was not present in *C. milleni*, and a fleshy seed coat which was thicker in the hybrid than in *C. milleni*. The seeds were much reduced in size, fragile, and embedded in a mass of colourless tissue. The miniature seed had two cotyledons and a tiny hairy embryo like the normal *C. milleni* seeds. The seeds obtained from the crosses of *C. nitida* (female) with *C. milleni* (male) were similar to the normal *C. nitida* except for a smaller size or a harder seed coat.

Some of the problems in kola production may be attributed to sterility/incompatibility and poor fruit/seed/cotyledon development. Morakinyo (1978) suggested the application of tissue culture techniques to the raising of seedlings from interspecific crosses when the embryo is well formed, but the problem seems to be lack of adequate food in the hybrid seed cotyledons. This may also be applicable to problems of poor seed germination. Successful shoot proliferation techniques need to be developed, using cotyledons, epicotyls, hypocotyl segments or any other plant parts that are suitable.

Future Research

Because of poor seed germination, it may be necessary to use somatic embryogenesis, possibly from immature zygotic embryos, to solve the problem. Polyploidization has been suggested as the most likely route for eliminating sterility in interhybrid crosses. Efficient techniques for isolation, culture, and fusion of protoplasts are available (Harms 1983; Gleba and Sytnic 1984; Horn et al. 1986) and their use in plant somatic cell hybridization might be worth investigating with respect to the incompatibility problem in *Cola* species.

References

Bodard, M. 1955. Contributions a lé tube de *Cola nitida* croissance et biologie florale. *Centre Recherches Agronomique Bingerville Bulletin* 11: 3-28.

Bodard, M. 1962. Contribution a l'étude systematique du genre *Cola* en Afrique Occidentale. *Annales de la Faculte des Sciences de l'Universite de Dakar* 7: 71-82.

Chevalier, A., and E. Perrot. 1911. *Les Vegetaux Utiles de l'Afrique Tropicale Française. VI. Les Kolatiers et les Noix de Kola.* Challamal, Paris.

Clay, D.W.T. 1964. Germination of the kolanut (*Cola nitida* (Vent.) Schott and Endl.). *Tropical Agriculture* (Trinidad) 41: 55-60.

Dublin, P. 1965. Le colatier (*C. nitida*) en Republique Centrafricaine, culture et amelioration. *Café, Cacao, The* 9: 97-115.

Eijnatten, C.L.M. van. 1964. Studies on the germination of kolanuts. *Cocoa Research Institute of Nigeria Memorandum* 4: 1-13.

Eijnatten, C.L.M. van. 1969. Kola: Its botany and cultivation. *Royal Tropical Institute Communication* 59: 1-20.

Eijnatten, C.L.M. van, and O.A. Odegbaro. 1966. Kolanut germination trials. Pages 93-94 in *Annual Report* 1964-65. Cocoa Research Institute of Nigeria, Ibadan, Nigeria.

Gleba, Y.Y., and K.M. Sytnik. 1984. Protoplast fusion and parasexual hybridization of higher plants. Pages 36-62 in *Protoplast Fusion: Genetic Engineering in Higher Plants* edited by R. Shoeman. Springer-Verlag, Berlin, Germany.

Harms, C.T. 1983. Somatic hybridisation by plant protoplast fusion. Pages 69-84 in *Protoplasts* edited by I. Potrykus, C.T. Harms, A. Hinnen, R. Huffer, P.J. King, and R.D. Shilitto. Birkhauser, Cambridge, Massachussetts, USA.

Horn, W., C.J. Jensen, W. Odenbach, and O. Schieder (eds). 1986. Genetic manipulations in plant breeding. In *Proceedings, International Symposium on Eucarpia, 8-13 September 1985, Berlin.* De Gruyter, Berlin, Germany.

Jacob, V.J., and L.K. Opeke. 1969. Interspecific hybridization in the genus *Cola.* Observations on pollinations and chromosome behaviour in F_1 hybrids of *Cola acuminata* x *Cola nitida.* In *Proceedings, 5th Conference of Nigeria Agricultural Society, Benin.*

Jacob, V.J., and M.B. Scott-Emuakpor. 1975. Use of UV-irradiated pollen in inducing pollen fertility in sterile interspecific hybrids of *Cola*. *Journal of Nuclear Agricultural Biology* 4(3): 57-58.

Keay, R.W.J. 1958. *Flora of West Tropical Africa* (2nd edn) Crown Agents, London, UK.

Knight, R., and H.H. Rogers. 1955. Incompatibility in *Theobroma cacao* L. *Heredity* 9: 77.

Morakinyo, J.A. 1978. Biosystematic studies in the genus *Cola* Schott and Endl. MSc dissertation, University of Ife, Nigeria.

Okoloko, G.E., and V.J. Jacob. 1971. Kolanut breeding. Pages 23-33 in *Progress in Tree Crop Research in Nigeria*. Cocoa Research Institute of Nigeria, Ibadan, Nigeria.

Olorode, O., and A.E. Olorunfemi. 1973. The hybrid origin of *Emilia practermissa* (Senecioneae; Compositae). *Annals of Botany* 37: 185-191.

Quarcoo, T. 1969. Development of kola and its future in Nigeria. *Proceedings, Agricultural Society of Nigeria* 6: 19-22.

Russell, T.A. 1955. The kola of Nigeria and the Cameroons. *Tropical Agriculture (Trinidad)* 32: 210-240.

Stebbins, G.L. 1971. *Chromosome Evolution in Higher Plants*. Edward Arnold, London, UK.

2.1

The use of wild species in crop improvement[1]

L.M. Monti

Department of Agronomy and Plant Genetics, University of Naples, Via Università 100, 80055 Portici, Italy

Abstract

The author discusses the importance of wild species as a source of useful traits for plant improvement, with particular emphasis on disease and pest resistance. The strategies used to overcome the barriers to interspecific crosses are outlined. Pre-fertilization barriers can be overcome by using in vivo techniques, among which the use of appropriate genotypes and of some meiotic mutants is most frequent. In vivo and in vitro techniques can be applied to circumvent post-fertilization barriers. Embryo rescue has been successfully used to overcome several interspecific barriers. Fusion between protoplasts has been used in several cases to produce symmetric and asymmetric hybrids and cytoplasmic hybrids (cybrids). New germplasm has also been produced through direct gene transfer using recombinant DNA techniques.

Wild species are known to be a source of useful genes, mainly for resistance or tolerance to diseases, pests, and abiotic factors not found in cultivated species. This higher level of resistance that characterizes wild species or ecotypes derives from their ability to withstand extreme environmental conditions.

Collection and Evaluation of Wild Species

While the importance of conserving resources is today widely recognized, the collection of wild species and the evaluation of their genetic variability for agronomically desirable traits is still in its infancy. Greater efforts should be made in collecting and screening wild species

---

1 Contribution No. 64 from the Research Centre for Vegetable Breeding, National Research Council, Via Università 133, 80055 Portici, Italy.

and ecotypes of the most important crops, in view of the fact that where systematic analyses have been carried out many useful traits have been found. Examples are provided by the work done on tomato (Figure 1) and *Phaseolus* species (Table 1).

Figure 1 Wild species of tomato and their contribution to the improvement of pest and disease resistance in tomato

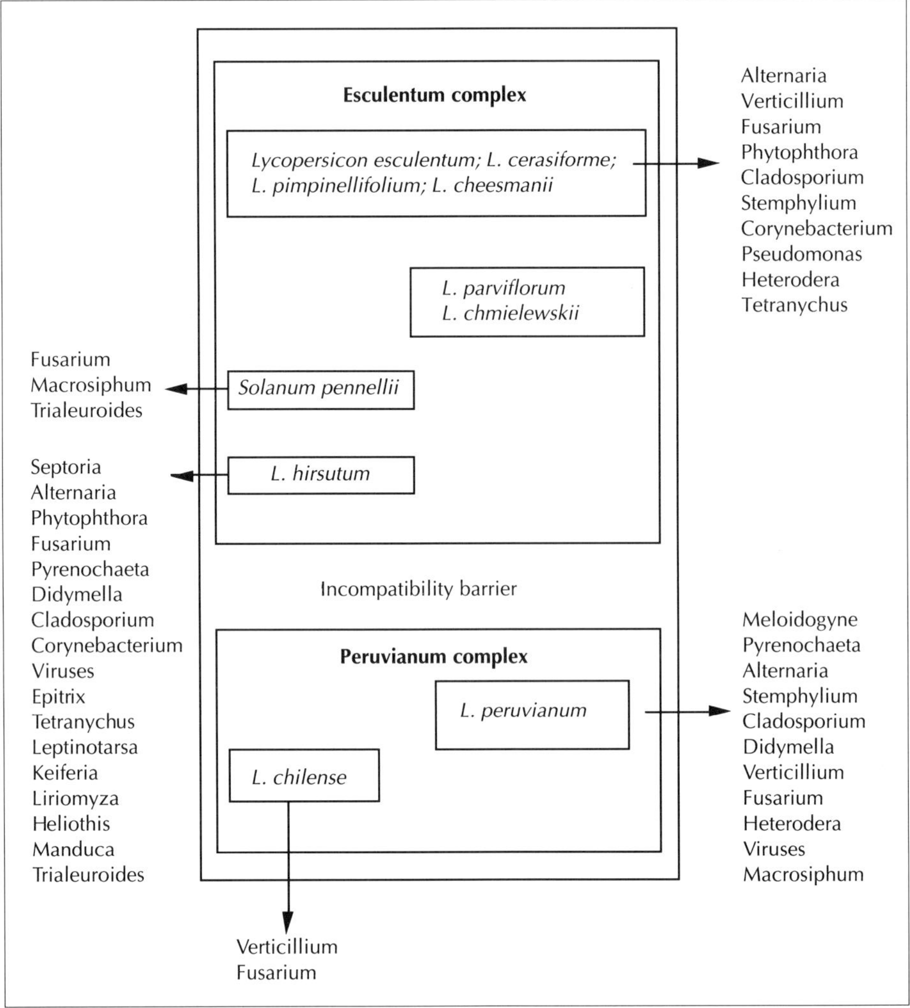

In a biological sense, the speciation process implies the development of isolation barriers which do not allow any gene flow between new species and the others. Nevertheless, interspecific crossings are designed in breeding programmes to transfer genes or relatively small blocks of foreign chromosomes from a wild species to a cultivated genotype of a related but distinct species. These experiments are successful either because of a misleading taxonomic classification of some species, which may not be different species at all, or

Table 1 Desirable traits identified in cultivated and wild *Phaseolus* species

Tolerance	Species
Xanthomonas blight	*P. acutifolius*
Root rots	*P. coccineus* *P. coccineus*
White mould	*P. coccineus*
Drought and heat	*P. coccineus*
Cold stress	*P. wrightii* *P. ritensis*
Frost	*P. wrightii*
Soil crusting	*P. coccineus*
Flooding	*P. polystachyus* *P. wrightii*

Source: Hucl and Scoles 1985.

because of appropriate experimental techniques. In the first case, there is no great difficulty in crossings; when distinct species are involved, the success or the failure of these crosses depends upon the genetic distance and the homology between the two parents.

After interspecific pollination, no seeds (or a few, often ill-formed, non-viable ones) can be obtained; sterile or semi-sterile F_1 hybrid plants are often obtained. To overcome the interspecific barriers, even between different genera, several techniques are used, aimed at zygotic embryo production through in vivo and in vitro techniques or at transformation through somatic hybridization and recombinant DNA techniques. Introgression (the introduction of alien variation into crops) is a constant objective of plant breeding, pursued through several techniques, including, most recently, genetic engineering. Table 2 (*overleaf*) gives some examples of useful traits transferred from wild species to cultivated species.

This paper will consider the strategies used to overcome the barriers to interspecific crosses, except the recombinant DNA technique, which is discussed elsewhere in this volume. The barriers that prevent fertillization will be considered separately from those arising after fertilization. It should be clear that a sound approach to overcoming these kinds of problems would involve appropriate histological and cytological analysis to study the reproductive biology of the species, the F_1 embryo development and the F_1 plants, as well as the use of good morphological, enzymatic, and molecular markers.

Pre-Fertilization Barriers

When no F_1 zygotes are formed, this indicates that allo-incompatibility exists between the two species. The barriers may be: at the level of the stigma, with no germination of the foreign pollen; at the level of the style, with no growth or retarded growth of the pollen tubes in the foreign styles; or at the level of the ovaries, with no fertilization of the egg cell nucleus

Table 2 Some disease and pest resistance traits transferred from wild species to crops

Crop	Diseases/pests	Donor species
Apple (*Malus domestica*)	apple scab	*M. floribunda*
Beet (*Beta vulgaris*)	leaf spot	*B. maritima*
Bread wheat (*Triticum aestivum*)	rust	*Ae. umbellulata*
	rust	*T. durum*
	rust	*T. timopheevi*
	wheat bunt	*Ae. elongatum*
	loose smut	*T. dicoccum*
Common bean (*Phaseolus vulgaris*)	viruses (BCMV, BYMV)	*P. coccineus*
	bean anthracnose	*P. acutifolius*
	bean blight	
Cotton (*Gossypium hirsutum*)	cotton bollworm	*G. barbadense*
Grape (*Vitus vinifera*)	downy mildew	*V. rotundifolia*
Lettuce (*Lactuca sativa*)	lettuce aphid	*L. virosa*
	downy mildew	*L. serriola*
Oat (*Avena sativa*)	powdery mildew	*A. barbata*
Potato (*Solanum tuberosum*)	late blight	*S. demissum*
Sunflower (*Helianthus annuus*)	nematodes	*H. tuberosus*
Tobacco (*Nicotiana tabacum*)	nematodes	*N. repanda*
		N. sylvestris
		N. tomentosiformis

Note: BCMV = bean common mosaic virus; BYMV = bean yellow mosaic virus.

by the sperm nucleus. When this incompatibility exists, it would be convenient to make crosses using different genotypes in an attempt to identify parents more effective at achieving interspecific fertilization.

The influence of the genotypes of the parents on interspecific crosses has been clearly demonstrated in several cases, such as in the cross between *Nicotiana repanda* and *N. tabacum* (Pittarelli and Stavely 1975), between *Trifolium repens* and *T. nigrescens* (Hoven 1962), and between *Tripsacum* species and *Zea mays* (Harlan and de Wet 1977). In wheat, two genes that have been identified and localized on chromosomes 5A and 5B control the fertilization with secale and other cereals (Riley and Chapman 1967).

When species with different levels of ploidy are present in a genus, crosses between species with the same ploidy level have more chance of success. In potato breeding, in order to cross diploid wild species ($2n = 2x = 24$) with tetraploid potatoes ($2n = 4x = 48$), haploid potato clones ($2n = 2x = 24$) are obtained through parthenogenesis and are crossed with wild species; the F_1 diploid clones obtained in this way are selected at the diploid level, which is relatively easy, and then taken back to the tetraploidy level (Peloquin et al. 1966).

Another possible way of crossing species with different levels of ploidy involves using particular meiotic mutants which form unreduced gametes. The formation of $2n$ gametes is widespread among plants; in potato breeding, wild diploid species forming $2n$ gametes can in this way be crossed with commercial tetraploid varieties (Mendiburu and Peloquin 1977).

To overcome interspecific incompatibility, the menthor pollen technique can be also used (Pandey and de Nettancourt 1975). Incompatible pollen is carried along the style together with compatible pollen that is unable to fertilize because it has been irradiated with high doses of gamma rays.

Post-Fertilization Barriers

A post-fertilization barrier can occur from incompatibility between the genome of one species and the cytoplasm of another species. Reciprocal crosses are used in an attempt to overcome this kind of barrier (Monti 1971). In many cases, zygotes can be formed but seeds are not developed because of retarded endosperm development, failure to transfer stored reserves to the embryo, or suspensor defects (reported in Ramsay et al. 1991). If these post-fertilization barriers are the causes of failure, embryo culture can be very valuable and has, indeed, determined the success of hundreds of difficult crosses (Pierik 1987). Histological analysis of F_1 embryo development has often led to the use of the embryo rescue technique.

Through embryo rescue after interspecific crosses, improvement was obtained in *Brassica* species (McNaughton and Ross 1978), groundnut (Singh and Moss 1984), common bean (Mok et al. 1986), cucurbits (Jelaska 1986), and tomato. Intergeneric crosses were also successful through embryo culture between barley and secale (Fedak 1986), between *Lycopersicon* and *Solanum* (Sink and Reynolds 1986), and in other cases.

In vitro culture of whole ovules and pistils followed by in vitro culture of the hybrid embryo was shown to be successful in some cases when isolated embryos failed to develop (McCoy and Walter 1984; Zenkteler 1973). To delay embryo abortion and obtain more developed embryos for in vitro culture, manipulating temperature (Pickering 1984) and applying growth substances (Al-Yasiri and Coyne 1964) may prove successful.

Through a study of the stage of embryo development reached before abortion, it was found that some genotypes of *Vigna vexillata* produced more developed embryos in crossings with *V. unguiculata* (Barone et al. in press). Heterozygotic parents were also shown to be more effective in interspecific crosses in *Phaseolus* (Pratt et al. 1985).

F_1 hybrids obtained after interspecific crosses can be highly sterile; this sterility usually results from reduced chromosome homology and the consequent lack of pairing of chromosomes at meiosis. As chromosomes are distributed at random in the gametes, only n gametes that include all or nearly all the chromosomes of one parent will be fertile. Doubling the number of chromosomes of F_1 sterile plants is the method usually tried to recover fertility; several allopolyploids have been bred in this way, among which triticale, first obtained by Rimpau just 100 years ago (Rimpau 1891), is the most famous and important because of its wide diffusion around the world.

In some experiments, breeders look for sterility in the F_1 plants after interspecific crosses. Male sterility, resulting from the interaction between cytoplasmatic and nuclear factors, is the basis for hybrid seed production; when both factors are absent in the cultivated species, interspecific crosses are made with this specific aim, as in wheat and sunflower. Another example of the utilization of F_1 interspecific sterility is that occurring after the pollination of barley with *Hordeum bulbosum*. In this cross, sterile F_1 plants are produced, but these plants are haploid because of the elimination of all *H. bulbosum* chromosomes in the mitoses from the zygote to the pro-embryo. Haploid production using this technique and the subsequent formation of dihaploids is used extensively in barley breeding (Kasha 1974).

Somatic Hybridization

Once obtained, protoplasts belonging to different species can be fused using two main techniques: the use of polyethelene glycol (PEG); and the use of electric pulses. After fusion, the hybrid cells must be able to regenerate plants. Fusion between the protoplasts of *N. glauca* and *N. langsdorfii* was the first successful case of somatic hybridization (Carlson et al. 1972) and was followed by several other interspecific fusions (Table 3).

Somatic hybridization can give rise to symmetric or asymmetric hybrids (according to the relative contributions of the nuclei of two parents), to cytoplasmic hybrids (cybrids) when the fusion is limited to the two cytoplasms and does not involve the nuclei, or to the combination of the nucleus of one parent with the cytoplasm of the other parent.

Table 3 Number of reports on somatic fusions

Type of fusion	Genus/family	Number of reports
Symmetric	Apocinacae	1
	Brassica	6
	Compositae	1
	Cucumis	1
	Dianthus	1
	Helianthus	1
	Lycopersicon	1
	Nicotiana	6
	Oryza	2
	Petunia	1
	Pyrus	1
	Solanum	11
Asymmetric	*Brassica*	3
	Lycopersicon	4
	Nicotiana	5
	Solanum	6
Cybridization	*Brassica*	3
	Lycopersicon	4
	Nicotiana	4
	Oryza	1
	Petunia	1
	Solanum	3

Source: Adapted from several papers presented at the 7th International Congress on Plant Tissue and Cell Culture, 24-29 June 1990, Amsterdam, The Netherlands.

Discussion

PADULOSI: There is a need to involve farmers closely in the conservation of different crops, especially the recalcitrant species according to the "Farmer's Rights" endorsed by many countries at FAO in 1990.

OLEMBO: There is a danger of loss of plant genetic resources in Africa where many hybrids are being released to farmers in favour of traditional crops and seeds. There is also a danger from "gene hunters", who may come to Africa, collect genetic material and, after some manipulation on them, may then patent them abroad, thereby making these materials inaccessible to the original owners in Africa.

MONTI: FAO has recognized the role of farmers in developing countries in saving the germplasm; farmers' rights have been undersigned by several countries and regulations on this act are in progress, probably by having a special fund through royalties on all the new varieties that are bred. This fund should be devoted to save other germplasm today.

FAUQUET: Concerning viruses that can be gene vectors, there is now a new system set up in Beachy's lab. It involves a cDNA infectious transcript of tobacco mosaic virus (TMV) where the movement protein has been deleted. It is then possible to insert in the TMV cDNA any protein gene which will be expressed in the transgenic plant during viral infection. After infection the virus containing the foreign gene is encapsidated and can be stored for years for another use. The system is also very secure because this TMV transcript cannot multiply on a wild type plant. This system has great potential but, so far, no application has been made.

MONTI: The importance of wild species is well known; numerous collections are in germplasm units and laboratories around the world. I think it is time to make an effort on a regional basis to start or coordinate a programme for evaluating these collections at least for the resistance to the most important biotic stresses.

HAMILTON: I would urge continued vigilance on the introduction of seed-transmitted viruses via wild species used as parents in breeding programmes. Caution should also be applied when using indigenous wild species, which may be suspected but symptomless carriers of seed-transmitted viruses.

References

Al-Yasiri, A., and D.P. Coyne. 1964. Effect of growth regulators in doubling pod abscission and embryo abortion in the interspecific crosses *Phaseolus vulgaris* x *P. acutifolius*. *Crop Science* 4: 433-435.

Barone, A., A. Del Giudice, and Q. Ng. (in press). Barriers to interspecific hybridization between *Vigna unguiculata* and *V. vexillata*. In *Sexual Plant Reproduction*. Wageningen Agricultural University, Wageningen, The Netherlands.

Carlson, P.S., H.H. Smith, and R.D. Dearing. 1972. Parasexual interspecific plant hybridization. *Proceedings of the National Academy of Science, USA* 69: 2292-2294.

Fedak, G. 1986. Hordecale (*H. vulgare* L. x *Secale cereale* L.). Pages 544-555 in *Biotechnology in Agriculture and Forestry. 2. Crop I* edited by Y.P.S. Bajaj. Springer-Verlag, Berlin, Germany.

Harlan, J.R., and J.M.J. de Wet. 1977. Pathways of genetic transfer from *Tripsacum* to *Zea mays*. *Proceedings of the National Academy of Science USA* 74: 3494-3497.

Hoven, A.W. 1962. Interspecific hybridization between *Trifolium repens* and *T. nigrescens* and analysis of hybrid meiosis. *Crop Science* 2: 251-254.

Hucl, P., and G.J. Scoles. 1985. Interspecific hybridization in the common bean: A review. *Hort. Science* 20: 352-357.

Jelaska, S. 1986. Cucurbits. Pages 371-386 in *Biotechnology in Agriculture and Forestry. 2. Crop I* edited by Y.P.S. Bajaj. Springer-Verlag, Berlin, Germany.

Kasha, K.J. 1974. Haploids from somatic cells. Pages 67-87 in *Haploids in Higher Plants: Advances and Potential* edited by K.J. Kasha. University of Guelph, Ontario, Canada.

McCoy, T.J., and K. Walter. 1984. Alfalfa. Pages 171-192 in *Handbook of Plant Cell Culture* edited by P.V. Ammirato, D.A. Evans, W.R. Sharp, and Y. Yamada. Macmillan, New York, USA.

McNaughton, I.H., and C.L. Ross. 1978. Interspecific and inter-generic hybridisation in the *Brassica* with special emphasis on the improvement of forage crops. Pages 75-105 in *Scottish Plant Breeding Station 57th Annual Report 1977/78.*

Mendiburu, A.P., and S.J. Peloquin. 1977. The significance of 2*n* gametes in potato breeding. *Theoretical and Applied Genetics* 43:53-61.

Mok, D.W.S., M.C. Mok, A. Rabakoarihanta, and C.T. Shii. 1986. *Phaseolus*: Wide hybridisation through embryo culture. Pages 309-318 in *Biotechnology in Agriculture and Forestry. 2. Crop I* edited by Y.P.S. Bajaj. Springer-Verlag, Berlin, Germany.

Monti, L.M. 1971. Genome-cytoplasmic interaction in *Pisum* species. *Pisum Newsletter* 3: 28.

Pandey, K.K., and D. de Nettancourt. 1975. Induced mutations and cross-compatibility. In *Induced Mutation in Cross-Breeding*. FAO/IAEA Advisory Group, Vienna, Austria.

Peloquin, S.J., R.W. Hougas, and A. Gabert. 1966. Haploidy as a new approach to the cytogenetics and breeding of *Solanum tuberosum*. Pages 21-28 in *Chromosome Manipulation and Plant Genetics* edited by R. Riley and K.R. Lewis. Oliver and Boyd, Edinburgh, UK.

Pickering, R.A. 1984. The influence of genotype and environment on chromosome elimination in crosses between *Hordeum vulgare* L. and *Hordeum bulbosum* L. *Plant Science Letter* 34: 153-164.

Pierik, R.L.M. 1987. *In Vitro Culture of Higher Plants*. Martinus Nijhoff, Dordrecht, The Netherlands.

Pittarelli, C.W., and J.R. Stavely. 1975. Direct hybridization of *Nicotiana repanda* x *N. tabacum*. *Journal of Heredity* 66: 281-284.

Pratt, R.C., R.A. Bressan, and P.M. Hasegawa. 1985. Genotypic diversity enhances recovery of hybrids and fertile backcrosses of *Phaseolus vulgaris* L. x *P. acutifolius* A. G. Gray. *Euphytica* 34: 329-344.

Ramsay, G., B. Pickersgill, and P.D.S. Caligari. 1991. The application of novel techniques in food, forage, and pasture legumes to overcome genetic barriers which prevent the use of potential germplasm resources. Pages 421-460 in *Legume Genetic Resources for Semi-Arid Temperate Environments* edited by A. Smith and L. Robertson. Proceedings, International Workshop on Genetic Resources of Cool-season Pasture, Forage, and Food Legumes for Semi-arid Temperate Environments, 19-24 June 1987, Cairo, Egypt. ICARDA, Aleppo, Syria.

Riley, R., and V. Chapman. 1967. The inheritance in wheat of crossability with rye. *Genetic Resources* 9: 259-267.

Rimpau, W. 1891. Krenzungsprodukte landwirtschaftlicher Kulturpflanzen. *Landwirtschafliche Jahrbuch* 20: 335-371.

Singh, A.N., and J.P. Moss. 1984. Utilisation of wild relatives in genetic improvement of *Arachis hypogea* L. V. Genome analysis in section *Arachis* and its implication in gene transfer. *Theoretical and Applied Genetics* 68: 1-10.

Sink, K.C., and F.J. Reynolds. 1986. Tomato (*Lycopersicon esculentum* L.). Pages 319-344 in *Biotechnology in Agriculture and Forestry. 2. Crop I* edited by Y.P.S. Bajaj. Springer-Verlag, Berlin, Germany.

Zenkteler, M. 1973. Test-tube fertilization and obtaining interspecific and intergeneric embryos. Abstract in *Proceedings, Eucarpia Meeting on Aseptic Culture Methods in Plant Breeding, Leeds, UK.*

2.2

Status of wide crosses in cassava and yam[1]

R. Asiedu, K.V. Bai, R.Terauchi, A.G.O. Dixon and S.K. Hahn

International Institute of Tropical Agriculture, PMB 5320, Ibadan, Nigeria

Abstract

Wild relatives of cassava (*Manihot* spp.) and yam (*Dioscorea* spp.) have many attributes of importance to the genetic improvement of the cultivated species. Some crosses between cassava and its wild relatives have led to polyploids (crosses to *M. glaziovii* and *M. epruinosa*) and resistance to two major diseases, cassava mosaic virus and cassava bacterial blight (crosses to *M. glaziovii*). No barriers to interspecific hybridization have been found in the genus *Manihot*, but hybridization between *Dioscorea* species is severely hampered by poor and erratic flowering. Much remains to be done in genome analysis in both genera.

Cassava (*Manihot esculenta* Crantz) and yams (*Dioscorea* spp.) are important staple foods for millions of people in the humid and subhumid tropics. Cultivated yams cover many species, but *M. esculenta* is the only species in the genus *Manihot* widely cultivated for food. Wild yams also serve as a valuable alternative food source in times of scarcity.

The high productivity of cassava and yams and their convenience as energy sources make them indispensable as integral components of the farming and food systems of many tropical countries, especially in Africa. One important avenue for enhancing the genetic ability of these crops to withstand the existing and emerging challenges in their environment has been the introduction of genes from their wild relatives.

The Genus *Manihot*

Cassava originated in Latin America, where it is still widely cultivated and from where it was introduced into Africa about the 16th century. It is difficult to state exactly how many

---

1 Contribution No. IITA/91/CP/20 from the International Institute of Tropical Agriculture, PMB 5320, Ibadan, Nigeria.

Manihot species exist, owing to the instability of species boundaries resulting from natural interspecific hybridization (Rogers and Appan 1973). Perry (1943) suggested 150 to 160 species, whereas Purseglove (1968) put the number between 100 and 200. Rogers and Appan (1973) provided a detailed description of 98 species, including perennial sub-shrubs and about 24 tall shrub or tree species.

Many useful traits have been reported in the wild *Manihot* species that are of potential benefit to cassava improvement (Bryne 1984). Rogers and Appan (1973) commented on the potential of many species for incorporating adaptation to abiotic stresses, resistance/ tolerance to biotic stresses, and quality traits. It is worth noting that the methodologies used for evaluating some of the traits, such as hydrocyanic acid (HCN) content of the tuberous roots, are open to criticism.

The wild species are not easily accessible to some of the breeding programmes that could benefit from their utilization. Knowledge about their reproductive biology, genomic constitution, and phylogenetic relationships is fairly limited. Also, it is argued that the wealth of variability in the cultigen and its capacity to generate further variation is adequate for many existing cassava improvement programmes. Indeed, it has been suggested that the available intraspecific variation arose partly through natural introgression from wild species into cassava (Rogers and Appan 1973; Bryne 1984; Grattapaglia et al. 1987).

M. glaziovii has been hybridized with cassava in many countries including Tanzania, Nigeria, India, Indonesia, Madagascar, and Brazil (Nichols 1947; Cours 1951; Magoon et al. 1966; Hahn et al. 1990). This might result partly from the relative availability of the species to the early breeding programmes. It was imported by the colonial authorities as a source of rubber into many countries where it now grows wild and is often used for nursing cash crops. It has proved to be the most useful cross-combination, as resistance to cassava common mosaic virus (CCMV) (Nichols 1947) and cassava bacterial blight (CBB) (Hahn et al. 1989) have been shown in the progenies. A selection of third backcross derivatives from Nichols' work, labelled 58308, is a parent of many cassava clones produced at the International Institute of Tropical Agriculture (IITA), including TMS 30572 and TMS 4(2)1425, which are popular varieties, particularly in Nigeria. Resistance to CCMV was also demonstrated in Nichols' (1947) hybrids between cassava and *M. dichotoma*. Hybrids between cassava and *M. tristis* ssp. *saxicola* were produced by Nichols (1947) and Bolhuis (1953). The anticipated transfer of the high protein content of the tuberous root from the wild species was not sustained in the progenies during backcrossing. Similar disappointment was reported by Jennings (1959) from his cross of *M. melanobasis* to cassava. The high fertility in the initial crosses of the two species to cassava raised questions about the classification of the respective parents as separate species.

IITA has carried out interspecific hybridization for over a decade (Hahn et al. 1990). The early hybrids involved crosses of cassava clones to wild species, including *M. epruinosa*, *M. chlorosticta*, *M. glaziovii*, *M. leptophylla*, and *M. brachyandra*. Recently, more accessions of these species, and additional species such as *M. tristis*, *M. tripartita*, *M. stricta*, *M. anomala*, *M. gracilis*, *M. catingae*, *M. pohlii*, and *M. neusana*, have been crossed as male to cassava (Asiedu et al. 1989). Numerous F_1 and backcross progenies are being evaluated for agronomic traits. Seeds are being collected from 19 paired combinations of wild *Manihot* species in an experiment conducted to investigate cross-compatibilities among the wild species *per se* and the elucidation of possible intergenomic epistatic interactions.

Polyploids have been isolated from the early hybrids which involved *M. glaziovii* and *M. epruinosa* as a result of the production of unreduced gametes by one or both parents

(Hahn et al. 1990). Some of the polyploids are undergoing performance trials at sites in Nigeria's three major agroecological zones — humid, subhumid, and semi-arid. Early observations are encouraging. The triploids also offer a unique opportunity to generate aneuploids systematically for the first time in the genus *Manihot*. Such genetic stock would be of great value in mapping the cassava genome and elucidating chromosome homoeologies.

Several F_1 hybrid progenies from crosses between IITA-improved cassava clones and seven accessions of local arborescent cassava clones are also undergoing evaluation for agronomic traits. Arborescent cassava clones grow wild in Nigeria and some other African countries, where they are sometimes used as shade trees or live fence posts. Nichols (1947) also hybridized cassava to one accession of arborescent cassava ("tree cassava"). He speculated that the arborescent cassava clone itself was probably a natural hybrid between cassava and *M. glaziovii*.

No systematic incompatibility has been reported that would prevent crossing within the genus *Manihot*. However, there are differences between genotypes in their performance as parents in crossing schemes (Bolhuis 1967). Seed set for 1989 crosses between seven cassava clones and several accessions of wild *Manihot* species at IITA ranged from about 1% to 30%, with a strong influence of the maternal cassava parent on the success. The edaphoclimatic conditions at the time of crossing, the efficiency of pollinators, and the level of pest damage on the developing fruits are additional factors. These factors, in addition to the wide range in the reported numbers of flowers pollinated per combination, call for caution in the comparison of results reported by different workers.

Most researchers have been more successful in using cassava as female parents in crosses to wild *Manihot* species (Bolhuis 1967, Nassar 1979) than in the reciprocal crosses. Koch (quoted by Nichols 1947), Abraham (1957), and Bolhuis (1967) were all unsuccessful in crossing *M. glaziovii* as female to cassava but they obtained limited success in the reciprocal crosses. Bolhuis (1967) reported 28% seed set from crossing cassava as female to *M. saxicola* and 8% from the reciprocal. Nichols (1947) succeeded in crossing *M. dichotoma* to cassava in reciprocal fashion, contrary to the failure recorded by Koch. Nassar (1979) reported 3-27% seed set in reciprocal crosses of cassava with *M. oligantha* ssp. *nesteli*, *M. anomala*, *M. gracilis*, and *M. zehntneri*.

Backcrossing to cassava as the recurrent parent has been the usual means of further exploitation of the F_1 interspecific hybrids. Inter-mating of backcross progenies has proved effective in inducing a stronger expression of characteristics, such as mosaic resistance, which are inherited in a predominantly additive and recessive fashion (Jennings 1976).

Few reports are available on the cytology of interspecific hybrids in genus *Manihot* (Bai 1987). Magoon et al. (1966, 1970) reported that there were generally 18 bivalents at meiotic metaphase 1 (M-I) in F_1 hybrids between cassava and *M. glaziovii*, although some precocious movement of chromosomes was observed. It was also observed that chromosomes with similar morphology in the two species did not necessarily pair with each other in the F_1 hybrid. The intimacy of synapsis in the F_1 hybrid, however, was considered to be an indication of the good potential for gene exchange and random chromosome assortment in backcross generations. Doughty (quoted by Bai 1987) found two univalents occasionally in F_1 hybrids of the same combination. Bai (1983) found variation in pollen fertility from zero to over 70% in F_1 hybrids between cassava and *M. epruinosa* (recorded as *M. dichotoma*). Plants which had complete pollen sterility exhibited complete asynapsis of chromosomes at M-I, whereas those which had more than 70% pollen fertility showed normal meiosis with 18 bivalents at M-I. Nichols (1947) recorded some problems with partial or complete pollen and ovule sterility in cassava x *M. dichotoma* F_1 hybrids and subsequent generations. Nevertheless, sterility of F_1 hybrids has usually not been a major obstacle in interspecific hybridization between *Manihot* species.

The Genus *Dioscorea*

The genus *Dioscorea* contains about 600 species but edible yams are found mainly in the following 10 species: *D. rotundata* Poir., *D. cayenensis* Lam., *D. alata* L., *D. dumetorum* Pax., *D. hispida* Dennst., *D. esculenta* (Lour.) Burk., *D. bulbifera* L., *D. opposita* Thunb., *D. japonica* Thunb., and *D. trifida* L. (Coursey 1967). Of these species, the first four species are economically the most important in Africa as a source of carbohydrate in human diets. By far the most important of these is white yam, *D. rotundata*, which is native to West Africa but is unknown in the wild state. It has been speculated that white yam probably developed in cultivation from *D. praehensilis* Benth.; some workers regard it as merely a subspecies of *D. cayenensis* (Coursey 1967). Akoroda and Chheda (1983) reviewed the published evidence of differences between *D. rotundata* and *D. cayenensis* and provided more data in support of the separate species concept. However, preliminary results from ongoing research at IITA, involving analysis of chloroplast DNA, do not support this finding (Terauchi, Paper 5.5, pp 255-260 this volume). *D. cayenensis* is also native to West Africa and can be found in the wild state as well as in cultivation. *D. alata* and *D. dumetorum* are widespread in distribution. The former is unknown in the wild but the latter has many wild forms that are extremely poisonous, probably because of the presence of the alkaloid dihydrodioscorine (Coursey 1967).

Chromosome counts on wild and cultivated species of *Dioscorea* have shown $2n = 20$ as normal for the primitive section, Stenophora, and $2n = 40$ as the most usual for the other sections, including the food yams (Coursey 1967). Higher ploidies occur, based on multiples of 10, in the Old World species. Furthermore, wide variations in chromosome number have been reported within species of the cultivated forms (Miege 1954; Martin and Ortíz 1963). Such variations are often attributed to the long history of yam cultivation through vegetative propagation. The few reports on the cytogenetics of yam have not yet established the genomes of the different species or what homologies or homoeologies might exist among their chromosomes.

In spite of the abundance of species in the genus *Dioscorea*, documentation of efforts at interspecific hybridization is limited. This is not too surprising when one considers the difficulties inherent even in intraspecific hybridization in yam. Poor and erratic flowering, coupled with asynchronization of male and female flowering, has frustrated many a breeder. Sexual seed production is influenced by climatic conditions and pollination technique (Akoroda 1983a), as well as a number of plant characteristics (Akoroda et al. 1984; Akoroda 1985a). Stagger planting at the Central Tuber Crops Research Institute (CTCRI) at Trivandrum, India has been partially successful in achieving synchronization of flowering (Bai and Jos 1986). At IITA, further success has been obtained through the siting of crossing blocks at locations with environmental factors conducive to flowering. Although storage of yam pollen for over a year has been demonstrated (Akoroda 1983b), the procedure has not been applied effectively in breeding programmes. Furthermore, the collection, characterization, evaluation, and documentation of wild yams has not been carried out sufficiently to enable their full potential to be appreciated.

Akoroda (1985b) reported 46% fruit set from a *D. rotundata* x *D. praehensilis* cross leading to viable seeds. His crosses, involving *D. rotundata* and different accessions of *D. cayenensis*, were less successful — fruit set of 2.0-8.6% and seedless except for the combination involving the use of pollen bulked from four accessions of *D. cayenensis*. He suggested a mismatch of chromosome numbers as a possible reason for this relatively low fruit set and seed fertility. From their crossing of 700 flowers of *D. pentaphylla* with *D. bulbifera*, Jos and Bai (1980) did not obtain any fruits. However, reciprocal crosses between *D. alata* and *D. rotundata* resulted in fruit set of 5.2-6.3% (CTCRI 1986). No information was provided on the viability of the resulting seeds.

The possibility of cross-incompatibilities among species has not been adequately tested This could easily result from differences in ploidies. The establishment of suitable genetic markers would facilitate confirmation of hybrids.

Conclusion

As increase in population and in competition for land call for innovations in food production technology, there is an urgent need to protect and broaden the genetic base of traditional staples, such as yam and cassava, in order to ensure food security well into the future. Interspecific hybridization in the genera *Manihot* and *Dioscorea* may seem futuristic to some researchers, but it is essential to place the right emphasis on it now so that the desired benefits will be available to meet the exigencies that are already too imminent to be ignored.

Discussion

ASIEDU: The wild relatives have advantages and disadvantages. Naturally, the cultivated species has been selected for certain traits and an addition of new genes could upset the balance. This is why we go through several cycles of backcrossing to get rid of the unwanted genetic material of the donor parent.

References

Abraham, A. 1957. Breeding of tuber crops in India. *Indian Journal of Genetics and Plant Breeding* 17: 212-217.

Akoroda, M.O. 1983a. Floral biology in relation to hand pollination of white yam. *Euphytica* 32: 831-838.

Akoroda, M.O. 1983b. Long-term storage of yam pollen. *Scientia Horticulturae* 20: 225-230.

Akoroda, M.O. 1985a. Sexual seed production in white yam. *Seed Science and Technology* 13: 571-581.

Akoroda, M.O. 1985b. Pollination management for controlled hybridization of white yam. *Scientia Horticulturae* 25: 201-209.

Akoroda, M.O., and H.R. Chheda. 1983. Agro-botanical and species relationships of Guinea yams. *Tropical Agriculture (Trinidad)* 60: 242-248.

Akoroda, M.O., J.E. Wilson, and H.R. Chheda. 1984. The association of sexuality with plant traits and tuber yield in white yam. *Euphytica* 33: 435-442.

Asiedu, R., S.K. Hahn, K.V. Bai, and A.G.O. Dixon. 1989. Introgression of genes from wild relatives into cassava. Paper presented at the 4th Triennial Symposium of the International Society for Tropical Root Crops—Africa Branch, 4-8 December 1989, Kinshasa, Zaire.

Bai, K.V. 1983. Cytogenetics of tuber crops. Pages 114-116 in *Annual Report for 1982*. IITA, Ibadan, Nigeria.

Bai, K.V. 1987. Recent advances in cassava genetics and cytogenetics. Pages 35-49 in *Cassva Breeding: A Multidisciplinary Review* edited by C.H. Hershey. Proceedings of a Workshop in the Philippines, 4-7 March 1985. CIAT, Cali, Columbia.

Bai, K.V., and J.S. Jos. 1986. Female fertility and seed set in *Dioscorea alata* L. *Tropical Agriculture (Trinidad)* 63: 7-10.

Bolhuis, G.G. 1953. A survey of some attempts to breed cassava varieties with a high content of proteins in the roots. *Euphytica* 2: 107-112.

Bolhuis, G.G. 1967. Intra and interspecific crosses in the genus *Manihot*. Pages 81-88 in *Proceedings, International Symposium on Tropical Root Crops (Vol. 1), St Augustine, Trinidad, 2-8 April 1967.*

Bryne, D.H. 1984. Breeding cassava. Pages 73-134 in *Plant Breeding Reviews* (Vol. 2) edited by J. Janick. AVI, Westport, Connecticut, USA.

Cours, G. 1951. Le manioc à Madagascar. *Mémoires de l'Institut Scientifique Madagascar Serie B. Biologie Vegetale* 3: 203-400.

Coursey, D.G. 1967. *Yams*. Longmans, London, UK.

CTCRI . 1986. *Annual Report 1986*. CTCRI, Trivandrum, India.

Grattapaglia, D., N.M.A. Nassar, and J.C. Dianese. 1987. Biossistematica de especies brasileiras do gennero *Manihot* baseada em padroes de proteina da semente. *Ciencia e Cultura* 39: 294-300.

Hahn, S.K., J.C.G. Isoba, and T. Ikotun. 1989. Resistance breeding in root and tuber crops at the International Institute of Tropical Agriculture (IITA), Nigeria. *Crop Protection* 8: 147-168.

Hahn, S.K., K.V. Bai, and R. Asiedu. 1990. Tetraploids, triploids and 2n pollen from diploid interspecific crosses with cassava. *Theoretical and Applied Genetics* 79: 433-439.

Jennings, D.L. 1959. *Manihot melanobasis* Muell. Agr. — A useful parent for cassava breeding. *Euphytica* 8: 157-162.

Jennings, D.L. 1976. Breeding for resistance to African cassava mosaic disease. Pages 39-44 in *Report of an Interdisciplinary Workshop, 19-22 February 1976, Mugaga, Kenya* edited by B.L. Nestel. IDRC-171e. IDRC, Ottawa, Canada.

Jos, J.S., and K.V. Bai. 1980. Cytogenetics of *Dioscorea*: Intervarietal hybridisation. Page 128 in *CTCRI Annual Report 1980*. CTCRI, Trivandrum, India.

Magoon, M.L., J.S. Jos, and S.G. Appan. 1966. Cytomorphology of the interspecific hybrid between cassava and ceara rubber. *Chromosome Information Service (Japan)* 7: 8-10.

Magoon, M.L., R. Krishnan, and K.V. Bai. 1970. Cytogenetics of the F_1 hybrid between cassava and ceara rubber and its backcross. *Genetica* 41: 425-436.

Martin, F.W., and S. Ortíz. 1963. Chromosome numbers and behavior in some species of *Dioscorea*. *Cytologia* 28: 96-101.

Miege, J. 1954. Nombres chromosomiques et repartition geographique de quelques plantes tropicales et equatoriales. *Revue de Cytologie et de Biologie Vegetale le Botaniste* 15: 312-348.

Nassar, N.M.A. 1979. Interspecific crosses between cassava and four wild *Manihot* species from central Brazil. *Egyptian Journal of Genetics and Cytology* 8: 175-179.

Nichols, R.F.W. 1947. Breeding cassava for virus resistance. *East African Agricultural and Forestry Journal* 12: 184-194.

Perry, B.A. 1943. Chromosome number and phylogenetic relationships in the *Euphorbiaceae*. *American Journal of Botany* 30: 527-543.

Purseglove, J.W. 1968. *Manihot*. Pages 171-180 in *Tropical Crops. I. Dicotyledons*. Longmans, London, UK.

Rogers, D.J., and S.G. Appan. 1973. *Manihot and Manihotoides* (*Euphorbiaceae*). Flora Neotropica Monograph No. 13. Hafner Press, New York, USA.

2.3

The use of simple biotechnological tools to facilitate plantain breeding[1]

R. Swennen[2], D. Vuylsteke[3] and S.K. Hahn[3]

Laboratory of Tropical Crop Husbandry, Katholieke Universiteit Leuven, Kardinal Mercierlaan 92, 3001 Heverlee, Belgium[2]; International Institute of Tropical Agriculture, PMB 5320, Ibadan, Nigeria[3]

Abstract

Plantains are a staple food crop in sub-Saharan Africa. Black sigatoka, a leaf spot disease, is the principal cause of current yield losses. Only 3 years after its inception, the plantain breeding programme at the International Institute of Tropical Agriculture (IITA) has produced four black sigatoka-resistant plantain hybrids by crossing the triploid plantains with a wild diploid. A few simple biotechnological tools, such as introduction of germplasm through in vitro meristem culture, in vitro micropropagation, and embryo rescue culture, were pivotal in this achievement.

Bananas are giant perennial herbs which can be grouped according to their fruit development into cultivated species and wild species. The wild species are diploids and produce small, almost completely pulp-free, fruits weighing between 30 g and 80 g. These fruits are usually filled with black or brown stony seeds 3-5 mm in diameter. The seeds germinate fairly easily and thus wild species of banana are well adapted to the jungle ecology of south-eastern Asia, where some 30 to 40 species thrive (Simmonds 1962). Parthenocarpy (that is, fruit development without the need for any stimulus by pollination) evolved in the species *Musa acuminata*, whose genomic constitution is designated by the letter 'A'. Because they are diploids, they are designated as AA bananas. Originally, AA parthenocarpic bananas produced seeds upon pollination; however, over time, man selected the ones that set only a few seeds, or none at all, after pollination and started to cultivate edible bananas which

1 Contribution No. IITA/91/CP/21 from the International Institute of Tropical Agriculture, PMB 5320, Ibadan, Nigeria.

were, and are, parthenocarpic and sterile, the result of two independent selection processes (Simmonds 1960).

The species *M. acuminata* is composed of several subspecies which co-exist in some regions (Simmonds 1962) and where they readily hybridize. As a result of the combined effects of hybridizations among the *M. acuminata* subspecies, the hybridizations between *M. acuminata* and *M. balbisiana* (which provides the B genome) and hybridizations with the parthenocarpic AA diploids, triploids of AAA, AAB, and ABB genomic constitution evolved (Simmonds and Shepherd 1955). Triploids developed because of restitution, followed by haploid pollination (Wilson 1946).

Triploids were favoured over diploids because of their higher productivity, induced by higher vigour and fruit growth rate (Simmonds 1948, 1953). Most importantly, triploidy conferred gametic sterility and thus these larger parthenocarpic fruits were seedless, an important asset (Simmonds 1962). Some tetraploids originated from restitution in these triploids, followed by haploid pollination. These tetraploids are very rare.

As a result of this selection process imposed by man, less than half the banana clones known today are diploid, the rest being primarily triploid. In cultivation, however, triploid plants outnumber diploid plants by 100 or more to 1, perhaps by several hundreds to one (Simmonds 1962). Thus, the bananas that provide one of the foremost basic staple foods to large populations in the humid and subhumid tropical regions of the world are triploids. The production of these triploids is now threatened by the spread of black sigatoka, a virulent fungal leaf spot disease caused by *Mycosphaerella fijiensis* (Meredith and Lawrence 1970).

Because black sigatoka reduces yield by 30-50% (Stover 1983), it poses a threat to all known plantain cultivars in Africa (Swennen and Vuylsteke 1991). About 60 million people in West Africa alone derive more than 25% of their carbohydrate intake from plantain (Wilson 1987). The bulk of these bananas are cultivated in compound gardens or backyards, and thus pesticide control of black sigatoka would impose a serious health hazard. In addition, pesticides are very expensive and beyond the reach of the subsistence farmer. Consequently, these farmers need access to resistant cultivars if the current decline in production is to be halted.

Breeding for Black Sigatoka Resistance in Plantain

All plantains are triploids of AAB genomic constitution and none is resistant to black sigatoka. Sources of resistance to this disease have been identified in diploids and other triploids (Swennen and Vuylsteke 1991). Because of gametic sterility, the susceptible but preferred bananas cannot be crossed with these resistant triploids. However, they can be crossed with the resistant diploids which are either parthenocarpic or wild. This cross is obviously a wide one.

At the International Institute of Tropical Agriculture (IITA) four black sigatoka-resistant plantains were produced recently from wide crosses of two African plantains and a wild banana (Swennen and Vuylsteke, in press) (Figure 1). This was achieved within 3 years of the initiation of the breeding programme. This is most surprising, since banana breeding started about 70 years ago and so far no hybrid bred by man had been found acceptable (Rowe 1984).

This paper discusses the considerable impact some simple biotechnological tools have had on the success and speed of the IITA plantain breeding programme.

Figure 1 **Black sigatoka-resistant plantain hybrid 548/4/78 (centre) obtained by crossing the plantain cultivar Obino l'ewai (left) with the wild diploid Calcutta 4 (right)**

Factors in the Improvement of Breeding Efficiency

Procurement of germplasm

IITA's plantain breeding programme is located in Nigeria where nearly 20 different plantain cultivars are cultivated (Swennen 1990). All these cultivars are susceptible to black sigatoka and until recently no sources of resistance to the disease were available. This germplasm was clearly insufficient to breed plantains for black sigatoka resistance. In addition, plantains were reputed to be highly sterile (De Langhe 1976). Hence, there was an urgent need to introduce sources of resistance to black sigatoka and as many plantain cultivars as possible that might show workable levels of female fertility.

Over the past 5 years, IITA has introduced 320 accessions of the *Musa* genus from all parts of the world, among which are 93 plantain cultivars and nearly 40 black sigatoka-resistant accessions. All accessions were introduced as in vitro propagules through the International Network for Improvement of Bananas and Plantains (INIBAP) Transit Centre located at the Katholieke Universiteit Leuven in Belgium (Vuylsteke et al. 1990a). The main advantages of in vitro propagules are ease of handling, highly reduced bulkiness, procurement of disease-free material, and the high speed of release of the material by the plant quarantine service.

Identification of seed producing plantains

The plantain collection was systematically screened for seed-producing ability (female-fertility). Freshly exposed female flowers of all plantain clones were pollinated in the morning with freshly exposed pollen of "Calcutta 4" (Figure 2). Calcutta 4 is a wild black sigatoka-resistant diploid belonging to *M. acuminata* ssp. *burmannicoides* (De Langhe and Devreux 1960). Thirty-one female-fertile plantains were identified (Swennen and Vuylsteke, submitted), of which most cultivars came from outside Nigeria. The plantain cultivar with the highest fertility was introduced from Cameroon.

Figure 2 Pollination of female plantain flowers at anthesis

Large-scale multiplication of parents

Screening for female fertility in plantain and for black sigatoka resistance in wild and edible diploids was carried out in the *Musa* species collection, where each accession was represented by five plants. Once a plantain cultivar that had a workable level of female fertility was identified, it was rapidly micropropagated using standard aseptic shoot-tip culture techniques (Vuylsteke and De Langhe 1985; Vuylsteke 1989). Similarly, black sigatoka-resistant diploids were quickly multiplied. Thus, within 1 year after discovering suitable germplasm, several thousand female and male parents were established in the field, allowing large-scale hybridizations.

Season-dependent seed production and planting schedule

Seed production from crosses between female fertile plantains and diploids varied with the season (Swennen and Vuylsteke 1990). In some crosses it varied up to 20-fold between pollinations done at the end of the dry season (maximum seed production) and those done early in the rainy season (minimun seed production). Thus, the schedule of successive plantings was adjusted to promote flowering during the peak of seed set. In contrast to the experience with conventional planting material, this could be readily achieved with in vitro plants as such material flowers very uniformly after a defined period of vegetative growth.

Seed germination

It is well known that seeds from artificial crosses in banana usually take several months to germinate and that the percentage of germinating seeds is very low. An abnormal embryo-endosperm relationship seems to be involved (Simmonds 1962). As thegermination rate in at Onne was about 1%, embryos were excised aseptically and cultured in vitro (Vuylsteke et al. 1990b). The embryo rescue technique increased the germination rate to 12%. Viable embryos responded within 2 weeks of culture initiation and developed into normal plantlets through the use of routine in vitro procedures (Vuylsteke and De Langhe 1985).

Distribution of resistant hybrids

Among the plantain hybrids obtained from the wide crosses mentioned, four showed very high levels of resistance to black sigatoka. Moreover, the yield of these hybrids was up to 125% higher than the fungicide-treated plantain from which they had originated (Swennen and Vuylsteke, in press). This promising material needs to be multiplied rapidly and distributed for clonal testing under a range of ecological conditions. In vitro micropropagation and distribution would obviously be the safest and fastest method.

Discussion

ENE-OBONG: It seems to me, from most papers presented during this session, that we are not making any attempt to control the agronomic quality of these crop (banana/plantain) hybrids. Is not the introduction of wild genepools into our crops, without evaluation, harmful? What about evaluation in terms of yield and quality performance, toxicity, etc.? The problem of acceptability has already been mentioned. How did IITA take care to avoid the introduction of viruses to Nigeria when introducing *Musa* germplasm?

SWENNEN: IITA relied on the following steps: (1) collection of apparently virus-free germplasm; (2) shipment of germplasm to the INIBAP Transit Centre at the Katholieke Universiteit Leuven in Belgium, a non-banana producing country; (3) germplasm sent by Transit Centre to a virus-indexing centre (in Australia, Philippines, etc); (4) when the virus-indexing centre reports that the material is virus-free, it is sent by the Transit Centre to the requester via the local plant quarantine service.

SHOYINKA: Most hybrids obtained from interspecific hybridization of wild species do not have the required consumer acceptability. I am particularly interested in the consumer response to the hybrid obtained from crossing plantain and banana, especially when it is fried; most Africans, particularly Nigerians, prefer the fried form to the cooking type.

References

De Langhe, E. 1976. Pourquoi l'amélioration génétique du plantain n'est-elle pas actuellement réalisable? *Fruits* 31: 537-539.

De Langhe, E. and M. Devreux. 1960. Une sous-espéce nouvelle de *Musa acuminata* Colla. *Bulletin du Jardin Botanique de Bruxelles* 30: 375-388.

Meredith, D., and J. Lawrence. 1970. Black leaf streak disease of bananas (*Mycosphaerella fijiensis*): Susceptibility of cultivars. *Tropical Agriculture (Trinidad)* 47: 275-288.

Rowe, P. 1984. Breeding bananas and plantains. Pages 135-155 in *Plant Breeding Reviews* edited by J. Janick. AVI Publishing Co., Westport, Connecticut, USA.

Simmonds, N. 1948. Genetical and cytological studies of *Musa*. X. Stomatal size and plant vigour in relation to polyploidy. *Journal of Genetics* 49: 57-68.

Simmonds, N. 1953. The development of banana fruit. *Journal of Experimental Botany* 4: 87-105.

Simmonds, N. 1960. Experiments on banana fruit development. *Annals of Botany* 24: 212-222.

Simmonds, N. 1962. *The Evolution of the Bananas.* Longmans Green, London, UK.

Simmonds, N., and K. Shepherd. 1955. The taxonomy and origins of cultivated banana. *Linnaean Society Botanical Journal* 55: 302-312.

Stover, R. 1983. Effet du cercospora noir sur les plantains en Amérique centrale. *Fruits* 38: 326-329.

Swennen, R. 1990. Limits of morphotaxonomy: Names and synonyms of plantain in Africa and elsewhere. Pages 172-210 in *Proceedings, International Workshop on Identification of Genetic Diversity in the Genus Musa, 5-10 September 1988, Los Baños, Philippines.* INIBAP, Montpellier, France.

Swennen, R., and D. Vuylsteke. 1990. Aspects of plantain breeding at IITA. Pages 252-266 in *Proceedings, International Workshop on Sigatoka Leaf Spot Diseases (Mycosphaerella spp.), 28 March-1 April 1989, San José, Costa Rica.* INIBAP, Montpellier, France.

Swennen, R., and D. Vuylsteke. 1991. Bananas in Africa: Diversity, uses, and prospects for improvement. Pages 151-159 in *Crop Genetic Resources of Africa* edited by N.Q. Ng, P. Perrino, F. Attere, and H. Zedan. IITA, Ibadan, Nigeria.

Swennen, R. and D. Vuylsteke. (in press). Breeding black sigatoka resistant plantains with a wild banana. *Tropical Agriculture (Trinidad).*

Vuylsteke, D. 1989. *Shoot-Tip Culture for the Propagation, Conservation, and Exchange of Musa Germplasm.* Practical Manuals for Handling Crop Germplasm In Vitro. 2. IBPGR, Rome, Italy.

Vuylsteke, D., and E. De Langhe. 1985. Feasibility of in vitro propagation of bananas and plantains. *Tropical Agriculture (Trinidad)* 62: 323-328.

Vuylsteke, D., J. Schoofs, R. Swennen, G. Adejare, M. Ayodele, and E. De Langhe. 1990a. Shoot-tip culture and third-country quarantine to facilitate the introduction of new *Musa* germplasm into West Africa. *FAO/IBPGR Plant Genetic Resources Newsletter* 81/82: 5-11.

Vuylsteke, D., R. Swennen, and E. De Langhe. 1990b. Tissue culture technology for the improvement of African plantains. Pages 316-337 in *Proceedings, International Workshop on Identification of Genetic Diversity in the Genus Musa, 5-10 September 1988, Los Baños, Philippines.* INIBAP, Montpellier, France.

Wilson, G. 1946. Cytological studies in the Musae. I. Meiosis in some triploid clones. *Genetics* 31: 241-258.

Wilson, G. 1987. Status of bananas and plantains in West Africa. Pages 29-35 in *Banana and Plantain Breeding Strategies: Proceedings of an International Workshop, Cairns, Australia, 13-17 October 1986* edited by G.J. Persley and E.A. De Langhe. ACIAR Proceedings No. 21. ACIAR, Canberra, Australia.

2.4

Wide crosses of *Vigna* food legumes[1]

N.Q. Ng

International Institute of Tropical Agriculture, PMB 5320, Ibadan, Nigeria

Abstract

Many successful wide crosses among Asiatic cultivated *Vigna* species and between them and a wild species, *V. trilobata*, have been reported However, despite many attempts to cross the African cultivated *Vigna* species with wild *Vigna* species and the Asiatic cultivated *Vigna* species, there has been no success as yet. Nevertheless, crosses between cultivated and wild subspecies of *V. unguiculata* were fairly easy and their hybrids are fertile. Studies conducted at the International Institute of Tropical Agriculture (IITA) have shown that crossing barriers between *V. unguiculata* and several African wild *Vigna* species were due to abortion of young embryos and pollen-pistil incompatibility. With more basic research on reproductive biology and tissue culture, and a larger number of crosses among species and genotypes, there will be a greater chance of success for wide crosses between *V. unguiculata* and wild *Vigna* species which possess some very desirable traits for the improvement of this cultivated species.

Seven species in the genus *Vigna* (Savi) Verdc. are cultivated as food legumes and constitute an important part of the human diet in many parts of Asia and Africa. Five of these originated in Asia and are classified under the subgenus *Ceratotropis*: mung bean, *V. radiata* (L.) R. Wilizek; black gram, *V. mungo* (L.) Hepper; rice bean, *V. umbellata* (Thunb.) Ohwi and Ohashi ; adzuki bean,*V. angularis* (Wild) Ohwi and Ohasi; and moth bean, *V. aconitifolia* (Jacq.) M.M. and S. The other two species are indigenous to Africa and are classified under the genus *Vigna* but placed in different sections: cowpea, *V. unguiculata* (L.) Walp, in the section *Catiang*; and bambara groundnut,*V. subterranea* (L.) Verdc., in the section *Vigna*.

Through hybridization between cultivars which differ genetically and subsequent selection from the hybrid populations, plant breeders have been very effective in increasing the production of cultivated food crops. Conventional plant breeding usually deals with

---

1 Contribution No. IITA/91/CP/22 from the International Institute of Tropical Agriculture, PMB 5320, Ibadan, Nigeria.

crosses between cultivars within a species or between a cultivar and its close relatives, which are easy to cross. However, if gene(s) conferring specific traits are not available within a cultivated species, breeders may explore the possibility of obtaining these from a wider genepool — other species or genera. This is the case with cowpea, which is highly susceptible to several flowering and post-flowering pests such as flower bud thrips, legume pod borer, and pod-sucking bug. Good sources of resistance to these pests have not been found in cultivated cowpea (IITA 1988b), but sources of resistance to the legume pod borer, pod-sucking bugs, and bruchid have been identified in many wild species of *Vigna* (N.Q. Ng and colleagues, IITA, unpubl.; Jackai and Singh 1990). Scientists at the International Institute of Tropical Agriculture (IITA) are investigating the possibility of exploiting the resistance in wild *Vigna* species for cowpea improvement. However, the introduction of resistant genes from wild species to cowpea is constrained by barriers to crossing between the species. The degree of genetic affinity between cowpea and wild *Vigna* species is unclear; a better understanding of this would enhance the use of the *Vigna* genepool.

It is in this context that the work on wide crosses within *Vigna* and experiments on interspecific hybridization between cowpea and wild *Vigna* species is reviewed here.

Wide Crosses: Definition and Constraints

The terms "wide cross" and "interspecific hybridization" are often used interchangeably with reference to crosses between different species. "Wide cross" is used here in a more restricted sense, that is, a cross between species or genera whose hybrids are sterile and in which the chromosomes from the parents do not pair (Shebeski 1983). Given this definition, crosses between a cultivated species and species belonging to its primary genepool, as defined by Harlan and de Wet (1971), are not considered as wide crosses because they can be intercrossed as easily as the cultivars within a species and their hybrids are fertile. Crosses between a cultivated species and another species in its secondary or tertiary genepool, which require considerable effort to overcome crossing barriers for the transfer of genes from those species to the cultivated species, are wide crosses.

The breeder's interest in making wide crosses has been limited by difficulties in introducing desirable genes from distant species to a recipient species, which involve crossing incompatibility, problems in raising the F_1 hybrids, and problems in restoring fertility and removing undesirable genes of the donor species from the hybrids. Pollination may be ineffective and fertilization may not occur. Several techniques, such as mutilation of the stigma, application of a hormone to flower buds, bud pollination, bridging species, and in vitro fertilization, may be effective in overcoming pollen-pistil incompatibility.

Even if fertilization occurs, the embryo may abort before a mature seed is produced. To raise a hybrid plant, embryo culture may be required. When a hybrid plant is produced, it may show the phenomenon of F_1 weakness — very weak plants which die prematurely or completely sterile plants. Fertility of F_1 hybrids can be restored by repeated backcrossing to the recipient parent, or by inducing the doubling of chromosomes.

Recent advances in biotechnology in the application of cell fusion techniques, followed by regeneration, have offered another opportunity for the recombination of genomes of a cultivated species with those of distantly related species and in transferring a fragment of chromosome from an alien species to a cultivated species. This area is not included in the present discussion.

Wide Crosses in Asiatic Cultivated *Vigna*

Several studies have been carried out on interspecific hybridization among the Asiatic cultivated *Vigna* species and between them and some wild *Vigna* species (Dana 1966a; De and Krishnan 1966; Biswas and Dana 1975; Chen et al. 1977, 1983; Chowdhury and Chowdhury 1977; Ahn and Hartmann 1978a, 1978b, 1978c). Successful wide crosses have been obtained for the following combinations (female x male):

V. radiata x *V. mungo* (Dana 1966a; Ahn and Hartmann 1978a; Chen et al. 1983)
V. radiata x *V. umbellata* (Dana 1967; Ahn and Hartmann 1978a; Chen et al. 1983)
V. radiata x *V. angularis* (Ahn and Hartmann 1978a, 1978b; Chen et al. 1983)
V. mungo x *V. angularis* (Ahn and Hartmann 1978a, 1978b; Chen et al. 1983)
V. mungo x *V. umbellata* (Biswas and Dana 1975; Ahn and Hartmann 1978c)
V. umbellata x *V. radiata* (Sawa 1974; Biswas and Dana 1975)
V. radiata x *V. trilobata* (Dana 1966b)
V. aconitifolia x *V. trilobata* (Biswas and Dana 1976)

From these studies it was evident that there was a maternal effect on crossing compatibility. All the crosses reported above, except *V. radiata* x *V. umbellata*, were successful only in one direction. Furthermore, almost all the crosses required embryo culturing to rescue the hybrids, and if the hybrids were successfully grown, they showed F_1 weakness. All those hybrids that grew to maturity were highly or completely male sterile, with pollen fertility (stainability) of less than 10%.

Restoring the Fertility of Hybrids

Hybrid sterility in wide crosses is usually the result of chromosomal, rather than genic, causes. Successive backcrossing of F_1 hybrids to either parent can, in most cases, restore the fertility of the hybrids, thus transferring the desired gene(s) or segments of chromosome to the recipient species. Amphidiploidy or inducing the doubling of chromosomes in the hybrids can also be effective in restoring the hybrid fertility. In *Vigna* species wide crosses, inducing the doubling of chromosomes by the application of colchicine effectively restored the fertility of hybrids, from less than 10% to about 80%, of the following crosses: *V. radiata* x *V. mungo* (Dana 1966a); *V. radiata* x *V. umbellata* (Sawa 1974); *V. radiata* x *V. trilobata* (Dana 1966b); and *V. aconitifolia* x *V. trilobata* (Biswas and Dana 1976).

Through allopolyploidy, scientists can increase the genetic diversity of a cultivated species with genetic material from another species. However, the results of these studies on Asiatic *Vigna* species have not been exploited for crop improvement.

Wide Crosses in African Cultivated *Vigna*

Many attempts have been made to cross *V. unguiculata* with other *Vigna* species, but no success has yet been reported (Faris 1963; Evan 1976; Mithen 1987; IITA 1988b; Ng 1990). Putative "hybrid seeds" have been obtained from crosses between *V. luteola* and cowpea (S.R. Schnapp, 1990, pers. comm.) but this has not been confirmed. However, successful crosses have been made by the author (unpubl.) between *V. luteola* and *V. oblongifolia* and have been reported by Schnapp et al. (1990).

In contrast, recent interspecific hybridization conducted at IITA has shown that, with few exceptions, most varieties of wild relatives of cowpea from diverse geographical regions and belonging to *V. unguiculata* ssp. *dekindtiana* (including var. *dekindtiana*, var. *pubescens* [syn. *V. pubescens* R. Wilczek], and var. *protracta*), ssp. *tenuis*, and ssp. *stenophylla* can readily intercrossed with cowpea (*V. unguiculata* ssp. *unguiculata*), and the hybrids are fertile (Ng 1990). This primary genepool of cowpea is large and very diverse and, when properly evaluated and exploited, it has the potential to revolutionize the production of the crop by conferring resistance to pests. In fact, potential sources of resistance to the legume pod borer and the brown cowpea coreid bug have been found in wild taxa (N.Q. Ng and colleagues, unpubl.; Jackai and Singh 1990); these should be exploited for cowpea improvement. IITA's breeders have already produced crosses between cultivated cowpea and wild cowpea (*V. unguiculata* ssp. *dekindtiana* var. *dekindtiana*) which are resistant to the pod borer and pod-sucking bugs.

IITA scientists are also interested in wide crosses between cultivated cowpea and wild *Vigna* species, particularly *V. vexillata* and others that have been shown in laboratory and field tests to exhibit a high level of resistance to pests (IITA 1988a). Crosses made at IITA and elsewhere between cowpea and *V. vexillata* and many other *Vigna* species have not been successful (IITA 1989; Ng 1990). F_1 pod setting of the crosses between *V. vexillata* and cowpea was high compared with other *Vigna* species that have been used. Some pods were retained for up to 12 days after pollination before abscissing.

Detailed histological studies on pollen tube and embryo development, following selfing in cowpea and *V. vexillata* and in interspecific crosses between them, have shown pollen-pistil incompatibility (Barone and Ng 1990; IITA 1989). Nevertheless, some pollen tubes did develop normally and reach the ovules, which were fertilized. Unfortunately, hybrid embryos of the crosses never developed to the heart-shaped or torpedo stage. Hybrid embryos developed only to the globular stage before collapsing (Barone and Ng 1990). In several other crosses between cowpea and wild *Vigna* (*V. leuteola* and *V. oblongifolia*) fertilization occurred, but the young hybrid pods aborted within 4 days of pollination (Agwaranze and Ng, unpubl.). Endosperm degeneration could have been caused by mitotic aberrations or other known factors.

Future Prospects for Wide Crosses in Cowpea

Given our present knowledge of wide crosses in *Vigna* species, the following generalizations can be made: first, there is a strong maternal influence on crossability; and second, there are species that can serve as bridge species, providing a means of transferring genes of interest between two species that normally do not cross. For example, *V. umbellata* and *V. trilobata* do not hybridize, but *V. radiata* can be successfully crossed with both *V. umbellata* and *V. trilobata*. Thus, *V. radiata* could serve as a bridge species to transfer genes in the wild species *V. trilobata* to *V. umbellata*. One should, therefore, seek genotypes that are more compatible with one another, thus identifying potential bridge species. A larger number of crosses should be made to provide a greater chance of success. For example, Thomas and Pratt (1982) reported that only one out of 10 000 hybrid embryos of crosses between cultivated tomato (*Lycopersicon esculentum*) and the wild species, *L. peruvianum*, could be rescued sufficiently to be developed into a plant on culture media. Similarly, Jan et al. (1982) found that only 22 out of 50 000 wheat ovaries pollinated by barley were true hybrids. Successful crosses were achieved, however, mainly because of advances in embryo culture techniques and the very large number of hybridizations made.

If different techniques are tried, more species and genotypes used, and large numbers of crosses made, it should be possible to find wild *Vigna* species other than wild *V. unguiculata* that are crossable with cultivated cowpea. In order to tap the wider genepool of wild *Vigna* for use in cowpea improvement, more basic research on reproductive biology, tissue culture, and biosystematics of *Vigna* is needed.

Discussion

HASEGAWA: What is your assessment of the potential of research on interspecific hybridization involving cowpea to incorporate insect resistance?

NG: A large diversity of wild *Vigna unguiculata* (including ssp. *dekindtiana*, ssp. *tenuis*, ssp. *steaophylla*, and other variants within this group) has been assembled and is available for use by breeders. Screening has shown that they are potential sources of resistance to some major cowpea pests. The diversity of this wild species should be thoroughly evaluated and the subspecies used in breeding for pest resistance, since they can hybridize easily with cowpea. Other wild *Vigna* species represent a wider genepool which can be tapped for cowpea improvement, because of their resistance to important pests such as cowpea bruchid, pod borers, and pod-sucking bugs. However, more basic research in cytology, embryo development following interspecific hybridization, embryo rescue, and other techniques needs to be continued and strengthened. More interspecific hybridization should be attempted. The potential of wild *Vigna* for cowpea improvement is high.

References

Ahn, C.S., and R.W. Hartmann. 1978a. Interspecific hybridization among four species of the genus *Vigna*. Pages 240-246 in *Proceedings, First International Mung Bean Symposium*. Asian Vegetable Research and Development Centre, Shunhua, Taiwan.

Ahn, C.S., and R.W. Hartmann. 1978b. Interspecific hybridization between mung bean (*Vigna radiata* (L.) Wilczek) and adzuki bean (*Vigna angularis* (Wild.) Ohwi and Ohashi). *Journal of the American Society for Horticultural Science* 103: 3-6.

Ahn, C.S., and R.W. Hartmann. 1978c. Interspecific hybridization between rice bean (*Vigna umbellata* (Thunb.) Ohwi and Ohashi) and adjuzuki bean (*V. angularis* (Wild.) Ohwi and Ohashi). *Journal of the American Society for Horticultural Science* 103: 435-438.

Barone, A., and N.Q. Ng. 1990. Embryological study on crosses between *Vigna unguiculata* and *V. vexillata*. Pages 151-160 in *Cowpea Genetic Resources* edited by N.Q. Ng and L.M. Monti. IITA, Ibadan, Nigeria.

Biswas, M.R., and S. Dana. 1975. Black gram x rice bean cross. *Cytologia* 40: 787-795.

Biswas, M.R., and S. Dana. 1976. *Phaseolus aconitifolius* x *P. trilobus*. *Indian Journal of Genetics and Plant Breeding* 36: 125-131.

Chen, N.C., J.F. Parrot, T. Jacobs, L.R. Baker, and P.S. Carlson. 1977. Interspecific hybridization of food legumes by unconventional methods of plant breeding. Pages 247-252 in *Proceedings, First International Mung Bean Symposium.* Asian Vegetable Research and Development Centre, Shanhua, Taiwan.

Chen, N.C., L.R. Baker, and S. Honma. 1983. Interspecific crossability among four species of *Vigna* food legumes. *Euphytica* 32: 925-937.

Chowdhury, R.K., and J.B. Chowdhury. 1977. Intergeneric hybridization between *Vigna mungo* (L.) Hepper and *Phaseolus calcaratus* Roxb. *Indian Journal of Agricultural Science* 47: 117-121.

Dana, S. 1966a. The cross between *Phaseolus aureus* Roxv. and *P. mungo* L. *Genetica* 37: 259-274.

Dana, S. 1966b. Species cross between *Phaseolus aureus* Roxb. and *P. trilobus* Ait. *Cytologia* 31: 176-187.

Dana, S. 1967. Hybrid from the cross *Phaseolus aureux* Roxb. x *P. calcaratus* Roxb. *Journal of Cytology and Genetics* 2: 92-97.

De, D.N., and R. Krishnan. 1966. Cytological studies of the hybrid *Phaseolus aureus* x *P. mungo. Genetica* 37: 588-600.

Evan, A.M. 1976. Species hybridization in the genus *Vigna.* Pages 31-34 in *Proceedings, IITA Collaborator's Meeting on Grain Legume Improvement* edited by R.A. Luse and K.O. Rachie. IITA, Ibadan, Nigeria.

Faris, D.G. 1963. Evidence for the West African origin of *Vigna sinensis* (L.) Savi. PhD thesis, University of California, USA.

Harlan, J.R., and J.M.J. de Wet. 1971. Towards a rational classification of cultivated plants. *Taxon* 20: 509-517.

IITA. 1988a. Host plant resistance of cowpeas to post-flowering insect pests. Pages 70-75 in *IITA Annual Report and Research Highlights 1987/88.* IITA, Ibadan, Nigeria.

IITA. 1988b. *Genetic Resources Unit Annual Report, 1987.* IITA, Ibadan, Nigeria.

IITA.1989. *Genetic Resources Unit Annual Report, 1988.* IITA. Ibadan, Nigeria.

Jackai, L.E.N., and S.R. Singh. 1990. Useful sources of pest resistance among wild *Vigna* species. Paper presented at the Cowpea Biotechnology Workshop, 16-20 July 1990, Purdue University, West Lafayette, USA.

Jan, C.C., C.A. Qualset, and J. Dvorak.1982. Wheat-barley hybrids. *California Agriculture* 36(8): 23-24.

Mithen, R.F. 1987. The African genepool of *Vigna.* I. *V. nervosa* and *V. unguiculata* from Zimbabwe. *FAO/IBPGR Plant Genetic Resources Newsletter* 70: 13-19.

Ng, N.Q. 1990. Recent developments in cowpea germplasm collection, conservation, evaluation and research at the Genetic Resources Unit, IITA. Pages 13-28 in *Cowpea Genetic Resources* edited by N.Q. Ng and L.M. Monti. IITA, Ibadan, Nigeria.

Sawa, M. 1974. On the interspecific hybridization between the adzuki bean, *Phaseolus angularis* (Wild.) W.F. Wight and the green gram, *Phaseolus radiatus* L. I. On the characteristics of amphidiploid in C_1 generation from the cross green gram x rice bean, *Phaseolus calcaratus* Roxb. *Japanese Journal of Breeding* 24: 282-286.

Schnapp, S.R., R.E. Shade, L.W. Kitch, L.L. Murdock, R.A. Bressan, and P.M. Hasegawa. 1990. Utilization of in vitro culture methods to facilitate introgession on insect resistance into cowpea (*Vigna unguiculata*). Page 200 in *Abstracts, 7th International Congress on Plant Tissue and Cell Culture, 24-29 June 1990, Amsterdam, The Netherlands.*

Shebeski, L.1983. The application of wide crosses to plants. Pages 569-576 in *Chemistry and World Food Supplies: The New Frontiers. CHEMRAWN II* edited by L.W. Shemilt. Pergamon Press, Oxford, UK.

Thomas, R., and D. Pratt. 1982. Embryo callus hybrids. *California Agriculture* 36(8): 27.

2.5

Cytogenetics and crop improvement[1]

K.V. Bai and S.K. Hahn

International Institute of Tropical Agriculture, PMB 5320, Ibadan, Nigeria

Abstract

In many crop species, the transfer of agronomically important genes from wild species has been possible through chromosomal manipulations. Haploids have been useful in the production of homozygous lines. Induced polyploids find direct use in crops in which the seed is not the economic produce. As illustrated by triticale and wheat, the development of amphiploids and addition and substitution lines is important in crop improvement. Meiotic modifications resulting in $2n$ gametes appear to be useful potential tools for unilateral/bilateral sexual polyploidization. Coupled with recent advances in several areas, including chromosome banding techniques and isozyme analysis, cytogenetics can play a significant role in crop improvement.

In the past, crop improvement has involved mainly conventional breeding methods that use the genetic variability existing in the cultivars. It is now clear that the enormous variability in the wild relatives of crop species should also be explored for improvement of crop plants. Hence, wide crosses (interspecific and intergeneric) have been attempted in various crop species, either by direct application of cytogenetic techniques or because of increased knowledge of the cytogenetic architecture of the species. Basic information on the cytogenetic architecture of cultivars and on cytogenetic relationships between a cultivated species and its wild relatives is vital to the successful introduction of genetic variation.

Chromosome structural changes, such as translocations, are potential tools for chromosome engineering; their significance in crop improvement has been reviewed by Burnham (1966). Increased knowledge of the cytogenetic aspects of crop plants and related species has resulted in several innovative approaches in crop breeding. The topics discussed in this

---

1 Contribution No. IITA/91/CP/23 from the International Institute of Tropical Agriculture, PMB 5320, Ibadan, Nigeria.

paper include wide crosses, ploidy manipulations, chromosome addition and substitutions, and structural changes in chromosomes.

Wide Crosses

Wide crosses for crop improvement are generally aimed at the transfer of desirable characters, such as resistance to pests and diseases or increase in yield or quality. As the characteristics to be transferred are controlled by gene(s) located on specific chromosomes, the successful integration of desirable characters into the hybrids and their derivatives depends largely on the genome homology and the nature of meiotic chromosome pairing of the species involved.

While numerous crossability barriers prevent successful gene transfer from wild species into cultivars (Stebbins 1958), many of these barriers have been overcome (Khush and Brar 1988) and successful transfer of alien genes into many of the cultivated crops achieved. These include the transfer of genes for mildew and crown rust resistance in oats (Zillinsky and Derick 1960; Aung and Thomas 1976), stem and leaf rust resistance in wheat (Knott 1971; Sharma and Gill 1983), nematode resistance in sugar beet (Savitsky 1975), leaf spot resistance in groundnut (Stalker 1985), increased grain yield in oats (Frey et al. 1984), and higher protein content in wheat (Kushmir and Halloran 1984).

Wild *Manihot* species have been used in cassava breeding programmes for transferring resistance to African cassava mosaic virus (ACMV) (Nichols 1947; Hahn et al. 1980) and cassava bacterial blight (CBB) (Hahn et al. 1989) and for enhancing tuber quality.

Ploidy Manipulations

The ploidy manipulations discussed here involve haploids, polyploids, amphiploids, meiotic modifications, and aneuploids.

Haploids

The term "haploid" is used to describe cells, tissues, and plants with the gametic (n) number of chromosomes. Haploids can arise through wide crosses, semigamy, chromosomal elimination, introduction of alien cytoplasm, and through anther and pollen culture. They occur spontaneously in *Zea mays* (Chase 1949). Wide crosses are the source of haploids in potatoes *(Solanum tuberosum)* wherein tetraploid potato cultivars are pollinated with the diploid species, *S. phureja* (Hougas and Peloquin 1957), and both the maternal and paternal genotypes influence the rate of haploid production (Hougas et al. 1964). In cotton, haploids have been produced through semigamy (Turcotte and Feaster 1974). Haploids have been obtained through elimination of alien chromosomes in interspecific crosses in *Hordeum* (Kasha and Kao 1970; Fedak 1985) and through the influence of alien cytoplasm, as in wheat x *Aegilops caudata* crosses (Tsunewaki et al. 1984). Other sources of haploids are anther culture in *Nicotiana tabacum* (Nitsch and Nitsch 1969; Tanaka and Nakata 1969), polyembryony and anther culture in *Asparagus* (Therenin 1974), and cultured, unfertilized ovaries in crops such as *Nicotiana*, *Hordeum*, *Oryza* (cited in Kasha and Seguin-Swartz 1983) and *Allium* (Campion and Alloni 1990).

Haploids have been successfully used for varietal improvement of barley (Choo et al. 1985) and *Brassica napus* (Thompson 1972). The uses of haploids in crop improvement have been adequately reviewed (Kasha 1974; Davies and Hopwood 1980; Wenzel 1980; Peloquin 1981; Kasha and Seguin-Swartz 1983).

Chromosome pairing at meiosis in haploids provides evidence on genome structure and evolution of plant species (Kimber and Riley 1963; Sadasivaiah 1974). Haploids have been used to produce aneuploid stocks, such as trisomics and monosomics, or interchange stocks, which may be utilized in crop improvement (Nitzsche and Wenzel 1977).

Polyploids

Polyploidy provides another route for crop improvement (Rana 1985). As doubling the chromosome number from diploid to tetraploid doubles the dosage of all genes, tetra- and other multi-allelic types can occur in polyploids and may confer new genetic potentialities on the polyploid variety. Polyploids are thus better buffered than their diploid parents and are capable of withstanding deeper and more diverse modifications and adaptability to extreme ecological conditions (Stebbins 1950).

Generally, crop improvement through increased ploidy consists of an overall enlargement in plant size, with broader and thicker leaves and larger cells, stomata, flower, fruits, and seeds. Slower growth and reduced fertility are the main drawbacks of autopolyploidy; in crops that suffer least from these drawbacks, polyploids can be successful. The crop must be low in chromosome number, cross-pollinated, and grown primarily for vegetative parts (Dewey 1980; Cauderon 1986). Many of the species have an optimum chromosome number or ploidy level beyond which additional chromosomes become detrimental to the plant.

Autopolyploidy induced by colchicine treatment has benefits, especially in forage, fruit, ornamentals, and root crops, in which fertility is not detrimental. Autotetraploids of Alsike clover (*Trifolium hybridum*), red clover (*T. pratense*), and Egyptian clover were superior to their diploid progenitors. Tetraploid grapes have larger berries and fewer seeds. Tetraploid zinnias, marigold, and snapdragons are popular for their flowers and foliage (Elliott 1958).

Highly irregular meiosis, resulting in high sterility and seedlessness associated with triploidy, is an economically valuable trait in fruits such as banana (Simmonds 1962), lime, apple, and watermelon (Kihara 1951). Sterility caused by triploidy in root and tuber crops is not decisive, as shown by triploid sugar beets, which are significantly superior to diploids or tetraploids and are commercially grown in Japan, Sweden, Denmark, and elsewhere.

In cassava (*Manihot esculenta*), sexual tetraploids and triploids originated from diploid interspecific crosses are very vigorous (Hahn et al. 1990). Reduced fertility and slower growth rates were noted in induced tetraploids of cassava (Abraham et al. 1964), which had higher water content than the diploids. However, sexual tetraploids of cassava (Hahn et al. 1990) gave higher (or equal) dry-matter percent than their diploid parents (unpubl.). Sexual tetraploid potatoes had higher total solids (specific gravity) than their parents (Masson and Peloquin 1987).

Amphiploids

In hybrids from wide crosses, a high degree of sterility is usually encountered. In such hybrids, fertility can be restored by chromosome doubling of the sterile hybrids through colchicine treatment, thus producing amphiploids. This treatment is significant in combining the genomes of wild and cultivated species. Such amphidiploids have been produced in several crops (*Triticum*, *Gossypium*, *Nicotiana*, and *Brassica)*. However, triticale produced from crosses between wheat and rye is the only successful amphiploid produced to date (Gupta and Priyadarshan 1982).

The amphiploids exhibit meiotic instability and other undesirable traits because of the lack of genomic harmony and nuclear and cytoplasmic interactions. However, stable and superior polyploid derivatives can be obtained from them, as in triticale (Gupta 1984); many triticale varieties so obtained have been released for commercial cultivation.

Meiotic modifications

A fascinating cytological phenomenon used recently in crop improvement is the *ps* (parallel spindle) mutant in cultivated potato (*Solanum tuberosum*). In these mutants, as a result of parallel spindles at M-II during microsporogenesis, $2n$ pollen are produced that are genetically equivalent to first division restitution (FDR) gametes (Mok and Peloquin 1974). Two other mechanisms of $2n$ pollen formation, premature cytokinesis-1 (*pc-1*) and premature cytokinesis-2 (*pc-2*), have also been reported in potato; these are equivalent to second division restitution (SDR) gametes. In both of these mechanisms, cytokinesis occurs after the first meiotic division and the second meiotic division is omitted. While in FDR all the heterozygous loci between the centrometer and first crossover on all chromosomes are also heterozygous and 75-80% of the heterozygosity is transmitted from parent to offspring, in SDR such loci will be homozygous in the gametes and only 40-45% of the heterozygosity of the parent is transmitted to the progeny. Tuber yields from $4x \times 2x$ (FDR) hybrids were 40-60% higher than from similar $4x \times 2x$ (SDR) hybrids (Mok and Peloquin 1975; Mendiburu and Peloquin 1977a, 1977b; Peloquin 1981). Thus, these simple cytological anomalies have become powerful breeding tools in potato.

In cassava, most interspecific diploid hybrids produce $2n$ pollen, and the origin of spontaneous sexual tetraploids and triploids is attributed to these $2n$ pollen (Hahn et al. 1990).

Aneuploidy

Aneuploidy can result from the addition of one (or more) entire chromosome to a genomic number (trisomics) or the loss of such chromosome material (monosomics). Aneuploids can be obtained from triploid x diploid crosses or other cross combinations, translocations, and asynaptic plants. They are useful for elucidating the cytogenetic architecture of plants. Primary trisomics (that is, plants with one extra chromosome — $2n + 1$) are useful in assigning genes to chromosomes and are used for characteristics which are relatively simply inherited. Each trisomic stock tests one entire chromosome. Monosomics ($2n - 1$) have been used to determine the marker genes carried by different chromosomes, and telocentric or isochromosomic stock helps identify the arm in which they are located (Khush 1973).

The development of a complete series of aneuploids, especially monosomics and trisomics in wheat (Sears 1954), has helped in identifying gene(s) that regulate chromosome pairing or control other characteristics, such as quality factors and disease resistance.

Cytogenetic stocks have also been used to assign RFLP linkage groups to specific chromosome arms. Monosomic stocks are useful as they permit detection of missing bands or alleles at an RFLP locus, which helps determine the chromosome upon which the particular locus resides (Helentjaris et al. 1986).

Chromosome Addition and Substitution Lines

Chromosome addition and substitution lines have been developed as an alternative to the addition of an entire alien genome into cultivated species, which is accompanied by many undesirable features. In alien chromosome substitution, one or more pairs of chromosomes of cultivated species are replaced by an equal number of pairs from the alien species. Unrau et al. (1956) and Khush (1973) described chromosome substitution methods in detail.

Different alien species of wheat, such as *Agropyron*, *Aegilops*, *Haynaldia*, *Secale*, *Elymus*, and barley, which are resistant to a variety of pathogens, have been used in wheat improvement through the transfer of a whole chromosome or a chromosome segment (Lacadena 1977; Islam et al. 1981; Mujeeb-Kazi and Jewell 1985). Alien addition lines have also been produced in cultivated oats, cotton, maize, rice, *Nicotiana*, and *Brassica* but they have not been successful as commercial varieties because of the instability of the alien chromosome and other undesirable characters associated with it. However, such chromosome addition lines have been used successfully in the transfer of alien genes.

Although substitution lines have been produced in *Avena*, *Nicotiana*, and cotton, only in wheat have they been as successful as commercial varieties. Alien substitution lines have been successful in wheat cultivars grown in Europe in which rye chromosome 1R is substituted for wheat chromosome 1B (Zeller 1973). Several wheat cultivars in which a segment of chromosome 1B of wheat is replaced by a segment of chromosome 1R of rye have wider adaptability and resistance to *Septoria tritici* (Rajaram et al. 1983). Substitution lines developed in other cultivated crops have not been successful because the alien chromosome does not fully compensate for the missing chromosome of the cultivated parent; it sometimes even carries undesirable genes along with desirable genes.

Chromosomal Structural Changes

As transfer of whole chromosomes was found to create meiotic instability, procedures for transferring only segments of alien chromosomes have been established. Such a transfer is considered as analogous to the transfer of a single gene or a gene cluster by conventional backcrossing between different varieties. Resistance to a number of wheat rusts has been achieved using this method (Swaminathan and Gupta 1983).

Induced chromosomal structural changes, such as translocations and deletions, are used to transfer desirable traits in crop plants. Sears (1956) used a bridging species, *Triticum dicoccoides*, and the polyploidy technique, together with radiation-induced chromosomal interchanges, to transfer leaf rust resistance from *Aegilops umbellulata* ($2n = 14$) to common wheat ($2n = 6x = 42$). Induced translocations have been used to transfer the gene for mildew resistance from *Avena barbata* to the cultivated oats (Aung and Thomas 1976). Induced deletions such as *phlb*, which permits homoeologous pairing and transfer of alien chromosomal segments (Sears 1977), and the male sterility mutant *mslc* (Driscoll 1977) are used in wheat improvement programmes.

Conclusion

Recent developments in the techniques of differential chromosome staining (banding), isozyme analysis, and restriction patterns of nuclear and cytoplasmic DNA, coupled with conventional cytogenetics, can play an important role in crop improvement.

References

Abraham, A., P.K.S. Panicker, and P.M. Mathew. 1964. Polyploidy in relation to breeding in tuber crops. *Journal of Indian Botanical Society* 43: 278-283.

Aung, T., and H. Thomas. 1976. Transfer of mildew resistance from the wild oat *Avena barbata* Pott. into the cultivated oat. *Euphytica* 27: 731-739.

Burnham, C.R. 1966. Cytogenetics in plant improvement. Pages 139-188 in *Plant Breeding* edited by K.J. Frey. Iowa State University Press, Ames, Iowa, USA.

Campion, B., and C. Alloni. 1990. Haploids from unpollinated ovules. *Rice Biotechnology Quarterly* 1 (4): 22.

Cauderon, Y. 1986. Cytogenetics in breeding programmes dealing with polyploidy, interspecific hybridization and introgression. Pages 83-104 in *Genetic Manipulation in Plant Breeding* edited by W. Horn, C.J. Jensen, W. Odenbach, and O. Schieder. De Gruyter, Berlin, Germany.

Chase, S.S. 1949. Monoploid frequencies in a commercial double cross hybrid maize and its component single cross hybrids and inbred lines. *Genetics* 34: 328-332.

Choo, T.M., E. Reinbergs, and K.J. Kasha. 1985. Use of haploids in breeding barley. *Plant Breeding Reviews* 3: 219-252.

Davies, D.R., and D.A. Hopwood (eds.). 1980. *The Plant Genome*. John Innes Charity, UK.

Dewey, D.R. 1980. Some applications and misapplications of induced polyploidy to plant breeding. Pages 445-470 in *Polyploidy — Biological Relevance* edited by W.H. Lewis. Plenum Press, New York, USA.

Driscoll, C.J. 1977. Registration of Cornerstone male-sterile wheat germplasm. *Crop Science* 17: 190.

Elliott, F.C. (ed.). 1958. *Plant Breeding and Cytogenetics.* McGraw-Hill, New York, USA.

Fedak, G. 1985. Wide crosses in *Hordeum.* Pages 155-186 in *Barley Monograph* (No. 26) edited by D.C. Rasmussan. *American Society of Agronomy*, Madison, Wisconsin, USA.

Frey, K.J.,T.S. Cox, D.M. Rodgers, and P. Bramel-Cox. 1984. Increasing cereal yields with genes from wild and weedy species. Pages 51-68 in *Genetics: New Frontiers (Vol. IV) Proceedings of the XV International Congress of Genetics, New Delhi, India* edited by V.L. Chopra, B.C. Joshi, R.P. Sharma, and H.C. Bansal. Oxford and IBH Publishing, New Delhi, India.

Gupta, P.K. 1984. Role of loss of rye heterochromation in the improvement of hexaploid triticale. *The Nucleus* 27: 100-107.

Gupta, P.K., and P.M. Priyadarshan. 1982. Triticale — Present status and future prospects. *Advances in Genetics* 21: 255-345.

Hahn, S.K., E.R. Terry, and K. Leuschner. 1980. Breeding cassava for resistance to cassava mosaic virus. *Euphytica* 29: 673-683.

Hahn, S.K., J.C.G. Isoba, and T. Ikotun. 1989. Resistance breeding in root and tuber crops at IITA, Ibadan, Nigeria. *Crop Protection* 8: 147-168.

Hahn, S.K., K.V. Bai, and R. Asiedu. 1990. Tetraploids, triploids and 2*n* pollen from diploid interspecific crosses with cassava. *Theoretical and Applied Genetics* 79: 433-439.

Helentjaris, T., D. Weber, and S. Wright. 1986. Use of monosomics to map cloned DNA fragments in maize. *Proceedings of the National Academy of Science, USA* 83: 6035-6039.

Hougas, R.W., and S.J. Peloquin 1957. A haploid plant of the potato variety Katahdin. *Nature (London)* 180: 1209-1210.

Hougas, R.W., S.J. Peloquin, and A.C. Gabert. 1964. Effect of seed parent and pollinator on frequency of haploids in *Solanum tuberosum*. *Crop Science* 4: 593-595.

Islam, A.K.R., K.W. Shepherd, and D.H.R. Sparrow. 1981. Isolation and characterization of euplasmic wheat-barley chromosome addition lines. *Heredity* 46: 161-164.

Kasha, K.J. (ed.). 1974. *Haploids in Higher Plants — Advances and Potential.* University of Guelph, Canada.

Kasha, K.J., and K.N. Kao. 1970. High frequency haploid production in barley *(Hordeum vulgare* L.). *Nature* 225: 874-876.

Kasha, K.J., and G. Seguin-Swartz. 1983. Haploidy in crop improvement. Pages 19-68 in *Cytogenetics of Crop Plants* edited by M.S. Swaminathan, P.K. Gupta, and U. Sinha. Macmillan India, New Delhi, India.

Khush, G.S. 1973. *Cytogenetics of Aneuploids.* Academic Press, New York, USA.

Khush, G.S., and D.S. Brar. 1988. Wide hybridization in plant breeding. Pages 140-188 in *Plant Breeding and Genetic Engineering* edited by A.H. Zakri. Society for the Advancement of Breeding Researches in Asia and Oceania, Bangi, Malaysia.

Kihara, H. 1951. Triploid watermelons. *Proceedings of American Society of Horticultural Science* 58: 217-230.

Kimber, G., and R. Riley. 1963. Haploid angiosperms. *Botanical Review* 29: 480-531.

Knott, D.R. 1971. The transfer of genes for disease resistance from alien species of wheat by induced translocations. Pages 67-77 in *Mutation Breeding for Disease Resistance.* IAEA, Vienna, Austria.

Kushmir, W., and G.M. Halloran. 1984. Transfer of high kernel weight and high protein from wild tetraploid wheat *(Triticum turgidum dicoccoides)* to bread wheat *(T. aestivum)* using homologous and homoeologous recombination. *Euphytica* 33: 249-256.

Lacadena, J. 1977. Interspecific gene transfer in plant breeding. Pages 45-62 in *Interspecific Hybridization in Plant Breeding* edited by E. Sanchez-Monge and F. Garcia-Olmedo. Eucarpia, Madrid, Spain.

Masson, M.F., and S.J. Peloquin, 1987. Heterosis for tuber yields and total solids content in 4x x 2x FDR-CO crosses in potato. Pages 213-217 in *The Production of New Potato Varieties: Technological Advances* edited by G.J. Jellis and D.E. Richardson. Cambridge University Press, Cambridge, UK.

Mendiburu, A.O., and S.J. Peloquin. 1977a. Bilateral sexual polyploidization in potatoes. *Euphytica* 26: 573-583.

Mendiburu, A.O., and S.J. Peloquin. 1977b. The significance of 2*n* gametes in potato breeding. *Theoretical and Applied Genetics* 49: 53-61.

Mok, D.W.S., and S.J. Peloquin. 1974. Genetic evidence of FDR and SDR diplandro-genesis in diploid potatoes and the mapping of S locus. *American Potato Journal* 5: 279-230.

Mok, D.W.S., and S.J. Peloquin. 1975. Three mechanisms of 2*n* pollen formation in diploid potatoes. *Canadian Journal of Genetics and Cytology* 17: 217-225.

Mujeeb-Kazi, A., and D.C. Jewell. 1985. CIMMYT's wide cross program for wheat and maize improvement. Pages 219-226 in *Biotechnology in International Agricultural Research.* IRRI, Los Banõs, Philippines.

Nichols, R.F.W. 1947. Breeding cassava for virus resistance. *East African Agricultural and Forestry Journal* 12: 184-194.

Nitsch, J.P., and C. Nitsch.1969. Haploid plants from pollen grains. *Science* 163: 85-87.

Nitzsche, W., and G. Wenzel. 1977. *Haploids in Plant Breeding.* Verlag Paul Parey, Berlin, Germany.

Peloquin, S.J. 1981. Chromosomal and cytoplasmic manipulations. Pages 117-150 in *Plant Breeding* (Vol. II) edited by K.J. Frey. Iowa State University Press, Ames, Iowa, USA.

Rajaram, S., E.E. Maan, G. Ortiz-Ferrara, and A. Mujeeb-Kazi. 1983. Adaptation, stability and high yield potential of certain 1B/1R CIMMYT wheats. Pages 613-621 in *Sixth International Wheat Genetics Symposium, Kyoto, Japan.* Plant Germplasm Institute, Kyoto, Japan.

Rana, R.S. 1985. Polyploidy in improvement of crop plants. Pages 1-21 in *Genetic Manipulation for Crop Improvement* edited by V.L. Chopra. Oxford and IBH Publishing, New Delhi, India.

Sadasivaiah, R.S. 1974. Haploids in genetic and cytological research. Pages 355-386 in *Haploids in Higher Plants — Advances and Potential* edited by K. Kasha. University of Guelph, Canada.

Savitsky, H. 1975. Hybridization between *Beta vulgaris* and *B. procumbens* and transmission of nematode *(Heterodera schachtii)* resistance to sugar beet. *Canadian Journal of Genetics and Cytology* 17: 197-209.

Sears, E.R. 1954. The aneuploids of common wheat. *Missouri Agricultural Experiment Station Research Bulletin* 572: 1- 58.

Sears, E.R. 1956. The transfer of leaf-rust resistance from *Aegilops umbellulata* to wheat. Pages 1-22 in *Genetics in Plant Breeding*. Brookhaven Symposia in Biology No. 9. Brookhaven National Laboratory, Upton, New York, USA.

Sears, E.R. 1977. An induced mutant with homoeologous pairing in common wheat. *Canadian Journal of Genetics and Cytology* 19: 585-593.

Sharma, H.C., and B.S. Gill. 1983. Current status of wide hybridization in wheat. *Euphytica* 32: 17-31.

Simmonds, N.W. 1962. *The Evolution of Banana*. Longmans Green, London, UK.

Stalker, H.T. 1985. Groundnut cytogenetics at North Carolina State University. Pages 119-123 in *Proceedings of an International Workshop on Cytogenetics of Arachis, 31 October -2 November 1983, ICRISAT Center, India.* ICRISAT, Patancheru, India.

Stebbins, G.L. 1950. *Variation and Evolution in Plants.* Columbia University Press, New York, USA.

Stebbins, G.L. 1958. The inviability, weakness and sterility of interspecific hybrids. *Advances in Genetics* 9: 147-215.

Swaminathan, M.S., and P.K. Gupta. 1983. Improvement of crop plants: Emerging possibilities. Pages 1-18 in *Cytogenetics of Crop Plants* edited by M.S. Swaminathan, P.K. Gupta, and U. Sinha. Macmillan India, New Delhi, India.

Tanaka, M., and K. Nakata. 1969. Tobacco plants obtained by anther culture and experiments to get diploid seeds from haploids. *Japanese Journal of Genetics* 44: 47-54.

Therenin, L. 1974. Haploids in asparagus breeding. Page 279 in *Haploids in Higher Plants — Advances and Potential* edited by K.J. Kasha. University of Guelph, Ontario, Canada.

Thompson, K.F. 1972. Oil seed rape. Pages 94-96 in *Report of the Plant Breeding Institute.* Plant Breeding Institute, Cambridge, UK.

Tsunewaki, K., Y. Mukai, and M. Yamamori. 1984. The salmon method of haploid production in common wheat. Page 21 in *Abstracts, International Symposium on the Genetic Manipulation of Crops. 2: Beijing, China.*

Turcotte, E.L., and C.V. Feaster. 1974. Methods of producing haploids: Semigametic production of cotton haploids. Pages 53-64 in *Haploids in Higher Plants — Advances and Potential* edited by K.J. Kasha. University of Guelph, Ontario, Canada.

Unrau, J., C. Person, and J. Kuspira. 1956. Chromosome substitution in hexaploid wheat. *Canadian Journal of Botany* 34: 629-640.

Wenzel, G. 1980. Recent progress in microspore culture of crop plants. Pages 185-213 in *The Plant Genome* edited by D. R. Davies and D.A. Hopwood. John Innes Charity, Norwich, UK.

Zeller, F.J. 1973. 1B/1R wheat-rye chromosome substitutions and translocations. Pages 209-221 in *Proceedings, Fourth International Wheat Genetics Symposium* edited by E.R. Sears and L.M.S. Sears. University of Missouri, Columbia, Missouri, USA.

Zillinsky, F.J., and R.A. Derick. 1960. Crown rust resistant derivatives from crosses between autotetraploid *Avena strigosa* and *A. sativa. Canadian Journal of Plant Science.* 40: 366-370.

2.6

Cytogenetics of cowpea

F. Saccardo[1], A. Del Giudice[1] and I. Galasso[2]

Department of Agronomy and Plant Genetics, University of Naples, Via Università 100, 80055 Portici, Italy[1]; Germplasm Institute, National Research Council, Via Amendola 165/A, 70100 Bari, Italy[2]

Abstract

This paper reports on the characterization of *Vigna unguiculata* chromosomes, performed in the mitotic prometaphase and metaphase and in pachytene using conventional techniques and C- and H-banding. Karyotypes of *V. unguiculata* and *V. vexillata* were obtained using an image analysis system. Meiotic analysis was used to investigate irregularities in the F_1 hybrids obtained from crosses between cultivated and wild *Vigna* species. The paper also reviews research on polyploidy and interspecific hybridization.

Cowpea, *Vigna unguiculata* (L.) Walp., is one of the most important grain legume crops in the tropics and subtropics, often constituting the main plant protein source in some agricultural regions. It has been the subject of genetic research since the beginning of the century (Fery 1985).

From a genetic point of view, several genes of this diploid species have been identified and characterized (Fery 1985). From a karyological viewpoint, however, little is known about *Vigna* species, mainly because of the extremely small size of their chromosomes.

Chromosome Number

Karpetchenko (1925) first reported chromosome numbers of $2n = 22$ for *V. unguiculata* and *V. catijang* (Burm.) Walp. Senn (1938) reported $n = 11$ for one cultivar of *V. catijang*, and Sen and Bhowal (1960a) reported $2n = 22$ for two cultivars each of *V. sinensis* (L.) Savi, *V. catijang*, and *V. sesquipedalis* (L.) Fruhw. Sen and Bhowal (1960b) found 44 chromosomes in artificial tetraploids of the same three species. For 192 *V. sinensis* cultivars and strains, Faris (1964) counted $2n = 22$ chromosomes in the root tips, and the pollen mother cells showed $n = 11$ chromosomes. In contrast, Kawakami (1930) found $n = 12$ for *V. sinensis* and Floresca et al. (1960) reported $2n = 24$ in one cultivar each of *V. sinensis* and *V. sesquipedalis*. More recently, Del Giudice et al. (1989), Barone and Saccardo (1990), and Pignone at al. (1990) found $2n = 22$ in *V. unguiculata*. Our own studies have also found $2n = 22$ for a considerable number of *V. unguiculata* genotypes (Table 1).

Table 1 Chromosome numbers of *Vigna unguiculata* genotypes, counted at the University of Naples, Italy

Variety or coultigroup	Genotype	2*n*
Coultigroup *unguiculata*	TVu 91	22
Coultigroup *unguiculata*	TVuu 662	22
Coultigroup *unguiculata*	TVu 946	22
Coultigroup *unguiculata*	TVu 1977	22
Coultigroup *unguiculata*	TVu 9062	22
Coultigroup *unguiculata*	TVu 9065	22
Coultigroup *unguiculata*	TVu 11953	22
Coultigroup *unguiculata*	TVu 11952	22
Coultigroup *unguiculata*	Vita 1	22
Coultigroup *unguiculata*	Vita 3	22
Coultigroup *unguiculata*	Vita 4	22
Coultigroup *unguiculata*	Vita 5	22
Variety *sesquipedalis*	TVu 21	22
Variety *sesquipedalis*	TVu 3652	22
Variety *biflora*	TVu 3657	22
Variety *biflora*	TVu 3658	22
Variety *dekintiana*	TVu 140	22
Variety *pubescens*	TVuu 112	22
Variety *pubescens*	TVuu 113	22

Karyomorphology

To overcome the limitations imposed by chromosome size, Mukherjee (1968) used pachytene chromosomes, which are longer. At this meiotic stage, the chromosomes are more easily characterized, as reported for other species: corn (McClintock 1929), tomato (Barton 1950), rice (Chu 1967), and *Phaseolus* species (Krishnan and De 1968).

Another way to study very small chromosomes is to make use of the mitotic prometaphase (Kurata and Omura 1978; Mouras et al. 1978; Pignone et al. 1990) because at this stage the chromosomes are much longer than they are at metaphase and the heterochromatin is already condensed. It is, therefore, possible to study its distribution along the chromosomes. Recently, chromosome banding techniques have also been used (Galasso et al. 1990; Pignone et al. 1990).

Besides conventional cytogenetic approaches, an automatic image analysis system for karyotyping *Vigna* chromosomes has been used. Systems for automated metaphase spread location and karyotyping were first developed for human cytogenetics (Casperson et al. 1971a, 1971b; Fleischman et al. 1971; Lubs and Ledley 1973; Marimuthu et al. 1974; Castelman and Melnyk 1976). Image analysis was first applied to plant chromosomes through the use of microdensitometric systems (Carlson et al. 1963; Mendelsohn et al. 1966; Filippone et al. 1983, 1984a, 1984b). Automated karyotyping has been carried out in barley (Fukui 1986), pea, wheat, and chickpea (Venora et al. 1991). The application of this method to *Vigna* chromosomes seems to be promising. The preliminary results have been used to compare *V. unguiculata* and *V. vexillata* karyotypes (Table 2).

Table 2 **Means and standard deviations of the relative lengths and short/long arm ratios of the chromosomes of the haploid set of *Vigna unguiculata* (TVu 9062) and *V. vexillata* (TVu 72), obtained using image analysis**

Chromosome pair number	*Vigna unguiculata* relative length (%)	short/long arm ratio	*Vigna vexillata* relative length (%)	short/long arm ratio
I	12.78 ± 1.54	0.76 ± 0.13[a]	11.07 ± 0.90	0.47 ± 0.12[a]
II	10.26 ± 0.49	0.59 ± 0.10	10.21 ± 0.60	0.59 ± 0.15
III	9.99 ± 0.42	0.73 ± 0.20	9.86 ± 0.29	0.67 ± 0.15
IV	9.63 ± 0.42	0.61 ± 0.14[b]	9.65 ± 0.39	0.78 ± 0.11[b]
V	9.23 ± 0.23	0.77 ± 0.26	0.08 ± 0.31	0.70 ± 0.12
VI	8.81 ± 0.30	0.68 ± 0.21	9.01 ± 0.35	0.80 ± 0.14
VII	8.44 ± 0.32[a]	0.71 ± 0.16	8.89 ± 0.27[b]	0.78 ± 0.12
VIII	8.46 ± 0.33	0.69 ± 0.19	8.70 ± 0.17	0.78 ± 0.11
IX	8.20 ± 0.39	0.84 ± 0.15	8.44 ± 0.30	0.69 ± 0.12
X	7.43 ± 0.73	0.78 ± 0.16	7.90 ± 0.47	0.76 ± 0.13
XI	6.78 ± 0.32	0.90 ± 0.06[b]	7.16 ± 0.72	0.72 ± 0.13[b]

Note: a Statistically significant difference ($P < 0.01$) between genotypes.
b Statistically significant difference ($P < 0.05$) between genotypes.

Using pachytene analysis on the Tabara cultivated strain of *V. sinensis*, Mukherjee (1968) identified 11 bivalents ranging from 44.3 μm to 19.4 μm which were classified, according to the criteria established for *Zea* by McClintock (1929), as three large chromosomes, seven medium-sized chromosomes and one small chromosome.

Barone and Saccardo (1990) used three parameters to identify each of the *V. unguiculata* TVu 9062 bivalents at pachytene stage: the total length of the chromosome; the arm ratio; and the pattern of chromomere distribution. The authors noted the presence of a nucleolar chromosome not mentioned by Mukherjee (1968). With respect to size, they found one very long chromosome (85.5 μm) and one very small one (14.1 μm); the chromosomes of intermediate length were allocated to three classes (51.5-45.8 μm; 39.2-30.5 μm; 22.6-22.0 μm) (Table 3, *overleaf*).

In the mitotic prometaphase study conducted by Pignone et al. (1990) the 22 chromosomes of *V. unguiculata* TVu 9062 were also characterized using barium-saline-giemsa (BSG), which was used in a modified form of the Vosa and Marchi (1972) technique. The

Table 3 **Average length, arm ratio, centromere position, and chromosome number of the 11 bivalents of *Vigna unguiculata* analysed in the pachytene stage**

Chromosome number	Total length (μm)	Long arm (μm)	Short arm (μm)	Short/long arm ratio	Centromere position	Chromomeres number
1	85.5	52.3	33.3	1.6	Sm	3
2	51.5	27.2	24.4	1.1	M	2
3	49.8	27.4	22.4	1.2	M-Sm	8
4	45.8	26.5	19.2	1.4	Sm	2
5	39.2	24.4	14.8	1.7	Sm	4
6	34.6	20.5	14.1	1.5	Sm	9
7	33.3	18.2	15.2	1.1	M	4
8	30.5	15.9	15.3	1.0	M	4
9	22.6	14.8	7.8	1.9	Sm-St	8
10	22.0	13.8	8.1	1.7	Sm	10
11	14.1	7.7	6.4	1.2	M-Sm	5

Note: a M = metacentric; Sm = submetacentric; St = subtelocentric.
Source: Barone and Saccardo 1990.

11 chromosome pairs were allocated to three groups based on size: five long, including satellited chromosome; five medium; and one short. Galasso et al. (1990) modified the above technique for use with the strain later chosen as a standard (Tvx 3236). In both studies, 27 C-bands were identified on the 22 chromosomes.

Polyploidy

Induced tetraploid forms of *V. unguiculata* have been studied by Sen and Bhowal (1960b). They reported marked cultivar differences in morphological traits, pollen sterility, and fruit setting. Meiotic irregularities, including the formation of quadrivalents, trivalents, and univalents, the presence of laggards, and unequal distribution in the anaphase stage were also observed. They concluded that pollen sterility was the most limiting factor in the development of commercially useful tetraploid cultivars.

Recently, a tetraploid accession of *V. glabrescens* ($2n = 4x = 44$), V1160 has been identified (Chen et al., 1989). A tetraploid *Phaseolus* species has been described by Dana (1964), who established that it belonged to the *Sigmoidotropis* section of the genus *Vigna*. Krishnan and De (1968) obtained colchicine-induced amphidiploids from the triploid hybrids of a cross between *V. mungo* and the above-mentioned tetraploid *Vigna* species.

Techniques for Handling *Vigna* Chromosomes

For mitotic analysis of *Vigna* species chromosomes, the squash technique for counting has been used extensively. It is a fairly conventional technique, consisting of pre-treatment of root tips in p-dichlorobenzene or other antimitotic substances, such as hydroxyquinoline, bromonaphalene, or colchicine; fixation in carnoy; staining in feulgen (after hydrolysis); and tolouidine blue, aceto-carmine, or aceto-orcein. Sometimes the technique differs in the presence or absence of pre-treatment or in the type and duration of pre-treatment, fixative, and stain. Frahm-Leliveld (1965) counted the chromosones for a number of wild species of *Vigna* using Navashin fluid as fixative and staining root tips with crystal violet.

Meiotic investigations in *Vigna* have been conducted by a few researchers. Mukherjee (1968) proposed the collection of floral buds in the morning, fixation in acetic alcohol (1:3) with a trace of ferric acetate, storage in 70% ethyl alcohol, keeping the suitable floral buds in 45% acetic acid for 15 min, and then squashing the anthers in a drop of 1% aceto-carmine. Barone and Saccardo (1990) obtained good results with propione alcohol (3:1) with a trace of ferric chloride as fixative and using 2% aceto-carmine, omitting the acetic acid stage.

For mitotic prometaphase analysis, Galasso et al. (1990) excised primary roots when 1.5-2.0 cm long and put them in a saturated aqueous solution of p-dichlorobenzene for 2 hours at 13°C. After maceration for 30 minutes in 3% pectinase at pH 4 and 35°C, the root tips were washed in distilled water and fixed in propione alcohol (1:1) for 24-48 hours at 4°C. Squashing was carried out in 45% propionic acid. To reveal C-bands on the prometaphase chromosomes, the slides were stained using the BSG technique.

Automated karyotyping, based on image analysis, has been used at the University of Naples. Further effort is needed to devise a single method which can be used as a standard.

Interspecific Hybrids

Interspecific hybridization is a useful tool in plant breeding for creating new genetic variations. Much research has been carried out on four oriental species of food legumes of the genus *Vigna*: mung bean, *V. radiata* (L.) Wilzcek; blackgram, *V. mungo* (L.) Hepper; rice bean *V. umbellata* (Thunb.) Ohwi and Ohashi; and adzuki bean *V. angularis* (Wild.) Ohwi and Ohashi. Varying degrees of success have been reported (Al-Yasiri and Coyne 1966; Dana 1966a, 1966b, 1966c; De and Krishnan 1966; Sawa 1973; Biswas and Dana, 1975; Ahn and Hartmann, 1977, 1978a; Chen et al., 1977; Chowdury and Chowdury, 1977). The overwhelming majority of the interspecific hybrids obtained are completely sterile or only partially fertile.

Biswas and Dana (1975) obtained hybrids from a *V. mungo* x *V. umbellata* cross through embryo rescue on a medium containing sucrose. The hybrids presented an irregular meiosis, with the presence of quadrivalents, monovalents, and bridging. Many authors have performed crosses between *V. radiata* and *V. mungo* but they were successful only when *V. radiata* was used as the female parent (Dana 1966a). Chowdhury and Chowdhury (1977) obtained hybrids that were sterile or prematurely aborted, depending on the*V. radiata* variety utilized. On the basis of morphological, genetic, and immunological features and of geographic distribution, De and Krishnan (1966) suggested that *V. mungo* and *V. radiata* are closely related species from a phylogenetic point of view. Meiosis of hybrids shows normal pairing of the 11 bivalents in the pachytene stage.

V. radiata and *V. umbellata* were used in interspecific hybridization by Dana (1966b) and by Machado et al. (1982), who obtained hybrids only by using *V. radiata* as the female parent. They reported an irregular meiosis in the hybrids. The species *V. radiata* was also used by Dana (1966c) in crosses with *V. trilobata*. Using the former as female, mature hybrids were obtained in 25% of the crosses, which showed irregular meiosis; an irregular pairing of the bivalents was observed in the pachytene stage, dicentric bridges, slower chromosomes, and the presence of micronuclei in subsequent stages.

Ahn and Hartmann (1978a) obtained a few hybrids in crosses involving *V. radiata* and *V. angularis* by rescuing the young embryos with appropriate in vitro medium. The meiosis of the hybrids was irregular. Crossing *V. umbellata* and *V. angularis*, these authors obtained hybrids directly by using *V. angularis* as female parent, or in the reciprocal cross by rescuing the embryos in vitro (Ahn and Hartmann 1978b). The meiosis of these hybrids was regular.

Other crosses have involved *V. umbellata* (Dana 1964), *V. radiata* (Dana 1965), and *V. mungo* (Krishnan and De 1968) with the tetraploid species of *Vigna* characterized by Dana (1964). Hybrids were obtained only when the diploid species were used as the female parent and in the cross involving *V. umbellata* only through embryo rescue. The meiosis of the triploid plants was highly irregular.

Krishnan and De (1968) analysed the meiosis of the colchicine-induced amphidiploids obtained from the triploid hybrids mentioned above, thereby initiating the discussion on the analogies between the polyploid, the hybrid, and the amphidiploid genomes.

Cowpea has not been successfully hybridized with any other *Vigna* or *Phaseolus* species. Evans (1976) failed in all attempts to cross *V. unguiculata* with *V. mungo*, *V. radiata*, *V. umbellata*, *V. aconitifolia* (Jacq.) Maréchal (moth bean), and *V. angularis*. Fatokun and Singh (1987) obtained hybrids between *V. unguiculata* and *V. pubescens* using embryo rescue. Cytological investigations showed the occurrence of quadrivalent associations during the meiosis of the hybrids, suggesting that the cowpea species share a structurally differentiated but basically homologous genome. In addition, lagging chromosomes were observed in anaphase I. No hybrid was obtained in an interspecific cross with *V. vexillata* (L.) A. Rich (Evans 1976; Barone and Ng 1990; Del Giudice et al. 1990; Saccardo 1990). Schnapp et al. (1990) reported successful hybridization between two wild *Vigna* species, *V. oblongifolia* and *V. luteola*.

The nature of pairing at diakinesis and pollen fertility of the hybrids mentioned above are summarized in Table 4.

Table 4 Meiotic analysis and pollen fertility in F_1 interspecific hybrids in the genus *Vigna*

Interspecific cross	Pairing	Pollen fertility	References diakinesis
V. mungo x *V. radiata*	abnormal	partial	Gosal and Bajaj (1983) Chen et al. (1983)
V. radiata x *V. umbellata*	abnormal	very low	Chen et al. (1983) Machado et al. (1982)
V. umbellata x *V. angularis*	regular	partial	Ahn and Hartmann (1978) Chen et al. (1983)
V. radiata x *V. angularis*	abnormal	not known	Ahn and Hartmann (1978)
V. pubescens x *V. unguiculata*	abnormal	low	Fatokun and Singh (1987)
V. luteola x *V. oblongifolia*	regular	high	Purdue and Portici (1990)

Conclusion

Knowledge of the cytogenetic structure of *Vigna* species can help to improve cultivated species and overcome barriers to interspecific compatibility. Many desirable traits, such as insect pest resistance, need to be introduced into cowpea from wild species. Bridge crosses and embryo rescue are promising techniques to ensure F_1 progeny.

Each chromosome pair of *V. unguiculata* has been described by the cytological techniques set up by the Italian laboratories collaborating in the cowpea improvement project. In addition to conventional cytogenetic approaches, chromosome banding techniques and an image analysis system for karyotyping have been used. Future studies concerning molecular cytogenetics could provide a new direction for investigating genetic similarities, evolutionary trends, and phylogeny of various *Vigna* species.

Discussion

QUESTION: What stain did you use for the pachytene analysis? Any silver nitrate stain used to detect nucleoli? There is a need to standardize your work with previous work on pachytene analysis and G-banding of cowpea chromosomes. For the fluorescence of pollen grain and tube, was the aniline blue technique used or was it fluorescein acetate?

SACCARDO: For pachytene analysis, we used (and are using) the classical staining with acetocarmine because we obtain good results. We tried the technique of overstaining, used in *Solanum*, but the results were not satisfactory. It is our plan to use a new technique based on the use of lacto-acetic orcein, but no attempts have been made yet. For nucleoli, we didn't use silver nitrate. We need to intensify our study, but I'd like to remind you that cowpea chromosomes are very small and not easy to characterize. Our presentation is not a definitive result. For the fluorescence analysis, we used aniline blue.

References

Ahn, C.S., and R.W. Hartmann. 1977. Interspecific hybridization among four species of the genus *Vigna*. Pages 240-246 in *Proceedings, First International Mung Bean Symposium*. Asian Vegetable Research and Development Centre, Shanhua, Taiwan.

Ahn, C.S., and R.W. Hartmann. 1978a. Interspecific hybridization between mung bean (*Vigna radiata* (L.) Wilczek) and adzuki bean (*Vigna angularis* (Willd.) Ohawi and Ohashi). *Journal of the American Society for Horticultural Science* 103: 3-6.

Ahn, C.S., and R.W. Hartmann. 1978b. Interspecific hybridization between rice bean (*Vigna umbellata* (Thunb.) Ohawi and Ohashi) and adzuki bean (*Vigna angularis* (Willd.) Ohawi and Ohashi). *Journal of the American Society for Horticultural Science* 103: 435-438.

Al-Yasiri, S.A., and D.P. Coyne. 1966. Interspecific hybridization in the genus *Phaseolus*. *Crop Science* 6: 59-60.

Barone, A., and N.Q. Ng. 1990. Embryological study on crosses between *Vigna unguiculata* and *V. vexillata*. Pages 151-160 in *Cowpea Genetic Resources* edited by N.Q. Ng and L.M. Monti. IITA, Ibadan, Nigeria.

Barone, A., and F. Saccardo. 1990. Pachytene morphology of cowpea chromosomes. Pages 137-143 in *Cowpea Genetic Resources* edited by N.Q. Ng and L.M. Monti. IITA, Ibadan, Nigeria.

Barton, D.W. 1950. Pachytene morphology of tomato chromosome complement. *American Journal of Botany* 37: 639-643.

Biswas, M.R., and S. Dana 1975. Black gram x rice bean cross. *Cytologia* 40: 787-795.

Carlson, L., T. Casperson, C.E. Faley, J. Kudynowski, G. Lomakka, E. Simonsson, and L. Soren. 1963. The application of quantitative cytochemical techniques to the study of individual mammalian chromosomes. *Experimental Cell Research* 31: 589-594.

Casperson, T., G. Lomakka, and A. Moller. 1971a. Computerized chromosome identification by aid of the quinacrine mustard fluorescent technique. *Hereditas* 67: 103-110.

Casperson, T., G. Lomakka, and L. Zech. 1971b. The 24 fluorescence patterns of the human metaphase chromosome — Distinguishing characters and variability. *Hereditas* 67: 89-102.

Castelmann, K.R., and J.H. Melnyk. 1976. *An Automated System for Chromosome Analysis: Final Report*. JPL Document 5040-30, Jet Propulsion Laboratory. California Institute of Technology, Pasadena, California, USA.

Chen, N.C., J.F. Parrot, T. Jacobs, L. Baker, and P.S. Carlson. 1977. Interspecific hybridization of food legumes by unconventional methods of plant breeding. Pages 247-252 in *Proceedings, First International Mung Bean Symposium*. Asian Vegetable Research and Development Centre, Shanhua, Taiwan.

Chen, N.C., L.R. Baker, and S. Honha. 1983. Interspecific crossability among four species of *Vigna* food legumes. *Euphytica* 32: 925-937.

Chen, H.K., M.C. Mok, S. Shanmugasundaram, and D.W.S. Mok. 1989. Interspecific hybridization between *Vigna radiata* (L.) Wilczek and *Vigna glabrescens*. *Theoretical and Applied Genetics* 78: 641-647.

Chowdhury, R.K., and J.B. Chowdhury. 1977. Intergeneric hybridization between *Vigna mungo* (L.) Hepper and *Phaseolus calcaratus* Roxb. *Indian Journal of Agricultural Science* 47: 117-121.

Chu, Y.E. 1967. Pachytene analysis and observations of chromosome association in haploid rice. *Cytologia* 32: 87-95.

Dana, S. 1964. Interspecific cross between tetraploid *Phaseolus* species and *P. ricciardianus* Ten. *Nucleus, Calcutta* 7: 1-10.

Dana, S. 1965. *Phaseolus aureus* Roxb. x tetraploid *Phaseolus* species cross. *Revista de Biologia, Lisboa* 5: 109-114.

Dana, S. 1966a. The cross between *Phaseolus aureus* Roxb. and *P. mungo* (L.). *Genetica* 37: 259-274.

Dana, S. 1966b. Cross between *Phaseolus aureus* Roxb. and *P. ricciardianus*. *Genetica Iberica* 18: 141-156.

Dana, S. 1966c. Species cross between *Phaseolus aureus* Roxb. x *P. trilobus* Ait. *Cytologia* 31: 176-187.

De, D.N., and R. Krishnan. 1966. Cytological studies of the hybrid *Phaseolus aureus* x *P. mungo*. *Genetica* 37: 588-600.

Del Giudice, A., A. Barone, and F. Saccardo. 1989. Caratterizzazione dei cromosomi pachitenici di *Vigna unguiculata*. Page 96 in *Atti del XXXIII Convegno Annuale della Società Italiana di Genetica Agraria, 23-26 October 1989, Porto Conte, Alghero, Italy*.

Del Giudice, A., A. Barone, and D. Consoli. 1990. Studio istologico dell'incompatibilità interspecifica tra *Vigna unguiculata* e *V. vexillata*. Page 93 in *Atti del XXXIV Convegno Annuale della Società Italiana di Genetica Agraria, 8-11 October 1990, Marina di Ugento (LE), Italy*.

Evans, A.M. 1976. Species hybridization in the genus *Vigna*. Pages 31-34 in *Proceedings, Collaborators' Meeting on Grain Legume Improvement, 9-13 June 1975, IITA, Ibadan, Nigeria*. IITA, Ibadan, Nigeria.

Faris, D.G. 1964. The chromosome number in *Vigna sinensis* (L.) Savi. *Canadian Journal of Genetics and Cytology* 6: 255-258.

Fatokun, C.A., and B.B. Singh. 1987. Interspecific hybridization between *Vigna pubescens* and *V. unguiculata* (L.) Walp. through embryo rescue. *Cell, Tissue and Organ Culture* 9: 229-233.

Fery, R.L. 1985. The genetics of cowpeas: A review of world literature. Pages 25-62 in *Cowpea Research, Production and Utilization* edited by S.R. Singh and K.O. Rachie. John Wiley, Chichester, UK.

Filippone, E., L.A. Smaldone, A. Errico, and L.N. Monti. 1983. Microdensitometric analysis of photographic chromosomes images of *Vicia faba* L. *Genetica Agraria* 37: 169-170.

Filippone, E., C. Conicella, A. Errico, and F. Saccardo. 1984a. Karyotype analysis of a mutant of *Vicia faba* L. by microdensitometric procedure. Pages 52-59 in *Systems of Cytogenetic Analysis in Vicia faba L.* edited by G.P. Chapman and S.A. Tarawali. Wye University College, University of London, UK.

Filippone, E., L.A. Smaldone, and L.M. Monti. 1984b. Microdensitometric analysis of *Vicia faba* L. chromosome image. Pages 225-230 in *Vicia faba L: Agronomy, Physiology and Breeding* edited by P.D. Hebbletwaite, T.C.K. Dawkins, M.C. Heath, and G. Lockwood. University of Nottingham, UK.

Fleischman, T., T. Gustaffson, and C.H. Hakansson. 1971. Computer display of the chromosomal fluorescence pattern. *Hereditas* 68: 325-328.

Floresca, E.T., J.M. Capinpin, and J.V. Pancho. 1960. A cytogenetic study of bush sitao and its parental types. *Philippines Agriculture* 44: 290-298.

Frahm-Leliveld, J.A. 1965. Cytological data on some wild tropical *Vigna* species and cultivars from cowpea and asparagus bean. *Euphytica* 14: 251-270.

Fukui, K. 1986. Standardization of karyotyping plant chromosomes by a newly developed chromosomes image analyzing system (CHIAS). *Theoretical and Applied Genetics* 72: 27-32.

Galasso, I., D. Pignone, N.Q. Ng, and S.R. Singh. 1990. Verso una standardizzazione del cariotipo di *Vigna unguiculata* (L.) Walp. Page 92 in *Atti del XXXIV Convegno Annuale della Società Italiana di Genetica Agraria, 8-11 October 1990, Marina di Ugento (LE), Italy.*

Gosal, S.A., and Y.P.S. Bajaj. 1983. Interspecific hybridization between *Vigna mungo* and *V. radiata* through embryo culture. *Euphytica* 32: 129-137.

Karpetchenko, G.D. 1925. On the chromosome of the Phaseolinae. *Bulletin of Applied Botany and Plant Breeding* 14: 143-148.

Kawakami, J. 1930. Chromosomes numbers in Leguminosae. *Botanical Magazine (Tokyo)* 44: 319-328.

Krishnan, R., and D.N. De. 1968. Cytological studies in *Phaseolus*. II. *Phaseolus mungo* x tetraploid species and the amphidiploid. *Indian Journal of Genetics and Plant Breeding* 28(1): 23-30.

Kurata, N., and T. Omura. 1978. Karyotype analysis in rice. A new method for identifying all chromosome pairs. *Japanese Journal of Genetics* 53: 251-255.

Lubs, H.A., and R.S. Ledley. 1973. Automated analysis of differentially stained human chromosome: A review of goals, problems and progress. *Nobel Symposium* 23: 61-76.

Machado, M., W. Tai, and L.R. Baker. 1982. Cytogenetic analysis of the interspecific hybrid *Vigna radiata* x *V. umbellata*. *Journal of Heredity* 73: 141-146.

Marimuthu, K.M., W.D. Selles, and P.W. Neurath. 1974. Computer analysis of giemsa banding patterns and automatic classification of human chromosomes. *American Journal of Human Genetics* 26: 369-377.

McClintock, B. 1929. Chromosome morphology of *Zea mays*. *Science* 69: 629-630.

Mendelsohn, M.L., T.I. Conway, D.A. Hungerford, W.A. Kolman, B.H. Perry, and I.M.S. Prewitt. 1966. Computer oriented analysis of human chromosomes. I. Photometric estimation of DNA content. *Cytogenetics* 5: 223-242.

Mouras, A., G. Salesses, and A. Lutz. 1978. Sur l'utilization des protoplasts en cytologie: Amelioration d'une methode recente en vue de l'identification des chromosomes mitotique des genres. *Caryologia* 31: 117-127.

Mukherjee, P. 1968. Pachytene analysis in *Vigna*: Chromosome morphology in *Vigna sinensis* (cultivated). *Science and Culture* 34: 252-253.

Pignone, D., S. Cifarelli, and P. Perrino. 1990. Chromosome identification in *Vigna unguiculata* (L.) Walp. Pages 144-150 in *Cowpea Genetic Resources* edited by N.Q. Ng and L.M. Monti. IITA, Ibadan, Nigeria.

Saccardo, F. 1990. Chromosome characterization in *Vigna* spp. In *Proceedings, Joint Biotechnology Workshop, Purdue University, 16-20 July 1990*. (Unpaged, informal publication). Purdue University, West Lafayette, Indiana, USA.

Sawa, M. 1973. On the interspecific hybridization between the adzuki bean *Phaseolus angularis* (Wild) W. F. Wright and the green gram, *Phaseolus radiatus* (L.). 1. Crossing between a cultivar of the green gram and a semi-wild relative of adzuki bean in endemic name "Bakaso". *Japanese Journal of Breeding* 23: 61-66.

Schnapp, S.R., R.E. Shade, L.W. Kitch, L.L. Murdock, R.A. Bressan, and P.M. Hasegawa. 1990. Utilization of in vitro culture methods to facilitate introgression of insect resistance into cowpea (*Vigna unguiculata*). In *Proceedings, Joint Biotechnology Workshop, Purdue University, 16-20 July 1990*. (Unpaged, informal publication). Purdue University, West Lafayette, Indiana, USA.

Sen, N.K., and J.G. Bhowal. 1960a. Cytotaxonomic studies on *Vigna*. *Cytologia* 25: 195-207.

Sen, N.K., and J.G. Bhowal. 1960b. Colchicine-induced tetraploids of six varieties of *Vigna sinensis*. *Indian Journal of Agricultural Science* 3: 149-161.

Senn, H.A. 1938. Chromosome number relationships in Leguminosae. *Bibliographical Genetics* 12: 175-336.

Venora, G., C. Conicella. A. Errico, and F. Saccardo. 1991. Karyotyping in plants by an image analysis system. *Journal of Genetics and Breeding* 45: 117-122.

Vosa, C.G., and P. Marchi. 1972. On the quinacrine and Giemsa staining patterns of the chromosomes of *Vigna faba*. *Giornale Botanico Italiano* 106: 151-159.

2.7

Plant genetic resources and their conservation

P. Perrino

Germplasm Institute, National Research Council, Via Amendola 165/A, Bari, Italy

Abstract

Of the more than 10 million species of living organisms, only about 10% have survived the ice ages (first narrowing), the spread of agriculture (second narrowing), and scientific plant breeding (third narrowing). Of the approximately 300 000 species of flowering plants, only 200 have been domesticated and today only eight crops supply the world's human population with about 75% of its food. However, in the second phase, the narrowing of plant diversity above species level was compensated for by an increase of diversity within crops. The spread of agriculture was largely due to this process of crop evolution under domestication. Genetic erosion started during the third phase. The need to conserve genetic diversity was recognized some 50 years ago. Today, over 1 million accessions of the most important crops are stored in nearly 100 seedbanks. However, more genebanks to store field collections and in vitro cultures are needed for species with recalcitrant seeds or those vegetatively propagated, for which in situ conservation is not practical. Despite all these efforts, it has been estimated that for some genera, especially for wild species, the representation of collected material is far from adequate. Urgent action is indicated for wild relatives in many ecosystems. Genetic resources are crucial to the development of biotechnology, and genebanks are a prime source of material in this field.

It has been estimated that of about 300 000 species of the higher plants originally extant in the world, only about 3000 (1%) are used by man. Of these only 200 have been brought into cultivation; and only 30 of these species supply nearly all the food consumed by the human population (*see* Bennet 1971; Scarascia Mugnozza 1984; Chang 1985). About 75% of this food is provided by only eight cereal species (wheat, rice, maize, barley, oats, sorghum, millet, and rye), with wheat, rice, and maize accounting for 75% of this amount. The first narrowing of the spectrum of plant species started with the ice ages, the second with the beginning of agriculture (neolithic revolution), and the third with the beginning of scientific plant breeding.

With the advent of scientific plant breeding early this century, the rapid spread of high-yielding varieties, characterized by a narrow genetic base, caused the displacement of traditional unimproved cultivars which had large genetic bases. The extinction of landraces because of the spread of improved varieties is called genetic erosion. This trend towards uniformity increases the genetic vulnerability of the major crops to diseases and insects (Anon. 1969). Narrowness of the genetic base of a crop may lead to disasters, as shown by historical examples (NAS 1972). Agronomists, now aware that genetic uniformity increases disease vulnerability, try to diversify their varieties within the constraints imposed by commercial breeding. New varieties are scientifically engineered to meet the tough requirements of machine harvesting, milling, brewing, and baking. While they may offer many advantages, their uniformity increases the danger of disease- and pest-related losses.

It is widely accepted that future plant breeding efforts will depend on a continuing and expanding supply of germplasm. Thus, an urgent task for the future is to conserve crops of major importance, such as wheat, rice, and maize, whose variability is experiencing increasing genetic erosion, and crops that may play a role in the development of new agricultural systems (Scarascia Mugnozza et al. 1988). Although it is impossible to conserve all available gene resources, a practical and successful strategy has been to identify germplasm categories and from each of them to collect and conserve a representative sample.

Methods of Conservation

Methods for the maintenance of genetic resources vary according to several factors, including the species, their geographical distribution, breeding system and seed behaviour (Perrino 1990). However, two main methods have so far been identified:

- *Ex situ conservation.* This is conservation in an artificial habitat or in a habitat different from the original one, depending on the kind of habitat (for example, preservation of seeds — seedbanks; preservation of plants — field collections, botanical gardens, parks, etc.; preservation of plant parts, such as cells, tissues, meristem culture, and even genes — in vitro culture and gene libraries). Collections of germplasm using any of these methods are often called genebanks.

- *In situ conservation.* This is conservation of a dynamic nature in natural or original habitats (for example, conservation of large tracts of land to protect plants and animals, both wild and domesticated; and conservation of landraces of cultivated species in their areas of cultivation).

Seedbanks

Seeds are by far the most convenient parts of plants for storage. With few exceptions, every seed has a different genetic constitution and thus a wide range of genetic variability may be included in a single sample stored in a small, sealed container.

The best storage temperature for base collections (long term) is about -20°C. Under these conditions some seed may survive for 100 years. Active collections are usually maintained at a temperature between 0°C and 5°C and a relative humidity of 35%. Under these conditions, seed viability may last for several years. In both cases, regular checking for viability is necessary. Samples must be grown out before the seeds begin to deteriorate so that a fresh generation of seeds can be obtained for continued storage.

A series of handbooks containing information on methods for conserving and testing seeds has been produced by the International Board for Plant Genetic Resources (IBPGR 1985, 1986). However, more studies on the regeneration of cross-pollinated species and other difficult material are needed for better management of genetic resources in seedbanks (Porceddu and Jenkins 1981).

Field genebanks

Species that do not easily produce seed and those with recalcitrant seeds may be conserved in field genebanks. These are pieces of land on which collections of plant have been assembled. In the past, they were limited to clones of banana, coconut, cacao, sugar cane, sweet potato, potato, cassava, and citrus. Today, one may add several others, many of them tropical and subtropical species. Some of the most valuable collections of banana, plantain, oil palm, and coffee are those established long ago in field genebanks.

Tissue culture and cryopreservation

In vitro methods are well suited for mass cloning of a single species or cultivar, as well as storage under "slow growth" conditions. However, each cultivar requires specially formulated techniques, and it is difficult and time-consuming to determine the best combination of nutrient and conditions for growing a particular species. Nevertheless, for plants that do not form seeds, such as those propagated from bulbs or rhizomes, in vitro techniques are the only option. The number of accessions that should be included in an active in vitro collection (at 0°C or 5°C) to represent the variability of a species has been estimated to be 500 for sugar cane, 1000 for banana, 2000 for sweet potato, and 5000 for cassava. It is estimated that between 25 to 625 cultures are necessary in order to cover the genetic diversity of each accession (IBPGR 1986).

The best answer for long-term conservation of germplasm in vitro seems to lie in cryopreservation — the storage of frozen tissue cultures at very low temperatures (for example, in liquid nitrogen at -196°C), which virtually stops all biologial processes. Cryopreservation of in vitro cultures of calli, suspensions, and protoplasts has proved successful for many herbaceous plants, including *Triticum*, *Glycine* and *Daucus*), as well as for forest trees, such as *Acer pseudoplatanus* and *Populus americana* and fruit trees (*Phoenix dactylifera*) (Stushnoff and Fear 1985). Apple shoot tips from in vitro cultures have shown a high degree of survival in liquid nitrogen. If this technique could be perfected so as to eliminate the damage caused by freezing and thawing, materials could be conserved almost indefinitely.

Much more work is needed to improve tissue culture techniques before in vitro conservation can be used widely and with confidence, and it may never match the efficiency of seed storage. However, for species where conservation as seeds is not possible, this method offers a promising alternative.

Gene libraries

Molecular biology may provide a new means of storing genetic resources. Gene and total genome libraries can now be readily produced (Peacock 1984). For most plant species whose seeds may be stored for a long time, gene libraries may have no particular advantage. Gene and genome libraries are simple collections of genes incorporated into some appropriate bacterial vector. They would preserve the total information but not coordinated information, as in a seedbank.

However, gene libraries may be useful as a first step in the acquisition of a given gene from a seedbank. In addition, a gene library may provide a new way for storing crop plants that have recalcitrant seeds or are vegetatively reproduced and for which at present there is no other satisfactory long-term storage technique.

Summing up, ex situ conservation cannot provide the opportunity for a wild relative to continue the evolutionary process that a species undergoes in its natural environment. There is no pressure to adapt to changing natural conditions nor to compete with other species. Moreover, genebanks, with large collections of germplasm in a small space, are vulnerable to a variety of mishaps, such as breaks in the power supply, incorrect labelling, or failure to regenerate the plants in time. Consequently, samples should be stored in more than one genebank to avoid the risk of total loss. The duplication of collections may be organized through the development of core collections (Frankel and Brown 1984).

However, ex situ methods do keep germplasm safe when plants are destroyed in their natural habitat and they have the advantage, for the user, of making material from widely scattered localities available in one place, ready for use. Although large genetic stocks have already been stored in existing genebanks (Plucknett et al. 1987), it has been estimated that for wild species the representation of some genera is far from adequate (Hoyt 1988). Thus, it is urgent that the collection of other germplasm continues in various centres of origin according to priorities at the regional and crop levels (IBPGR 1981, 1986; Chang 1985).

Conserving germplasm in natural ecosystems

A great advantage of in situ conservation is that species can continue to evolve, allowing the appearance of new recombinant forms. For tropical and perennial species with recalcitrant seeds, in situ conservation may be an essential complement to in vitro conservation. In recent years, the objectives in the conservation of natural areas, particularly in the tropics, have been to protect a representative sample of each ecosystem and to establish a world network of biosphere reserves to maintain communities of plants and animals in their natural ecosystems. In this way, landraces of important crops are protected in their natural habitat.

Genetic Resources and Biotechnology

Germplasm is a basic resource for biotechnology. The best expression of this relationship was given by Lewis (1985): "An adequate gene resource conservation programme is to genetic engineering as a library is to knowledge. They both are sources of the past, the present, and the future, and both are equally essential to the development of their counterpart." Germplasm collections of crops, their wild relatives, and unrelated plants are essential if biotechnology is to move ahead (Witt 1985).

There are two good reasons why biotechnologists need germplasm collections. First, they need models to synthesize genes because, at the moment, they cannot invent genes. Second, biotechnologists will continue to depend heavily on naturally occurring genes in their experiments to influence the future. Genebanks are thus the mecca for gene hunters.

Conclusion

Every year the world continues to lose forests and over 20 million hectares become desert or economically unproductive. Recently, efforts have begun to deal with these problems from technical, economic, and political perspectives (Esquinas-Alcàzar 1987).

Technically, the development of new methods for genetic resources conservation have opened up new possibilities, but further studies to overcome theoretical and practical difficulties are needed. Economically, resources for ex situ and in situ conservation are insufficient. In March 1987, under a proposal supported by 77 countries, the Food and Agriculture Organization of the United Nations (FAO) created the International Fund of Plant Genetic Resources; more extensive funding is now necessary for it to be effective. Politically, the recognition of plant genetic resources as a common heritage of all mankind has been a major step in giving germplasm preservation its rightful status. Nevertheless, many countries have not yet decided to participate in these common efforts. Legislation must have an important role in controlling genetic erosion, protecting genetic resources, and promoting free access to them. The concept of "farmers' rights" includes recognition of the fact that farmers have developed and continue to develop genetic diversity. This subject, still under discussion, may represent another way to create and perfect useful germplasm.

Awareness of the narrowing of genetic resources, like other problems such as lead poisoning, fertilizer runoff, industrial water pollution, massive solid waste production, and the greenhouse effect, must become part of the thinking of every human being.

Discussion

MURDOCK: Is there (or should there be) a plan to collect and preserve any genetically transformed plant materials at some location so that it will be available to all for purposes other than those originally intended?

PERRINO: Yes, there should be a plan to collect and preserve transformed plants but this is less urgent than the need to conserve landraces and wild relatives, either in situ or ex situ.

References

Anon. 1969. Genetic dangers in the Green Revolution. *Ceres* 2 (5): 35-37.

Bennet, E. 1971. The origin and importance of agroecotypes in South-West Asia. Pages 219-234 in *Plant Life in South-West Asia* edited by P.H. Davis, P.C. Harper, and I.C. Hedge. Edinburgh Botanical Society, Edinburgh, UK.

Chang, T.T. 1985. Principles of genetic conservation. *Iowa State Journal of Research* 59(4): 325-348.

Esquinas-Alcàzar, J.T. 1987. Plant genetic resources: A base for food security. *Ceres* 20(4): 39-45.

Frankel, O.H., and A.H.D. Brown. 1984. Plant genetic resources today: A critical appraisal. Pages 249-257 in *Crop Genetic Resources: Conservation and Evaluation* edited by J.H.W. Holden and J.T. Williams. Allen and Unwin, London, UK.

Hoyt, E. 1988. *Conserving the Wild Relatives of Crops*. IBPGR, Rome, Italy.

IBPGR. 1981. *Revised Priorities among Crops and Regions*. IBPGR, Rome, Italy.

IBPGR. 1985. *Ecogeographical Surveying and In Situ Conservation of Crop Relatives*. IBPGR, Rome, Italy.

IBPGR. 1986. *Design, Planning and Operation of "In Vitro" Genebanks: Report of a Sub-Committee of the IBPGR Advisory Committee on "In Vitro" Storage*. IBPGR, Rome, Italy.

Lewis, L.N. 1985. Genetic engineering: One leg of three-legged stool. *California Agriculture* 39: 2.

NAS. 1972. *Genetic Vulnerability of Major Crops*. National Academy of Sciences, Washington DC, USA.

Peacock, W.J. 1984. The impact of molecular biology on genetic resources. Pages 268-276 in *Crop Genetic Resources: Conservation and Evaluation* edited by J.H.W. Holden and J.T. Williams. Allen and Unwin, London, UK.

Perrino, P. 1990. Germplasma e ambiente. *Biologi Italiani* 7-8: 19-30.

Plucknett, D.L., N.J.H. Smith, J.T. Williams, and N. Anishetty Murthy. 1987. *Genebanks and the World's Food*. Princeton University Press, Princeton, New Jersey, USA.

Porceddu, E., and G. Jenkins. 1981. Seed regeneration in cross-pollinated species. Page 293 in *Proceedings, CEC Eucarpia Seminar*, Nyborg, Denmark, 15-17 July 1981.

Scarascia Mugnozza, G.T. 1984. La fame nel mondo, il problema scientifico e organizzativo, non emotivo. Pages 209-242 in *Agricoltura e Ambiente, il Problema del XXI Secolo*. REDA, edizioni per l'agricoltura, Rome, Italy.

Scarascia Mugnozza, G.T., E. Porceddu, and C. De Pace. 1988. Genetic resources and modern agriculture. Pages 701-738 in *Agricolo e Sicurezza Alimentare* edited by G.T. Scarascia Mugnozza. Fondazione Degli Studi Tropicalisti e Cotonieri, Firenze, Italy.

Stushnoff, C., and C. Fear. 1985. *The Potential Use of "In Vitro" Storage for Temperate Fruit Germplasm*. IBPGR, Rome, Italy.

Witt, S.C. 1985. *Briefbook: Biotechnology and Genetic Diversity*. California Agricultural Lands Project, San Francisco, USA.

3.1

Recent advances in cell and tissue culture[1]

E. Filippone[2], M. Leone[2] and R. Penza[3]

Research Centre for Vegetable Breeding, National Research Council, Via Università 133, Portici, Italy[2]; Department of Agronomy and Plant Genetics, University of Naples, Via Università 100, Portici, Italy[3]

Abstract

In the past, plant cell and organ cultures were seen simply as a tool for mass propagation or virus eradication. Today, cell and tissue cultures are playing a central role in applied plant biotechnology for mutant selection, gene transfer, artificial seed and secondary metabolites production, to mention just a few applications, whilst robotics is being applied on order to increase the productivity in mass propagation of ornamental crops. The focus of this paper is on two topics involving tissue and cell culture: the maintenance of genetic stocks; and the introduction of variability to widen the genetic base of a species.

Since the first successful attempt to grow plant tissue in a fully controlled environment, or in vitro, and to regenerate a complete plant from a single cell (Vasil and Hildebrandt 1965), there have been tremendous improvements in techniques and knowledge. In vitro tissue culture techniques are not only a powerful system to study plant development, from a single cell to a complete formed plant, and to study the physiology of growth in the absence or presence of stress, but also now represent a way to introduce foreign genetic material directly in the nucleus of a cell, overcoming sexual barriers (Evans et al. 1983). However, as reported by Robertson (1990), this genetic and technological revolution is leaving millions of people in developing countries far behind. The simplest application of in vitro technology, meristem culture, and virus eradication could produce thousands of new clones of staple food plants ready to be transferred in the field, with an obvious potential for increased production. This kind of technology is available worldwide and, if laboratory

1 Contribution No. 72 from the Research Centre for Vegetable Breeding, National Research Council, Via Università 133, 80055 Portici, Italy.

facilities are provided, several developing countries could, within a few years, increase the productivity of economically important plants. In Mexico, for example, where micropropagation of selected plants, isolation of secondary metabolites from cell cultures, and the discovery of new sources of germplasm and its preservation have only recently been achieved by Mexican agribusiness, there is now the prospect of a new Green Revolution through the application of these technologies (Medina-Gomez 1990). This paper discuss some of these technologies in terms of their application in developing countries in the medium term as well as the long term.

Germplasm Preservation and Propagation

Meristem culture and cryopreservation

Meristem cultures are widely applied to obtain virus-free clones. The success of Morel and Martin (1952) in eliminating virus from infected dahlia plants, and subsequently from potato plants, stimulated meristem culture in many horticultural species. The biological basis of virus eradication via meristem culture is founded on the observation that the titre of a virus is highest in leaves or stems of plants and is low or absent in newly developed meristem tissue. The smaller the meristem-tip explant cultured, the higher the expected recovery of virus-free plants is, but explant survival and plant regeneration will be reduced compared to regeneration from larger shoot-tip explants. Regenerated plantlets need to be indexed for virus presence before they are given a virus-free designation (Walkey et al. 1980; Poehlman 1987). Once isolated, virus-free clones can be maintained in vitro to provide disease-free stocks for multiplication through conventional or in vitro propagation. Commercial applications of virus eradication via meristem cultures have been reported for *Camellia* species, *Euphorbia pulcherrima*, *Pelargonium zonale*, *Pisum sativum*, *Narcissus* species and *Solanum tuberosum* (Manette 1983). Some of the plant species submitted to cryopreservation are listed in Table 1.

The conservation of plant material in vitro is necessary when seed storage or other conventional genetic-stock conservation is inadequate or not possible. Basically, there are two ways to preserve genetic stocks: with or without freezing. The latter system, at an

Table 1 Some plant species submitted to cryopreservation

Arachis hypogaea	*Malus pumilia*
Cicer arietinum	*Manihot esculenta*
Citrus sinensis	*Nicotiana tabacum*
Cocos nucifera	*Palma* spp.
Coffea arabica	*Petunia hybrida*
Dactylis spp.	*Picea glauca*
Daucus carota	*Pisum sativum*
Dianthus caryophyllus	*Pyrus communis*
Elaeis guineensis	*Saussurea lappa*
Fragaria x *ananassa*	*Solanum tuberosum*
Lycopersicon esculentum	*Triticum aestivum*

elementary level, consists in subculturing in vitro nodes or other vegetative organs. This method could introduce some genetic variability; for example, in the legume *Stylosanthes* about 40% of the plants coming from leaf tissues were found to be variants (Godwin et al. 1987). Hence, it is necessary to avoid any de-differentiation and subsequent re-differentiation of new buds (Towill 1989).

When cells or little pieces of tissue are conserved at a very low temperature (about -196°C using liquid nitrogen), every genetic modification process is greatly reduced (Scowcroft 1985). Protocols have been established for meristems, somatic embryos, shoot tips, seedlings, winter buds, anthers, pollen embryos, suspension cell culture, protoplasts, and calluses of various crops (Table 2), although this process is still empirical (Kartha 1982). Viability of cells after the cryogenic storage depends upon various factors, such as age of the plant material, length of the conservation period, and composition of the cryogenic medium. Above all, the recovery stage is the most important and delicate one. Basic cryopreservation procedures are described in Kartha (1985). It is necessary to point out that in vitro germplasm conservation is the major task for international institutes (Withers et al. 1990).

Table 2 **Viability after recovery following cryopreservation of some plant species**

Species	Type of material	Viability (%)
Arachis hypogaea	meristems	23-31
Cicer arietinum	meristems	27-36
Daucus carota	somatic embryos	100
Dianthus caryophyllus	shoot tips	80
Dactylis sp.	seedlings	30
Fragaria x *ananassa*	meristems	95
Lycopersicum esclulentum	seedlings	30
Malus pumilia	winter buds	100
Manihot esculenta	meristems	13
Nicotiana tabacum	anthers	2
Petunia hybrida	anthers	5
Pisum sativum	meristems	60
Solanum tuberosum	meristems	60

Source: Kartha 1982.

Mass propagation

Mass propagation by means of in vitro tissue culture has attracted increasing interest as an alternative to conventional asexual propagation of economically important plant species. It offers the following potential advantages: thousands of cloned plants can be regenerated from a small amount of plant tissue; in vitro propagation could provide an alternative method for species that are resistant to conventional bulk propagation; and international exchange of plant stocks could be safer when in vitro tissue cultures are involved. In fact,

Table 3 Basic procedures for in vitro propagation of plants

Stage I	Selection of small pieces of conventionally grown plants, their sterilization and inoculation onto medium
Stage II	Transfer to a medium or sequence of media that promote induction and/or proliferation of shoots
Stage III	Transfer of shoots to media or conditions that induce root formation, followed by planting out in the greenhouse

Source: Hu and Wang 1983.

the sterile status of the cultures reduces the likelihood of spreading diseases, and thus quarantine requirements are reduced. In vitro plant propagation procedures are summarized in Table 3.

In some species of high value, meristems and bud culture have achieved propagule multiplication rates which are competitive with conventional asexual propagation based on vegetative cuttings (Beversdorf 1990). Recently, Levin and Vasil (1989) reviewed the progress and potential for reducing the costs associated with the multiplication phase of in vitro propagule production, in which the labour of manual shoot separation and transfer represents 60-80% of the production costs. Improved bioreactor technology and an automated or robotic handling system may reduce propagule costs significantly (Levin et al. 1988).

Periodic inspection of stocks is necessary to check for the presence of somaclones or chimaeric shoots.

Somatic embryogenesis and artificial seeds

Somatic embryogenesis is the development of embryos from cells that are not the product of gamete fusions; in fact, they develop directly from somatic cells without gametogenesis. Somatic embryos closely resemble their zygotic counterparts from the structural and biochemical points of view (Ammirato 1987).

This type of adventitious embryony has been reported in an ever-increasing number of plants, including cereals (Vasil 1988), woody plants (Von Arnold and Wallin 1988), grain or large-seeded legumes and even recalcitrant tropical crops such as banana (Novak et al. 1989), cassava, and mango (Table 4). Somatic embryogenesis creates a number of opportunities not available when plants are regenerated via organogenesis: in one step it is possible to obtain a bipolar structure bearing both roots and shoot apices; large-scale propagation is achieved compared with vegetative propagation; and artificial seed production technology can be applied.

The availability of somatic embryos, or microspore-derived embryos, has prompted investigation into ways of delivering them to the planting site. One approach is the realization of synthetic or artificial seeds consisting of somatic embryos coated or incapsulated to create analogs to botanic seeds. The most recent advances in the production of artificial

Table 4 Ex-plant source, culture medium, and growth regulators inducing somatic embryogenesis in some monocotyledonous and dicotyledonous plants

Species	Ex-plant source	Medium and hormone[a]
Acrocomia aculeata	zygotic embryos	MS; 2,4D
Apium graveolens	cell culture	MS; ABA
Asparagus officinalis	hypocotyl	MS; 2,4D
Carica papaya	mature zygotic embryo	LS; NAA, KIN, GA
Cicer arietinum	hypocotyl	MS; BAP, IAA
Cycad sp.	zygotic embryo	B5; 2,4D, KIN
Elaeis guinnensis	leaf	MS; 2,4D,KIN
Euterpe edulis	immature embryo	MS; 2,4D
Glycine max	immature cotyledon	MS; 2,4D
Nicotiana tabacum	cotyledonary leaf	MS; 2,4D
Picea abies	mature zygotic embryo	MS; NAA/2,4D, BAP
Pisum sativum	epicotyl	MS; picloram
Solanum melongena	leaf	MS; NAA
Solanum tuberosum	immature embryo	MS; 2,4D
Trifolium alexandrinum	hypocotyl	MS; 2,4D, NAA, KIN
Triticum aestivum	root or leaf	MS; NAA, BAP

Note: a LS = Lindsmaier and Skoog 1965; MS = Murashige and Skoog 1962; B5 = Gamborg et al. 1968; 2,4D = 2,4-dichlorophenoxy-acetic acid; ABA = abscisic acid; BAP = 6-benzyl-aminopurine; GA = giberellic acid; IAA = indolacetic acid; KIN = 6-furfurylaminopurine; NAA = 1-naphthalene acetic acid.
Source: IAPTC 1990.

seeds are those relating to *Medicago sativa* (Senaratna et al. 1989), *Hordeum vulgare* (Datta and Potrykus 1989), and *Apium graveolens* (Kim and Janick 1989). Recently, a review on application of artificial seed technology to tropical crops was published (Redenbaugh 1990).

Induction and Expression of Genetic Variability

Pollen, ovary culture, and in vitro fertilization

Interest in haploids stems largely from their considerable potential in plant breeding. As haploids possess only one set of alleles at each locus, it is possible for recessive mutants to be detected without awaiting segregation. Furthermore, doubling of the chromosome number of haploids offers a method for the rapid production of homozygous plants useful in obtaining inbred lines for hybrids production. In the same way, the recovery of unique recombinants is made considerably easier. Haploids may occur spontaneously as a result of apomixy or parthenogenesis, or they can be induced experimentally by anther, pollen, or ovule culture (D'Halluin and Keimer 1986; Beversdorf 1990).

The culture of isolated pollen offers the following advantages over the anther culture: uncontrolled effects of the anther wall and other associated tissues are eliminated; pollen is ideal for uptake, transformation, and mutagenic studies, as pollen may be uniformly exposed to chemicals or physical mutagens; unlike single callus cells, pollen is transformed directly into an embryo, and thus is suitable for understanding the physiology and biochemistry of androgenesis; and a higher yield of plants per anther can be expected.

Fertilized ovules are sometimes cultured to rescue embryos from wide crosses without excision of the embryo from the ovule.With ovule culture, the embryos may be cultured at an earlier stage than with an excised embryo. Embryos developing within the ovule may have a more favourable chemical and physical environment for growth and development than embryos cultured outside the ovule. This technique enables us to obtain a great number of haploids, due to a genetic mechanism which controls chromosome elimination.

Among the reviews published on anther, pollen, and ovary culture of herbaceous as well as tree plants are Williams et al. (1987), Bonga et al. (1988), and Gobel and Lorz (1988).

Somaclonal variation

Micropropagation is often simplistically perceived as the cloning of selected genotypes. In reality, the process is far more complicated. The selection of a mother plant is often based on the phenotype. The micropropagation procedure may cause phenotypic (physiological) or genotypic (somaclonal) changes, which result in some plants being different from the mother plant. The process itself is highly dependent on:

- The genotype — in fact, some cultivars have been shown to give rise to more somaclonal variation than others (Larkin 1987)
- The ploidy of the source material, which influences the frequency and nature of the variation obtained, although, in general, polyploids exhibit greater somaclonal variation than diploids (Fish and Karp 1986)
- The tissue culture procedure (excluding media composition); for example, protoplast regeneration is associated with more somaclonal variation than regeneration from cultured immature embryos and explants (Armstrong and Phillips 1988)
- The tissue sources; for example, in potato plants regenerated from different tissues, a higher frequency of aneuploidy was seen in regenerants from tuber pieces (Wheeler et al. 1985)
- The hormones in the media, which seem to influence variation in tissue cultures (Karp et al. 1984)

Furthermore, it is well known that variation is already present in plant tissues, as spontaneous mutation rates in plants have been estimated at between 10^{-4} to 10^{-7} per locus (Bhatia et al. 1985). Together with the fact that endopolyploidy and polyteny occur during the differentiation of some plant tissues, this indicates that some variation is already present in the source tissue. Different tissue sources should, therefore, be tested for their capacity to generate instability.

Culture conditions constitute another source of variation. In normal plant development, any mutations arising during DNA replication will be subjected to selective constraints during growth or reproduction. Cells in culture, however, are separated by many cycles of cell division from the original source plant. Each cycle of DNA replication will not only propagate any variation inherited from the ancestral cell but also generate new aberrations, which will accumulate if selective constraints are removed. An alternative explanation for the role of tissue culture phase on somaclonal variation is that it imposes stress. This stress could then be viewed as enhancing the spontaneous mutation rate disproportionately from the rate of DNA repair. There is evidence indicating that particular DNA sequences have a role in generating instability in culture (Johnson et al. 1987).

Whatever the process causing somaclonal variation, induced variation has been used as a basis for selection of new genotypes, showing better adaptation to biotic and abiotic stresses (Evans and Sharp 1986).

Gene transfer

Agrobacterium-mediated transformation

Agrobacterium tumefaciens is frequently used for plant transformation, becoming a good flexible gene delivery system. The host range of *Agrobacterium* includes soybean, cotton, rapeseed, tomato, alfalfa, potato, and many other dicotyledonous crops. A list of some species stably transformed by co-cultivation with *Agrobacterium* is reported in Paper 3.3 (pp 127-134 this volume).

The four major indirect techniques for the transformation of plant cells by *Agrobacterium tumefaciens* are:

- *Wounding and inoculation of intact or decapitated plants.* In this case, wounding stems or leaves induces a response that plays a role in the specific bacterium : tissue attachment.

- *Inoculation of explants in vitro.* This technique has several advantages for the production of transformed plants. An antibiotic, such as cefotaxime, allows the elimination of the bacteria after transformation has occurred. Some strains of *Agrobacterium* have been used to induce growth of transformed shoots that can be grown to maturity by grafting them to normal plants.

- *Co-cultivation of growing protoplast-derived cells with the bacteria.* This permits very stringent selection for resistance to antibiotics and has several advantages over explant infection. In fact, protoplasts show a decreased tendency to clump, permitting the isolation of a clone of cells originating from a single transformation event.

- *Leaf disc transformation.* This avoids the mutations or chromosomal abnormalities induced in cell culture, and it is appropriate for those species that do not show well-developed protoplast regeneration. It consists of the co-cultivation of leaf mesophyll cells that are still within the tissue of a leaf slice or disc. Leaves provide an abundant source of genetically uniform cells with the capacity to regenerate whole plants for a wide range of species.

Direct gene transfer

Agrobacterium cannot yet be used to transform cereals or other species that are recalcitrant to regeneration; hence, the need has arisen for a generally effective gene transfer technique.

One possible solution is the development of direct gene transfer techniques, the most promising of which are:

- *Macroinjection*, using needles with diameters far greater than cell diameters, allows the delivery of DNA into wound sites within tissues. The DNA solution is further absorbed by cells surrounding the wounded site. Most experiments involving macroinjection of ovules, embryos, and other tissues have not attracted much attention. Experiments have been carried out on floral meristems of rye (De la Pena et al. 1987).

- *Microinjection* allows the delivery of DNA directly into the nucleus of a cell. Microinjection has been applied to rapeseed microspore-derived embryos (Neuhaus et al. 1987). Plants derived from macro- or microinjection may be mostly chimaeric; an appropriate selection procedure is necessary to recover transformed tissues.

- *Liposome fusion* with protoplasts is an established method for the production of transgenic plants. The rationale behind this technique is that liposomes help the DNA to enter through the plasmodesmata or that lipids impregnate the cell walls, making it easier for DNA to penetrate (Gad et al. 1986). Again, this method works only in species that can regenerate from a single cell; however, regenerated plants are totally transformed.

- *Microlaser-driven genetic transformation* is based on a laser beam that is focused into the light path of a microscope. The focused laser beam can be used to burn holes in the walls of pollen and cells in tissues, and can also open membranes, including those of chloroplasts, for short intervals. Attempts to use these openings for the entry of DNA molecules have yielded ambiguous data (Weber et al. 1990).

- *Electrodiffusion* is another physical method used to drive DNA into cells. There is evidence to indicate uptake and expression of DNA applied electrophoretically to barley cariopses. However, the data cannot be interpreted as proof of integrative transformation (Ahokas 1989).

- *Biolistics* is a recently developed physical technique that is very effective, at least for transient expression. This method uses acceleration of small metal particles covered with DNA to deliver DNA into plant tissue. It is seen as a technique applicable for gene transfer into every desired variety of plant species (Sanford 1990). The advantages of biolistics are: it is relatively easy to handle; one shot yields many hits; the cell survives the intrusion of particles; and it is unnecessary to regenerate new plants from tissues, so that problems caused by somaclonal variation are avoided. Some stable transformants rescued from transgenic chimaeric plants have been obtained in soybean (Christou et al. 1989) and maize (Gordon-Kamm et al. 1990), after the recovery of solid mutants in the progeny of plants obtained after embryo bombardment.

Conclusion

There are a number of techniques available for conservation against gene erosion or preservation of germplasm, for inducing somatic embryogenesis, for widening the genetic base of a species through somaclonal variation, and for transferring new genes with or without the need of in vitro regeneration. This wide variety of techniques allows those involved in breeding programmes, as well as in research, to follow strategies with a good chance of reaching predetermined goals in a short time. Unfortunately, problems still arise when genetic uniformity is required in regenerants, or when a reliable method of transferring DNA is needed to shorten breeding programmes. Most importantly, however, some basic techniques, such as micropropagation and virus eradication, are already available for widespread application in laboratories in developing countries, where even a limited effort to propagate virus-free clones or new varieties could be of great benefit to agriculture.

Discussion

ADU-AMPOMAH: Since factors controlling somaclonal variation are not well understood, is it not dangerous to mass propagate plants via somatic embryogenesis?

FILIPPONE: Although some somaclonal variants are found in plants from somatic embryogenesis, the proportion is still very low compared with the proportion introduced by regeneration via organogenesis. In fact, somatic embryogenesis is a rather complex phenomenon that appears to occur only in cells that are not physiologically or genetically disturbed. In short, it seems that diplontic selection acts to avoid the entrance of a variant cell into somatic embryogenesis. In view of this, somatic embryogenesis could be used to obtain artificial seeds and thereby propagate healthy plants, although control of the material produced is essential.

OKORO: It is interesting to note that increasing numbers of species are lending themselves easily to somatic embryogenesis. Has the germination of encapsulated seeds become possible?

FILIPPONE: The technology for producing somatic embryos in bioreactors is now being developed in many laboratories. A major problem is to find a substance that can satisfactorily mimic the normal endosperm. Some resins have been tested, mainly in Janick's laboratory at Purdue University, to control moisture content and thereby extend storage life, as well as to ensure recovery of a complete plant following sowing. Until now, encapsulated somatic embroys have been able to survive for only a few weeks.

References

Ahokas, H. 1989. Transfection of germinating barley seed electrophoretically with exogenus DNA. *Theoretical and Applied Genetics* 77: 469-472.

Ammirato, P.V. 1987. Organizational events during somatic embryogenesis. Pages 57-81 in *Plant Tissue and Cell Culture* edited by C.E. Green, D.A. Somers, W.P. Hackett, and D.D. Biesboer. Alan R. Liss, New York, USA.

Armstrong, C.L., and R.L. Phillips. 1988. Genetics and cytogenetics variation in plants regenerated from organogenic and friable embryogenic tissue cultures of maize. *Crop Science* 28: 363-369.

Bhatia, C.R., D.C. Joshua, and H. Mathews. 1985. Somaclonal variation: A genetic interpretation based on the rates of spontaneous chromosomal aberrations and mutations. *Trends in Plant Research* 5: 317-326.

Beversdorf, W.D. 1990. Micropropagation in crop species. Pages 3-12 in *Progress in Plant Cellular and Molecular Biology: Current Plant Science and Biotechnology in Agriculture* (Vol. 9) edited by H.J.J. Nijkamp, L.H.W. Van Der Plas, and J. Van Aartrijk. Kluwer Academic Publishers, Dordrecht, The Netherlands.

Bonga, J.M., P. Von Aderkas, and D. James. 1988. Potential application of haploid cultures of tree species. Pages 57-77 in *Genetic Manipulation of Woody Plants* edited by J.M. Hanover, and D.E. Keathley. Plenum Press, New York, USA.

Christou, P., W.F. Swain, and D.E. McCabe. 1989. Inheritance and expression of foreign genes in transgenic soybean plants. *Proceedings of the National Academy of Science, USA* 86: 671-674.

Datta, S.K., and I. Potrykus. 1989. Artificial seeds in barley: Encapsulation of microspore-derived embryos. *Theoretical and Applied Genetics* 77: 820-824.

D'Halluin, K., and Keimer, B. 1986. Production of haploid sugar beets (*Beta vulgaris* L.) by ovule culture. Pages 307-309 in *Genetic Manipulation in Plant Breeding* edited by W. Horn, C.J. Jensen, W. Odenbach, and O. Shieder. Walter de Gruyter, Berlin, Germany.

De la Pena, A., H. Lorz, and J. Schell. 1987. Transgenic plants obtained by injecting DNA into young floral tillers. *Nature* 325: 274-276.

Evans, D.A., and W.R. Sharp. 1986. Somaclonal and gametoclonal variation. Pages 97-132 in *Handbook of Plant Cell Culture* edited by D.A. Evans, W.R. Sharp, and P.V. Ammirato (Vol. 4) Macmillan, New York, USA.

Evans, D.A., W.R. Sharp, P.V. Ammirato, and Y. Yamada (eds). 1983. *Handbook of Plant Cell Culture* (Vol. 1). Macmillan, New York, USA.

Fish, N., and A. Karp. 1986. Improvements in regenerations from protoplast of potato and studies on chromosome stability. 1. The effect of initial culture media. *Theoretical and Applied Genetics* 72: 405-412.

Gad, A. E., G. Elyashiv, and N. Rosenberg. 1986. The induction of large unilamellar vesicle fusion by cationic polypepttides: The effects of mannitol, size, charge density and hydrophobicity of the cationic polypeptides. *Biochimica et Biophysica Acta* 860: 314-324.

Gamborg, O.L., R.A. Miller, and K. Ojima. 1968. Plant cell cultures. I. Nutrient requirements of suspension cultures of soybean root cells. *Experimental Cell Research* 50: 151-158.

Gobel, E., and H. Lorz. 1988. Genetic manipulation of cereals. Pages 1-22 in *Oxford Surveys of Plant Molecular and Cell Biology* (Vol. 5.) Oxford University Press, Oxford, UK.

Godwin, I.A., G.H. Gordon, and D.F. Cameron. 1987. Callus culture-derived somaclonal variation in the tropical pasture legume *Stylosanthes guianensis* (Aubl.) Sw. *Plant Breeding* 98: 220-227.

Gordon-Kamm, W.J., T.M. Spencer, M.L. Mangano, T.R. Adams, R.J. Daines, W.G. Start, J.V. O'Brien, S.A. Chambers, W.R. Adams Jr., N.G. Willetts, T.B. Rice, C.J. Mackey, R.W. Krueger, A.P. Kausch, and P.G. Lemaux. 1990. Transformation of maize cells and regeneration of fertile transgenic plants. *The Plant Cell* 2: 603-618.

Hu, C.Y., and P.J. Wang. 1983. Meristem, shoot tip and bud culture. Pages 177-227 in *Handbook of Plant Cell Culture* (Vol. 1) edited by W.R. Sharp, D.A. Evans, P.V. Ammirato, and Y. Yamada. Macmillan, New York, USA.

IAPTC. 1990. *Abstracts, Seventh International Congress on Plant Tissue and Cell Cultures, 24-29 June 1990, Amsterdam, The Netherlands.*

Johnson, S.S, R.L. Philips, and H.W. Rines. 1987. Possible role of heterochromatin in chromosome breakage induced by tissue culture in oats (*Avena sativa* L.). *Genome* 29: 439-446.

Karp, A., R. Risiott, M.G.K. Jones, and S.W.J. Bright. 1984. Chromosome doubling in monohaploid and dihaploid potatoes by regeneration from cultured leaf explants. *Plant Cell, Tissue and Organ Culture* 3: 363-373.

Kartha, K.K. 1982. Cryopreservation of germplasm using meristem and tissue culture. Pages 139-161 in *Application of Plant Cell and Tissue Culture to Agriculture and Industry* edited by D.T. Tomes, B.E. Ellis, P.M. Harney, K.J. Kasha, and R.L. Peterson. University of Guelph, Ontario, Canada.

Kartha, K.K. (ed). 1985. *Cryopreservation of Plant Cells and Organs*. CRC Press, Boca Raton, Florida, USA.

Kim, Y.H., and J. Janick. 1989. ABA and Polyox-encapsulation or high humidity increases survival of desiccated somatic embryos of celery. *HortScience* 24: 674-676.

Larkin, P. J. 1987. Somaclonal variations, history, methods and meaning. *Iowa State Journal of Research* 61: 393-434.

Levin, R., and I.K. Vasil. 1989. Progress in reducing the cost of micropropagation. *IAPTC Newsletter* 59: 2-12.

Levin, R., G. Gaba, B. Tal, S. Hirsch, D. De-Nola, and I.K. Vasil. 1988. Automated plant tissue culture for mass propagation. *Bio/technology* 6: 1035-1040.

Lindsmaier, E.M., and F. Skoog. 1965. Organic growth factor requirements of tobacco tissue cultures. *Physiologia Plantarum* 18: 100-127

Manette, P.L. 1983. Virus eradication through in vitro techniques. In *Combined Proceedings, International Plant Propagators' Society* (Vol. 33). Saanichton Research and Plant Quarantine Station, Sidney, Canada.

Medina-Gomez, F. 1990. Impact of tissue culture technique on the development of agribusiness in Mexico. Pages 795-800 in *Progress in Plant Cellular and Molecular Biology: Current Plant Science and Biotechnology in Agriculture* (Vol. 9) edited by H.J.J. Nijkamp, L.H.W. Van Der Plas, and J. Van Aartrijk. Kluwer Academic Publishers, Dordrecht, The Netherlands.

Morel, G.M., and C. Martin. 1952. Guerison de dahlias atteints d'une maladie a virus. *Comptes Rendus l'Academie Sciences (Paris)* 235: 1324-1325.

Murashige, T., and F. Skoog. 1962. A revised medium for rapid growth and bioassays with tobacco tissue cultures. *Physiologia Plantarum* 15: 473-497

Neuhaus, G., G. Spangenberg, O. Mittelsten Scheid, and H.G. Shweiger. 1987. Transgenic rapeseed plants obtained by microinjection of DNA into microspore-derived embryoids. *Theoretical Applied Genetics*. 75: 30-36.

Novak F.J., R. Afza, M. Van Duren, M. Perea-Dallos, B.V. Conger, and T. Xiaolang. 1989. Somatic embryogenesis and plant regeneration in suspension cultures of dessert (AA and AAA) and cooking (ABB) bananas (*Musa* spp.). *Bio/technology* 7: 154-159.

Poehlman, J.M. 1987. Plant cell and tissue culture applications in plant breeding. Pages 149-170 in *Breeding Field Crops*. AVI Publishing, New York, USA.

Redenbaugh, K. 1990. Application of artificial seed to tropical crops. *HortScience* 25: 251-255.

Towill, L.E. 1989. Biotechnology and germplasm preservation. *Plant Breeding Reviews* 7: 159-182.

Robertson, A.I. 1990. Pages 783-788 in *Progress in Plant Cellular and Molecular Biology: Current Plant Science and Biotechnology in Agriculture* (Vol. 9) edited by H.J.J. Nijkamp, L.H.W. Van Der Plas, and J. Van Aartrijk. Kluwer Academic Publishers, Dordrecht, The Netherlands.

Sanford, J.C. 1990. Biolistic plant transformation. *Physiologia Plantarum* 79: 206-209.

Scowcroft, W.R. 1985. *Genetic Variability in Tissue Culture: Impact on Germplasm Conservation and Utilization*. IBPGR, Rome, Italy.

Senaratna, T., B.D. McKersie, and S.R. Bowley. 1989. Desiccation tolerance of alfalfa (*Medicago sativa* L.) somatic embryos. Influence of abscisic acid, stress pretreatments and drying rates. *Plant Science* 65: 253-259.

Towill, L.E. 1989. Biotechnology and germplasm preservation. *Plant Breeding Reviews* 7: 159-182.

Vasil, I.K. 1988. Progress in the regeneration and genetic manipulation of cereal crops. *Bio/technology* 6: 397-402.

Vasil, V., and A.C. Hildebrandt. 1965. Differentiation of tobacco plants from single, isolated cells in microculture. *Science* 150: 889-892.

Von Arnold, S., and A. Wallin. 1988. Tissue culture methods for clonal propagation of forest trees. *IAPTC Newsletter* 56: 2-13.

Walkey, D.G.A., H.A. Neely, and P. Crisp. 1980. Rapid propagation of white cabbage by tissue culture. *Scientia Horticulturae* 12: 99-107.

Weber, G., S. Monajembashi, K.O. Greulich, and J. Wolfrum. 1988. Injection of DNA into plant cells with a UV laser microbeam. *Naturwissenschaften* 75: 35-36.

Wheeler, V.A., N.E. Evans, D. Foulger, K.J. Webb, A. Karp, J. Franklin, and S.W.J. Bright. 1985. Shoot formation from explant cultures of fourteen potato cultivars and studies of the cytology and morphology of regenerated plants. *Annals of Botany* 55: 309-320.

Williams, E.G., G. Maheswaran, and J.F. Hutchinson. 1987. Embryo and ovule culture in crop improvement. *Plant Breeding Reviews* 5: 181-236.

Withers, L.A., S.K. Wheelans, and J.T. Williams. 1990. In vitro conservation of crop germplasm and the IBPGR databases. *Euphytica* 45: 9-22.

3.2

Application of cell culture techniques to cereal improvement

A.M. Casas and P.M. Hasegawa

Center for Plant Environmental Stress Physiology, Department of Horticulture, Purdue University, West Lafayette, Indiana 47907, USA

Abstract

Over the past few years, efficient protocols for the regeneration of various cereal crops via somatic embryogenesis have been established. Embryogenic cereal cell suspensions have been developed and are used mainly as a source for protoplasts. Regeneration of plants from isolated protoplasts has been attained in maize and rice. Stable transformation by electroporation or DNA uptake mediated by polyethylene glycol (PEG) in protoplasts, followed by recovery of transgenic plants, has been achieved in maize and rice. Also, transgenic maize plants have been obtained after high velocity microprojectile bombardment of cell suspensions.

Plant breeding has been useful in the past as a means of increasing yield, disease resistance, nutritional quality, and other agronomic traits in cereals, including barley, maize, rice, and wheat. Further efforts are now needed to provide food for an increasing world population. In the past decade, in vitro technology was expected to increase the efficiency of the process of improving crop productivity. Special attention has been paid to somaclonal variation and in vitro selection. Plants with useful agronomic traits such as disease and herbicide resistance, and which are high yielding and rich in protein or sugar content have been obtained (Bajaj 1990) and are being incorporated into crop improvement programmes.

In the past few years it has been suggested that recombinant DNA methodology combined with cell and tissue culture techniques has considerable potential to facilitate crop improvement. Theoretically, this methodology could increase the genepool for agronomically desirable traits by identifying new sources of genetic variation and by circumventing reproductive affinity barriers and allowing the direct introgression of a single trait into the genetic background of a genotype with agronomic utility or importance as a parental line.

Substantial progress has been made on techniques to isolate and modify potentially useful genes, to make appropriate gene constructions for expression in plants, and to develop vectors and methods for the transformation of plants with these gene constructions.

A few years ago it was considered, perhaps rather optimistically, that biotechnology would allow the transfer of genes to any given species in a short period of time. Certainly, advances have been made, but, as far as cereals are concerned, regeneration is still a problem. Morphogenesis is limited to few genotypes of the various cereal crops and only in a few cases has the stable insertion, integration, and expression of characterized alien genes into the host plant genome been proven. In this paper, we review recent advances in various aspects of cereal tissue culture leading to the regeneration and stable transformation of plants. Additional information can be found in several recent reviews (Vasil 1987; Gobel and Lorz 1988; Lorz et al. 1988; Bhaskaran and Smith 1990; Potrykus 1990).

Cereal Regeneration via Somatic Embryogenesis from Callus

Plant morphogenesis in vitro can be attained through organogenesis of shoots and roots or by somatic embryogenesis. Embryogenesis is the preferred method of morphogenesis since the single cell origin of somatic embryos should, in theory, give uniformly transgenic regenerants. In contrast, multicellular regeneration of shoots or roots may result in chimaeric plants since each transformed cell represents a unique transgenic genotype. Somatic embryogenesis has also been postulated to result in less somaclonal variation in regenerated plant populations (Vasil 1987). Systematic evaluations of progeny from plants produced by embryogenesis do indicate the occurrence of heritable variation (Cai et al. 1990). The literature indicates that almost all the important cereal crops can be manipulated in vitro to produce somatic embryos (Table 1).

To induce embryogenesis, the explants are cultured on media with a high concentration of auxin (2-5 mg/L). Cytokinins are not strictly required at this stage. The induction of embryogenesis takes place in a short period of time after auxin treatment. The embryogenic callus cultures are then maintained in a medium with lower auxin and higher cytokinin. The development of embryos occurs in the absence of hormones. The most common growth regulator used in establishing regenerative cultures is the auxin 2,4D (Vasil 1987; Bhaskaran and Smith 1989, 1990). Low levels of cytokinin have been used in addition to high 2,4D for inducing embryogenic callus in different species, such as sorghum (Bhaskaran and Smith 1988; Cai and Butler 1990; Wernicke et al. 1982) and rice (Raghava Ram and Nabors 1984).

A major factor affecting the induction of embryogenic callus is the type of explant used. In many dicotyledonous species, leaf or shoot segments are used as explants, but in cereals these explants have limited use. Immature embryos with meristematic and undifferentiated cells have been successfully used to establish regenerative cultures of a great variety of cereal species. Immature inflorescences or young leaves have also been used successfully. Somatic embryos have been obtained from mature embryos of barley (Lupotto 1984), rice (Oono 1985), and sorghum (Cai et al. 1987), but the efficiency of regeneration is lower.

Embryogenic cultures often lose their morphogenic ability after 4-8 months. If somatic embryogenesis can be maintained, plant regeneration would be possible for a longer period. It is important to select visually embryogenic parts of the culture. Embryogenic calli are characteristically compact, rather organized, and white to pale yellow (Heyser and Nabors 1982; Armstrong and Green 1985; Vasil and Vasil 1986; Bhaskaran and Smith 1988).

Table 1 Reports on the regeneration of cereals via somatic embryogenesis

Avena sativa	Heyser and Nabors 1982
Hordeum vulgare	Thomas and Scott 1985, Luhrs and Lorz 1987
Oryza sativa	Chen et al. 1985
Panicum miliaceum	Rangan and Vasil 1983
Pennisetum americanum	Chandler and Vasil 1984
Secale cereale	Lu et al. 1984
Sorghum bicolor	Cai and Butler 1990, Ma et al. 1987, Wernicke and Brettell 1980, 1982
Triticum aestivum	Maddock et al. 1983
Triticale	Stolarz and Lorz 1986
Zea mays	Armstrong and Green 1985, Duncan et al. 1985, Gobel et al. 1986, Hodges et al. 1986, Tomes and Smith 1985, Vasil and Vasil 1986

There is an indication that heritable genetic variation for the expression of totipotency exists in different cereal species (Tomes and Smith 1985; Hodges et al. 1986). It is not clear if the variation is a response to the culture media or conditions used to facilitate the proliferation of embryogenic cells, or if it is a reflection of the percentage of potentially totipotent cells in the explant. Regardless, it appears that isolation of embryogenic cultures can be achieved with most genotypes with the use of the appropriate explant. However, the efficiency of obtaining embryogenic cultures may be genetically controlled.

Embryonic Cell Suspension Cultures

Embryogenic callus is the primary source of material for establishing cell suspensions capable of somatic embryogenesis. Although considerably more difficult to establish and maintain than callus, embryogenic cell suspensions, from which plants can be regenerated, have been obtained for the most of the important cereal crops (Table 2). Cell cultures can

Table 2 Regeneration of cereals from cells in suspension culture

Species	Regenerant	Reference
Hordeum vulgare	albino plantlets	Kott and Kasha 1984
	albino plantlets	Luhrs and Lorz 1988
Oryza sativa	plants	Fujimura et al. 1985
	plants	Zimny and Lorz 1986
Sorghum vulgare	plants	Wei and Xu 1990
Triticale	plants	Stolarz and Lorz 1986
Triticum aestivum	plants	Maddock 1987
	plants	Wang and Nguyen 1990
Zea mays	plants	Green et al. 1983
	plants	Kamo and Hodges 1986

be used in physiological studies, to select mutant cell lines for tolerance to salt, herbicides, or pathogenic toxins, and mainly as a source for the production of protoplasts.

Regeneration from Protoplasts

The efficient isolation of protoplasts and their regeneration to plants has been considered as a pre-requisite for the development of genetic engineering methods in cereals. A summary of the culture and regeneration of plants from protoplasts in cereals is provided in Table 3.

Table 3 Regeneration of cereals from protoplasts

Species	Regenerant	Reference
Hordeum vulgare	albino plantlets	Luhrs and Lorz 1988
Oryza sativa		
O. japonica	plants	Abdullah et al. 1986
	plants	Yamada et al. 1986
	plants	Li and Murai 1990
O. indica	plants	Kyozuka et al. 1988
	plants	Datta et al. 1990
Panicum maximum	plantlets	Lu et al. 1981
Pennisetum americanum	plantlets	Vasil and Vasil 1980
Sorghum vulgare	plants	Wei and Xu 1990
Triticum aestivum	plants	Vasil et al. 1990
Zea mays	sterile plants	Rhodes et al. 1988a, 1988b
	plants	Prioli and Sondahl 1989
	plants	Shillito et al. 1989

In species belonging to the Solanaceae, protoplasts have usually been isolated from leaves, with considerable success. However, until now, all culture experiments with mesophyll protoplasts isolated from cereal species have been unsuccessful (Vasil 1987). An alternative approach is the use of embryogenic cell cultures for the isolation of protoplasts. In this case, regeneration can be expected if plants can be regenerated efficiently from the callus or suspension cultures, which are subsequently used for the protoplast preparation. To increase the rate of regeneration and the plating efficiency of the protoplasts, the most critical factor is the development of rapidly dividing and highly embryogenic cell suspensions.

In recent years, there have been encouraging results with the regeneration of rice (Fujimura et al. 1985; Yamada et al. 1986), maize (Rhodes et al. 1988a), and wheat plants from protoplasts (Vasil et al. 1990). However, success with the direct transformation of organized meristems has shown that gene transfer in difficult cereal species does not require the regeneration of plants from protoplasts (Fromm et al. 1990; Gordon-Kamm et al. 1990; Raineri et al. 1990).

Status of In Vitro Culture for the Production of Transgenic Cereal Plants

In summary, most of the important cereal crops can be regenerated from embryogenic callus and cell suspensions via somatic embryogenesis, although manipulation of suspension cultures is considerably more difficult. The efficiency of obtaining embryogenic cultures is affected by the choice of explant and perhaps genotype, and it can be influenced by refinements to culture media and protocols. Isolation and culture of protoplasts for regeneration of major cereal crops has been demonstrated. However, these protocols and media are complex and the efficiency of regeneration is low relative to callus and cell suspension, associated with the sterility of the regenerated plants.

For many years, protoplasts have been considered the most likely material for use in the transformation of cereals. Stable transformation of protoplasts with DNA whose intracellular influx has been facilitated by electroporation (Fromm et al. 1985) or direct DNA uptake, which is usually chemically stimulated (Lorz et al. 1985; Paszkowski et al. 1984), has been achieved with barley (Lazzeri et al. 1991), maize (Fromm et al. 1986, Lyznik et al. 1989), and rice (Uchimiya et al. 1986). Recent advances in tissue culture have contributed to the regeneration of the first transgenic plants of maize (Rhodes et al. 1988b) and rice (Toriyama et al. 1988; Zhang et al. 1988; Hayashimoto et al. 1990; Peng et al. 1990) from protoplasts.

Agrobacterium, the most common vector used to transform dicotyledons, does not seem to infect most monocots, in particular cereals. Nevertheless, it has been used to introduce viral DNA by agroinfection in wheat (Dale et al. 1989) and barley (Creissen et al. 1990), although the efficiency of the transformation was very low. Also, Raineri et al. (1990) reported the transformation of rice embryos with a virulent strain of *Agrobacterium*. Other methods that have been used with variable results include direct injection of DNA in rye tillers (De la Peña et al. 1987) or electrophoretic transfection of barley seeds (Ahokas 1989).

In 1987 Klein et al. published the first report of the use of particle discharge bombardment to deliver DNA to plant cells. Since then, it has become the most promising technique for transforming cereals, overcoming the problems associated with regeneration of protoplasts. In fact, the rapidly increasing use and success with particle bombardment of organized meristems will probably eliminate most efforts using protoplasts for transformation. This delivery system has been effectively used to achieve transient expression in cell suspensions, callus, and embryos of numerous cereals, including barley, maize, rice, and wheat (Klein et al. 1988; Wang et al. 1988; Kartha et al. 1989; Mendel et al. 1989; Lonsdale et al. 1990). Microprojectile bombardment has been used successfully to obtain stably transformed maize cells (Klein et al. 1989; Spencer et al. 1990) and transgenic maize plants (Fromm et al. 1990; Gordon-Kamm et al. 1990, 1991).

Discussion

ROBERTSON: My impression is that there is a particular age at which it is best to excise embryos in order to get regeneration of plants, yet you indicate that the best excision time is day 1. Am I wrong?

HASEGAWA: Experimental evidence can be interpreted in this manner. Perhaps the inconsistency arises from the difference between practice and hypothesis.

NDEBERI: In cereal regeneration through callus culture, it seems that there is varietal difference in organogenesis. My question is: When working on immature embryos of cereals, especially rice, have you encountered varietal differences in behaviour?

HASEGAWA: Yes, somatic embryogenesis has been shown to be heritable.

LADEINDE: Proline is usually associated with stress. Why do you explain your results in terms of the importance of threonine in stress expression rather than proline?

HASEGAWA: I believe that there is a misunderstanding here. I was referring to an amino acid change in the QB protein, serine to threonine, which is the basis for atrazine resistance of photoautotropic cell suspensions. However, I am aware of research that confirms the relationship between proline accumulation and yield stability in drought environments.

References

Abdullah, R., E.C. Cocking, and J.A. Thompson. 1986. Efficient plant regeneration from rice protoplasts through somatic embryogenesis. *Bio/Technology* 4: 1087-1090.

Ahokas, H. 1989. Transfection of germinating barley seed electrophoretically with exogenous DNA. *Theoretical and Applied Genetics* 77: 469-472.

Armstrong, C.L., and C.E. Green. 1985. Establishment and maintenance of friable, embryogenic maize callus and the involvement of proline. *Planta* 164: 207-214.

Bajaj, Y.P.S. (ed). 1990. *Biotechnology in Agriculture and Forestry* (Vol. II). Somaclonal variation in crop improvement. I. Springer-Verlag, Berlin/Heidelberg, Germany.

Bhaskaran, S., and R.H. Smith. 1988. Enhanced somatic embryogenesis in *Sorghum bicolor* from shoot tip culture. *In Vitro Cellular and Developmental Biology* 24: 65-70.

Bhaskaran, S., and R.H. Smith. 1989. Control of morphogenesis in sorghum by 2,4 dichlorophenoxyacetic acid and cytokinins. *Annals of Botany* 64: 217-224.

Bhaskaran, S., and R.H. Smith. 1990. Regeneration in cereal tissue culture: A review. *Crop Science* 30: 1328-1336.

Cai, T., and L. Butler. 1990. Plant regeneration from embryogenic callus initiated from immature inflorescences of several high-tannin sorghums. *Plant Cell Tissue and Organ Culture* 20: 101-110.

Cai, T., B. Daly, and L. Butler. 1987. Callus induction and plant regeneration from shoot portions of mature embryos of high tannin sorghums. *Plant Cell Tissue and Organ Culture* 9: 245-252.

Cai, T., G. Ejeta, J.D. Axtel, and L.G. Butler. 1990. Somaclonal variation in high tannin sorghums. *Theoretical and Applied Genetics* 79: 737-747

Chandler, S.F., and I.K. Vasil. 1984. Optimization of plant regeneration from long term embryogenic callus cultures of *Pennisetum purpureum* Schum. (Napier grass). *Journal of Plant Physiology* 117: 147-156.

Chen, T.H., L. Lam, and S.C. Chen. 1985. Somatic embryogenesis and plant regeneration from cultured young inflorescences of *Oryza sativa*. *Plant Cell, Tissue and Organ Culture* 4: 51-54.

Creissen, G., G. Smith, R. Francis, H. Reynolds, and P. Mullineaux. 1990. *Agrobacterium* — and microprojectile — mediated viral DNA delivery into barley microspore-derived cultures. *Plant Cell Reports* 8: 680-683.

Dale, P.J., M.S. Marks, M.M. Brown, C.J. Woolston., H.V. Gunn., P.M. Mullineaux, D.M. Lewis, J.M. Kemp, D.F. Chen, D.M. Gilmour, and R.B. Flavell. 1989. Agroinfection of wheat: Inoculation of in vitro grown seedlings and embryos. *Plant Science* 63: 237-245.

Datta, S.K., K. Datta, and I. Potrykus. 1990. Genetically engineered fertile indica-rice recovered from protoplasts. *Bio/Technology* 8: 736-740.

De la Peña, A., H. Lorz, and J. Schell. 1987. Transgenic rye plants obtained by injecting DNA into young floral tillers. *Nature* 325: 274-276.

Duncan, D.R., M.E. Williams, B.E. Zehr, and J.M. Wildhom. 1985. The production of callus capable of plant regeneration from immature embryos of numerous *Zea mays* genotypes. *Planta* 165: 322-332.

Fromm, M.E., L.P. Taylor, and V. Walbot. 1985. Expression of genes transferred into monocot and dicot plant cells by electroporation. *Proceedings of the National Academy of Science, USA* 82: 5824-5828.

Fromm, M.E., L.P. Taylor, and V. Walbot. 1986. Stable transformation of maize after gene transfer by electroporation. *Nature* 319: 791-793.

Fromm, M.E., F. Morrish, C. Armstrong, R. Williams, J. Thomas, and T.J. Klein. 1990. Inheritance and expression of chimeric genes in the progeny of transgenic maize plants. *Bio/Technology* 8: 833-839.

Fujimura, T., M. Sakurai, T. Negishi, and A. Hirose. 1985. Regeneration of rice plants from protoplasts. *Plant Tissue Culture Letters* 2: 74-75.

Gobel, E., and H. Lorz. 1988. Genetic manipulation of cereals. *Oxford Surveys of Plant Molecular and Cell Biology* 5: 1-22.

Gobel, E., P.T.H. Brown, and H. Lorz. 1986. In vitro culture of *Zea mays* and analysis of regenerated plants. Pages 21-27 in *Nuclear Techniques and In Vitro Culture for Plant Improvement*. IAEA, Vienna, Austria.

Gordon-Kamm, W.J., T.M. Spencer, M.L. Mangano, T.R. Adams, R.J. Daines, W.G. Start, J.V. O'Brien, S.A. Chambers, W.R. Adams Jr., N.G. Willetts, T.B. Rice, C.J. Mackey, R.W. Krueger, A.P. Kausch, and P.G. Lemaux. 1990. Transformation of maize cells and regeneration of fertile transgenic plants. *The Plant Cell* 2: 603-618.

Gordon-Kamm W.J., T.M. Spencer, J.V. O'Brien, W.G. Start, R.J. Daines, T.R. Adams, M.L. Mangano, S.A. Chambers, S.J. Zachwieja, N.G. Willetts, W.R. Adams Jr., C.J. Mackey, R.W. Krueger, A.P. Kausch, and P.G. Lemaux. 1991. Transformation of maize using microprojectile bombardment: An update and perspective. *In Vitro Cell Dev Biol* 27P: 21-27.

Green, C.E., L. Armstrong, and P.C. Anderson. 1983. Somatic cell genetic systems in corn. Pages 147-157 in *Advances in Gene Technology* edited by K. Downey, R.W. Voellmy, F. Ahmad, and J. Schultz. Academic Press, New York, USA.

Hayashimoto, A., Z. Li, and N. Murai. 1990. A polyethylene glycol-mediated protoplast transformation system for the production of fertile transgenic rice plants. *Plant Physiology* 93: 857-863.

Heyser, J.W., and M.W. Nabors. 1982. Long term plant regeneration, somatic embryogenesis and green spot formation in secondary oat (*Avena sativa* L) callus. *Zeitschrift für Pflanzenphysiologie* 107:153-160.

Hodges, T.K., K.K. Kamo, C.W. Imbrie, and M.R. Becwar. 1986. Genotype specificity of somatic embryogenesis and regeneration in maize. *Bio/Technology* 4: 219-223.

Kamo, K.K., and T.K. Hodges. 1986. Establishment and characterization of long-term embryogenic maize callus and cell suspension cultures. *Plant Science* 45:111-117.

Kartha, K.K., R.N. Chibbar, F. Georges, N. Leung, K. Caswell, E. Kendall, and J. Qureshi. 1989. Transient expression of chloramphenicol acetyltransferase (CAT) gene in barley cell cultures and immature embryos through microprojectile bombardment. *Plant Cell Reports* 8: 429-432.

Klein, T.M., E.D. Wolf, R. Wu, and J.C. Sanford. 1987. High-velocity microprojectiles for delivering nucleic acids into living cells. *Nature* 327: 70-73.

Klein, T.M., M. Fromm, A. Weissinger, D. Tomes, S. Schaaf, M. Sletten, and J.C. Sanford. 1988. Transfer of foreign genes into intact maize cells with high-velocity microprojectiles. *Proceedings of the National Academy of Science, USA* 85: 4305-4309.

Klein, T.M., L. Kornstein, J.C. Sanford, and M.E. Fromm. 1989. Genetic transformation of maize cells by particle bombardment. *Plant Physiology* 91: 440-444.

Kott, L.S., and K.J. Kasha. 1984. Initiation and morphological development of somatic embryos from barley cell cultures. *Canadian Journal of Botany* 62: 1245-1249.

Kyozuka, J., E. Otoo, and K. Shimamoto. 1988. Plant regeneration from protoplasts of indica rice: Genotypic differences in culture response. *Theoretical and Applied Genetics* 76: 887-890.

Lazzeri, P.A., R. Brettschneider, R. Luhrs, and H. Lorz. 1991. Stable transformation of barley via PEG-induced direct DNA uptake into protoplasts. *Theoretical and Applied Genetics* 81: 437-444.

Li, Z., and N. Murai. 1990. General medium for efficient plant regeneration from rice (*Oryza sativa* L) protoplasts. *Plant Cell Reports* 9: 216-220.

Lonsdale, D., S. Onde, and A. Cuming. 1990. Transient expression of exogenous DNA in intact, viable wheat embryos following particle bombardment. *Journal of Experimental Botany* 41: 1161-1165.

Lorz, H., B. Baker, and J. Schell. 1985. Gene transfer to cereal cells mediated by protoplast transformation. *Molecular and General Genetics* 199: 178-182.

Lorz, H., E. Gobel, and P. Brown. 1988. Advances in tissue culture and progress toward genetic transformation of cereals. *Plant Breeding* 100: 1-25.

Lu, C.Y., V. Vasil, and I.K. Vasil. 1981. Isolation and culture of protoplasts of *Panicum maximum* (Guinea grass): Somatic embryogenesis and plantlet formation. *Zeitschrift für Pflanzenphysiologie* 104: 311-318.

Lu, C., S.F. Chandler, and I.K. Vasil. 1984. Somatic embryogenesis and plant regeneration in cultured immature embryos of rye (*Secale cereale* L). *Journal of Plant Physiology* 115: 237-244.

Luhrs, R., and H. Lorz. 1987. Plant regeneration in vitro from embryogenic cultures of spring and winter-type barley (*Hordeum vulgare* L.). *Theoretical and Applied Genetics* 75:16-25.

Luhrs, R., and H. Lorz. 1988. Induction of morphogenic cell suspension and protoplast cultures of barley. *Planta* 175: 71-81.

Lupotto, E. 1984. Callus induction and plant regeneration from barley mature embryos. *Annals of Botany* 54: 523-529.

Lyznik, L.A., R.A. Ryan, S.W. Ritchie, and T.K. Hodges. 1989. Stable cotransformation of maize protoplasts with *gusA* and *neo* genes. *Plant Molecular Biology* 13: 151-161.

Ma, H., M. Gu, and G.H. Liang. 1987. Plant regeneration from cultured immature embryos of *Sorghum bicolor* (L.) Moench. *Theoretical and Applied Biology* 73: 389-394.

Maddock, S.E. 1987. Suspension and protoplast culture of hexaploid wheat (*Triticum aestivum* L.). *Plant Cell Reports* 6: 23-26.

Maddock, S.E., V.A. Landaster, R. Risiott, and J. Franklin. 1983. Plant regeneration from cultured immature embryos and inflorescences of 25 cultivars of wheat (*Triticum aestivum* L.). *Journal of Experimental Botany* 34: 915-926.

Mendel, R.R., B. Muller, J. Schulze, V. Kolesnikov, and A. Zelenin. 1989. Delivery of foreign genes to intact barley cells by high-velocity microprojectiles. *Theoretical and Applied Genetics* 78: 31-34.

Oono, K. 1985. Putative homozygous mutation in regenerated plants of rice. *Molecular and General Genetics* 198: 377-384.

Paszkowski, J., R.D. Shillito, M. Saul, V. Mandak, T. Hohn, B. Hohn, and I. Potrykus. 1984. Direct gene transfer to plants. *EMBO Journal* 3: 2717-2722.

Peng, J., L.A. Lyznik, L. Lee, and T.K. Hodges. 1990. Cotransformation of indica rice protoplasts with gusA and neo genes. *Plant Cell Reports* 9: 168-172.

Potrykus, I. 1990. Gene transfer to cereals: An assessment. *Bio/Technology* June: 535-542.

Prioli L.M., and M.R. Sondahl. 1989. Plant regeneration and recovery of fertile plants from protoplasts of maize (*Zea mays* L.). *Bio/Technology* 7: 589-594.

Raghava Ram, N.V., and M.W. Nabors. 1984. Cytokinin mediated long term, high frequency plant regeneration in rice tissue culture. *Zeitschrift für Pflanzenphysiologie* 113: 315-323.

Raineri, D.M., P. Bottino, M.P. Gordon, and E.W. Nester. 1990. *Agrobacterium*-mediated transformation of rice (*Oryza sativa* L.). *Bio/Technology* 8: 33-38.

Rangan, T.S., and I.K. Vasil. 1983. Somatic embryogenesis and plant regeneration in tissue cultures of *Panicum miliaceum* L. and *Panicum miliare* Lamk. *Zeitschrift für Pflanzenphysiologie* 109: 49-53.

Rhodes, C.A., K.S. Lowe, and K.L. Ruby. 1988a. Plant regeneration from protoplasts isolated from embryogenic maize cell cultures. *Bio/Technology* 6: 56-60.

Rhodes, C.A., D.A. Pierce, I.J. Metler, D. Mascarenhas, and J.J. Detmer. 1988b. Genetically transformed maize plants from protoplasts. *Science* 240: 204-207.

Shillito, R.D., G.K. Carswell, C.M. Johnson, J.J. DiMaio, and C.T. Harms. 1989. Regeneration of fertile plants from protoplasts of elite inbred maize. *Bio/Technology* 7: 581-587.

Spencer, T.M., W.J. Gordon-Kamm, R.J. Daines, W.G. Start, and P.G. Lemaux. 1990. Bialaphos selection of stable transformants from maize cell culture. *Theoretical and Applied Genetics* 79: 625-631.

Stolarz, A., and H. Lorz. 1986. Somatic embryogenesis, in vitro multiplication and plant regeneration from immature embryos of hexaploid *Triticale* (x Triticosecale Wittmack). *Z Pflanzenzücht* 96: 353-362.

Thomas, M.R., and K.J. Scott. 1985. Plant regeneration by somatic embryogenesis from callus initiated from immature embryos and immature inflorescences of *Hordeum vulgare* L. *Journal of Plant Physiology* 121: 159-169.

Tomes, D.T., and O.S. Smith. 1985. The effect of the parental genotype on initiation of embryogenic callus from elite maize (*Zea mays* L.) germplasm. *Theoretical and Applied Genetics* 70: 505-509.

Toriyama, K., Y. Arimoto, H. Uchimiya, and K. Hinata. 1988. Transgenic rice plants after direct gene transfer into protoplasts. *Bio/Technology* 6: 1072-1074.

Uchimiya, H., T. Fushimi, H. Hashimoto, H. Harada, K. Syono, and Y. Sugawara. 1986. Expression of a foreign gene in callus derived from DNA-treated protoplasts of rice (*Oryza sativa* L). *Molecular and General Genetics* 206: 204-207.

Vasil, I.K. 1987. Developing cell and tissue culture systems for the improvement of cereal and grass crops. *Journal of Plant Physiology* 128: 193-218.

Vasil, V., and I.K. Vasil 1980. Isolation and culture of cereal protoplasts. Part 2: Embryogenesis and plantlet formation from protoplasts of *Pennisetum americanum. Theoretical and Applied Genetics* 56: 97-99.

Vasil, V., and I.K. Vasil. 1986. Plant regeneration from friable embryogenic callus and cell suspension cultures of *Zea mays* L. *Journal of Plant Physiology* 124:399-408.

Vasil, V., F. Redway, and I.K. Vasil. 1990. Regeneration of plants from embryogenic suspension culture protoplasts of wheat (*Triticum aestivum* L). *Bio/Technology* 8: 429-434.

Wang, W.C., and H.T. Nguyen. 1990. A novel approach for efficient plant regeneration from long-term suspension culture of wheat. *Plant Cell Reports* 8: 639-642.

Wang, Y.C., T.J. Klein, M. Fromm, J. Cao, J.C. Sanford, and R. Wu. 1988. Transient expression of foreign genes in rice, wheat and soybean cells following particle bombardment. *Plant Molecular Biology* 11: 433-439.

Wei, Z.M., and Z.H. Xu. 1990. Regeneration of fertile plants from embryogenic suspension culture protoplasts of *Sorghum vulgare*. *Plant Cell Reports* 9: 51-53.

Wernicke, W., and R. Brettell.1980. Somatic embryogenesis from *Sorghum bicolor* leaves. *Nature* 287: 138-139.

Wernicke, W., and R. Brettell. 1982. Morphogenesis from cultured leaf tissue of *Sorghum bicolor*. Culture initiation. *Protoplasma* 111: 19-27.

Wernicke, W., I. Potrykus, and E. Thomas. 1982. Morphogenesis from cultured leaf tissue of *Sorghum bicolor* (L.). The morphogenetic pathways. *Protoplasma* 111: 53-62.

Yamada, Y., Z.Q. Yang, and D.T. Tang. 1986. Plant regeneration from protoplast derived callus of rice (*Oryza sativa*). *Plant Cell Reports* 5: 85-88.

Zhang, H.M., Yang H., Rech E.L., Golds T.J., Davis A.S., Mulligan B.J., Cocking E.C., and Davey M.R. 1988. Transgenic rice plants produced by electroporation-mediated plasmid uptake into protoplasts. *Plant Cell Reports* 7: 379-384.

Zimny, J., and H. Lorz. 1986. Somatic embryogenesis and plant regeneration from meristematic tissue of *Secale cereale* (rye). Pages 503-505 in *Genetic Manipulation in Plant Breeding* edited by W. Horn, C.J. Jensen, W. Odenbach, and O. Schieder. Walter de Gruyter, Berlin, Germany.

Acknowledgements

We wish to thank A.K. Kononowicz and R.A. Bressan for helpful suggestions. A.M. Casas was supported by a fellowship of the Ministerio de Educación y Ciencia (Spain).

3.3

Isolation of protoplasts and genetic manipulation[1]

E. Filippone[2], F. D'Ambrosio[3] and T. Cardi[2]

Research Centre for Vegetable Breeding, National Research Council, Via Università 133, 80055 Portici, Italy[2]; Department of Agronomy and Plant Genetics, University of Naples, Via Università 100, 80055 Portici, Italy[3]

Abstract

Protoplasts have been isolated from many plant sources. This paper summarizes the present status of protoplast isolation, culture, and regeneration. It outlines techniques developed for the genetic manipulation of protoplasts so that they can be used to transfer and express genes stably.

The term protoplast refers to a cell without a cell wall. To release protoplasts, many techniques using physical or chemical treatments have been devised. Among these is enzymatic digestion of the cell wall, which is now used for a number of plant species. The use of protoplasts in plant breeding is currently restricted by the difficulty experienced in routinely regenerating plants from single cells and calluses in some economically important crops, such as cereals. However, many techniques have been developed to transfer and express genes stably in plant protoplasts.

Protoplant Isolation, Culture, and Regeneration

Generally, the entire cycle from plant to protoplast, and back again to plant, is a rather complex phenomenon which involves several intermediate steps. Each of these steps is

1 Contribution No. 73 from the Research Centre for Vegetable Breeding, National Research Council, Via Università 133, 80055 Portici, Italy.

influenced by various factors, which may be dependent on or independent of plant genotype.

Protoplasts have been isolated from a variety of sources, such as leaves, petals, petioles, shoot apices, roots, fruits, coleoptiles, hypocotyls, stems, embryos, aleurone layers of cereal grains, root nodules, microspore mother cells, microspore tetrads, pollen grains, pollen, callus cultures, and suspension cultures (Eriksson 1985). The quality and amount of isolated protoplasts are influenced primarily by the kind of tissue and by the physiological state of the donor plant. In most dicotyledons, young leaves, preferably excised from aseptic shoot cultures (Bengochea and Dodds 1986) and sometimes pre-treated (Eriksson 1985), are generally used for protoplast isolation. In monocotyledons and woody crops, however, embryogenic callus or suspension cultures, which have been derived from immature embryos, are preferred (Roest and Gilissen 1989). Whatever the origin, donor tissues are generally digested with a one-step treatment in which pectinases dissolve the middle lamella, while cellulases and/or hemicellulases degrade the cell wall (Nagata and Ishii 1979; Eriksson 1985). Methods for the purification of protoplasts include filtration and centrifugation steps, with or without gradient, for removing undigested tissues and cellular debris and eventually to select protoplast populations. The viability of freshly isolated protoplasts can be assessed by means of fluoresceine diacetate (FDA) or phenosaphranin (Bengochea and Dodds 1986).

To achieve regeneration of the cell wall and sustained division, purified protoplasts must be plated at an optimal population density, usually ranging between 5×10^3 and 10^5 protoplasts/mL (Bengochea and Dodds 1986). According to Eriksson (1985), protoplasts can be plated in liquid or in semi-solid media, or in a combination of both. For media solidification, better results are generally obtained with agarose than with agar (Lörz et al. 1983; Shillito et al. 1983). Feeder techniques are necessary for special purposes (Eriksson 1985), while with the recently developed microculture technology it is possible to culture individual protoplasts in nanoliter drops (Potrykus 1988). The composition of the culture medium, especially to achieve hormone balance, varies with the genotype used. Generally, dim light or darkness are required at the beginning of culture, but some exceptions are known for legume species (Eriksson 1985). A temperature between 20°C and 28°C is generally used. Under optimal conditions, protoplasts give rise to cell colonies that produce a callus. Regeneration of shoots can generally be achieved by modifying the composition of the medium and varying the environmental conditions.

According to a recent review (Roest and Gilissen 1989), more than 200 plant species, belonging to almost 100 genera in over 30 families, have been regenerated to plants, or in some cases to embryoids or shoots. Regeneration protocols are now available for staple food crops, such as maize, rice, wheat, potato, cassava, and others (Roest and Gilissen 1989). Among factors affecting protoplast culture and regeneration, the genotype is outstandingly important. The family Solanaceae, with almost 30% of the listed regenerable species, appears to be very responsive, whereas species of Graminaceae and Papilionaceae and woody crops are considered recalcitrant to regenerate in vitro. However, genetic variability for this characteristic has been reported even between varieties in cultivated crops (Webb 1988) or accessions of wild species (Cardi et al. 1990). It has been suggested that genetic control of in vitro response is under the control of a few dominant genes (Koornnef et al. 1987). Regeneration of plants can be achieved either via organogenesis, as happens almost exclusively within the Solanaceae, or via embryogenesis, as has been reported for the species belonging to the Graminaceae (Roest and Gilissen 1989).

Regenerated plants coming from protoplast culture can display some variation, which can be used in breeding programmes (Larkin and Scowcroft 1981). Some species in which regeneration via protoplast culture has been achieved are listed in Table 1.

Table 1 Some plant species in which protoplast regeneration has been achieved

Actinidia arguta	*Hyoscyamus muticus*
Actinidia deliciosa	*Limonium perezii*
Beta vulgaris	*Oryza sativa*
Brassica juncea	*Pogostemon cablin*
Brassica napus	*Primula malacoides*
Brassica oleracea	*Rauvolfia vomitoria*
Chrysanthemum morifolium	*Rosa persica* x *xanthina*
Cucumis melo	*Sesbania bispinosa*
Digitalis lanata	*Solanum tuberosum*
Eucalyptus spp.	*Triticum aestivum*
Glycine max	*Zea mays*
Hordeum vulgare	

Source: IAPTC 1990.

Genetic Manipulation of Protoplasts

Somatic fusion

The first report on somatic hybrid plants obtained through induced fusion mediated by iso-osmotic solutions of sodium nitrate dates back to only 1972 (Carlson et al. 1972). Soon afterwards, the use of protoplast fusion became a reliable technology in cell biology and plant breeding when other procedures were developed, such as the high pH-high Ca^{++} method and the polyethylene glycol (PEG) method (Bengochea and Dodds 1986). In the early 1980s, a physical procedure based on the use of an electric AC field to align protoplasts and a DC field to induce fusion was developed by Zimmermann and his co-workers (Gaynor 1986).

Applications of protoplast fusion include the combination of two complete nuclear genomes and cytoplasms in so-called "symmetric fusion" and the transfer of only a part of the nuclear genome, in "asymmetric fusion". Partial nuclear genome transfer has been attempted mainly by fusing strongly irradiated (with gamma or X-rays) donor protoplasts with untreated recipient ones (Bates 1990). However, chromosome-mediated gene transfer (Dudits 1988) and microcell-mediated cell transfer (Ramulu et al. 1988) have been proposed.

Whatever the final goal or the technique employed for fusing plant protoplasts, a major task is the identification and/or selection of fusion products. Several methods have been adopted, including complementation of specific traits, resistance to drugs, and differential growth rates of hybrids and parental calluses. Further investigation, such as morphological, cytological, biochemical, and genetic analyses, is necessary to find evidence of the hybridity of selected material.

Besides bypassing crossing barriers, protoplast fusion can lead to the production of a wide range of hybrid types, as a consequence of new combinations between nuclei and cytoplasms, recombination between mitochondria or chloroplasts, and somaclonal variation (Pehu et al. 1989). As far as applications to crop improvement are concerned, during the past few years several interesting results have been reported in Brassicaceae and Solanaceae, where traits such as cytoplasmic male sterility, drought and cold tolerance, resistance to pathogens and insects, and specific fatty acid composition were investigated in somatic fusion experiments. Interspecific hybrids are now under evaluation (Table 2). Somatic fusion for the transfer of cytoplasmic male sterility and resistance to both biotic and abiotic stresses has been recently extended to important new crops, such as rice (Finch 1990), pear (Ochatt et al. 1989), and ornamental plants belonging to the family Compositae (Malaure et al. 1990).

Organelle transfer

About 5-10% of expressed DNA in plant cells is present in organelles such as mitochondria and plastids. Several traits of agronomic importance, such as plastid development, male fertility, leaf morphology, and flower differentiation, are carried out, at least in part, by these organelles (Sager and Lane 1972), giving particular importance to the development of a reliable technique for transferring organelles from one species to another.

It has been shown that protoplasts can take up viruses, bacteria, chloroplasts, and mitochondria by means of endocytosis. This phenomenon is related to the size of particle to be taken up; special stress conditions are necessary to promote endocytosis of particles with a size greater than 0.5 μm (Cocking 1977). Experiments have been performed in various species to produce "cybrids" (cytoplasmic hybrids) by irradiating donor protoplasts, treating them chemically or actually removing (in cytoplasts) the nucleus of one of the fusion partners. The transfer of chloroplasts has been attempted mainly in species of *Datura, Nicotiana,* and *Petunia* (Whitaker and Evans 1986), but no evidence has been observed for successful chloroplast take up by recipient protoplasts.

Gene transfer

Gene transfer into protoplasts is performed using a variety of methods: chemical helpers, such as poly-L-ornithine, polyethylene glycol, and calcium phosphate microcrystals; liposomes, which have been shown to be effective in organelle transfer; electroporation, that differs from protoplast electrofusion in the absence of an AC field and the decay of a DC pulse that is in the range of tenths of milliseconds (Bates et al. 1988); and *Agrobacterium tumefaciens* co-cultivation, which has not yielded good results till now, although it shows several advantages over explant infection (for example, the possibility of isolating single transformed clones) (Wullems et al. 1986). In each case except the last, DNA is added to the protoplast medium as plasmids, in circular (coiled or supercoiled) or linear form. Traits to be transferred into protoplasts must have at least a selectable marker gene at the cellular level, to avoid regenerating thousands of untransformed plants.

Plant species have been transformed using a variety of methods (Table 3, *overleaf*). Many dicotyledonous and some monocotyledonous species have been transformed by using

Table 2 Some crops in which somatic hybridization via protoplast culture has been achieved

Recipient species		Donor species
Symmetric	*Brassica campestris*	*B. oleracea*
	Brassica napus	*Barbarea vulgaris*
	Busicca napes	*Thlaspi perfoliatum*
	Busicca napes	*B. oleracea*
	Catharantus roseus	*Vinca minor*
	Chrysanthemum morifolium	*Felicia berguiana*
	Cucumis melo	*Cucurbita pepo*
	Cucumis melo	*Cucurbita moschata*
	Dianthus chinensis	*D. barbatus*
	Helianthus spp.	*H. petiolaris*
	Helianthus spp.	*H. rigidus*
	Helianthus spp.	*H. debilis*
	Lycopersicon esculentum	*L. hirsutum*
	Lycopersicon esculentum	*L. cheesmanii*
	Lycopersicon esculentum	*L. pennellii*
	Lycopersicon esculentum	*L. peruvianum*
	Nicotiana sylvestris	*N. tomentosiformis*
	Nicotiana tabacum	*N. rusica*
	Nicotiana tabacum	*N. debneyii*
	Oryza sativa	*Porteresia coarctacta*
	Oryza sativa	*O. officinalis*
	Oryza sativa	*O. punctata*
	Pyrus communis	*Prunus avium* x *pseudocerasus*
	Rauwolfia serpentina	*Catharantus roseus*
	Rauwolfia serpentina	*Vinca minor*
	Rauwolfia serpentina	*Rhazya stricta*
	Solarnum tuberosum	*S. brevidens*
	Solarnum tuberosum	*S. commersonii*
	Solanum tuberosum	*Nicotiana plumbaginifolia*
	Solanum tuberosum	*S. phureja*
	Solanum melongena	*S. aethiopicum*
	Solanum lycopersicoides	*Lycopersicon esculentum*
Asymmetric	*Beta vulgaris*	*B. vulgaris*
	Brassica napus	*Arabidopsis thaliana*
	Nicotiana tabacum	*N. glutinosa*
	Nicotiana tabacum	*N. repanda*
	Nicotiana plumbaginifolia	*Petunia hybrida*
	Lycopersicon esculentum	*L. peruvianum*
	Lycopersicon esculentum	*L. pennellii*
	Lycopersicon esculentum	*L. hirsutum*
	Lycopersicon esculentum	*Solanum commersonii*
	Lycopersicon esculentum	*Solanum nigrum*
	Lycopersicon esculentum	*Solanum rickii*
	Lycopersicon esculentum	*Solanum tuberosum*
	Solanum tuberosum	*Lycopersicon pimpinellifolium*
	Solanum tuberosum	*S. brevidens*

Source: IAPTC 1990.

Table 3 Methods used for protoplast transformation and genes transferred in some plant species

Species	Method	Gene(s)[a]
Arabidopsis thaliana	PEG[b]	*NptII*
Beta vulgaris	*Agrobacterium tumefaciens*	*NptII* + *Gus*
Brassica napus	electroporation	*NptII* + *Gus*
	PEG	*Hpt*
Citrus spp.	PEG	*NptII, cat*
Daucus carota	electroporation	*Cat*
Eucalyptus gunii	electroporation, PEG	*Gus*
Eucalyptus saligna	electroporation	*NptII*
Fragaria x *ananassa*	electroporation	*Hpt* + *Gus*
Helianthus annuus	electroporation, PEG	*Gus*
Linum suffriticosum	PEG	*Hpt*
Lycopersicon esculentum	liposome	*Cat* + TYLCV
Malus pumilia	electroporation	*Cat*
Nicotiana tabacum	liposome	*Cat* + TYLCV
	electroporation, PEG	*Gus*
Oryza sativa	electroporation, PEG	*Lux, Gus, Hpt*
Solanum brevidens	electroporation	*Gus*
Solanum dulcamara	electroporation, PEG	*NptII*
Solanum tuberosum	electroporation	*NptII*
Triticum aestivum	electroporation	*Gus*
Zea mays	electroporation, PEG	*Gus, Bar* + *Pat*

Note: a *Bar* = phosphinotricin acetyltransferase; *Cat* = chloramphenicol acetyltransferase; *Gus* = ß-glucuronidase; *Hpt* = hygromycin phosphotransferase; *Lux* = firefly luciferase; *NptII* = neomycin phosphotransferase II; *Pat* = patatin; TYLCV = tomato yellow leaf curl virus coat protein.

b PEG = polyethylene glycol method.

Source: IAPTC 1990.

physical and chemical helpers other than polyethylene glycol, and plants showing stable integration of the transferred gene have been regenerated.

In summary, the major constraints to protoplast transformation are related to difficulties in regeneration in vitro for some species. Transformants, however, will be true mutants as they derive from a single cell.

Discussion

MASONA: What is the difference between the cell constituents of embryonic and non-embryonic cell suspensions (cell walls) as regards the isolation of protoplasts? Why is it easier to isolate protoplasts from embryonic cells as compared to nonembryonic cells?

FILIPPONE: I have only a few suggestions about how to isolate protoplasts from embryogenic cells that seem to be difficult to digest. Generally speaking, it is necessary to use two or even more enzymes together: cellulases, pectinases, and even hemicellulases. There are enzymes, such as pectolyase, which are more "aggressive" than others, that need to be tested in combination with physical treatments (low or high intensity, temperature, shocking) and chemical treatments (pH, mineral composition of the medium).

EARLE: I have two points. First, perhaps the fact that embryogenic suspensions are usually growing rapidly results in their having thinner cell walls, which are easier to digest. Second, many non-regenerable suspensions, such as black Mexican sweet maize, are excellent for protoplast isolation.

References

Bates, G.W. 1990. Transfer of tobacco mosaic virus resistance by asymmetric protoplast fusion. Pages 293-298 in *Progress in Plant Cellular and Molecular Biology: Current Plant Science and Biotechnology in Agriculture* (Vol. 9) edited by H.J.J. Nijkamp, L.H.W. Van Der Plas, and J. Van Aartrijk. Kluwer Academic Publishers, Dordrecht, The Netherlands.

Bates, G.W., W. Piastuch, C.D. Riggs, and D. Rabussay. 1988. Electroporation for DNA delivery to plant protoplasts. *Plant Cell, Tissue and Organ Culture* 12: 213-218.

Bengochea, T., and J.H. Dodds. 1986. *Plant Protoplasts: A Biotechnological Tool for Plant Improvement*. Chapman and Hall, New York, USA.

Cardi, T., K.J. Puite, and K.S. Ramulu. 1990. Plant regeneration from mesophyll protoplasts of *Solanum commersonii* Dun. *Plant Science* 70: 215-221.

Carlson, P.S., H.H. Smith, and R.D. Dearing. 1972. Parasexual interspecific plant hybridization. *Proceedings, National Academy of Science, USA* 69: 2292.

Cocking, E.C. 1977. Uptake of foreign genetic material by plant protoplasts. *International Review of Cytology* 48: 323-343.

Dudits, D. 1988. Increase of genetic variability by asymmetric cell hybridization and isolated chromosome transfer. Pages 275-281 in *Progress in Plant Protoplast Research* edited by K.J. Puite, J.J.M. Dons, H.J. Huizing, A.J. Kool, M. Koornnef, and S.A. Kreus. Kluwer Academic Publishers, Boston, Massachussetts, USA.

Eriksson, T.R. 1985. Protoplast isolation and culture. Pages 1-20 in *Plant Protoplasts* edited by L.C. Fowke and F. Constabel. CRC Press, Boca Raton, Florida, USA.

Finch, C.H. 1990. Protoplast fusion between rice (*Oryza sativa* L.) and its salt tolerant wild relative, *Porteresia coarctata*. Page 209 in *Abstracts, Seventh International Congress on Plant Tissue and Cell Cultures, 24-29 June 1990, Amsterdam, The Netherlands.*

Gaynor, J.J. 1986. Electrofusion of plant protoplasts. Pages 149-171 in *Handbook of Plant Cell Culture* (Vol. 4) edited by D.A. Evans, W.R. Sharp and P.V. Ammirato. Macmillan, New York, USA.

Koornnef, M., C.J. Hanhart, and L. Martinelli. 1987. A genetic analysis of cell culture traits in tomato. *Theoretical and Applied Genetics* 74: 633-641.

IAPTC. 1990. *Abstracts, Seventh International Congress on Plant Tissue and Cell Cultures, 24-29 June 1990, Amsterdam, The Netherlands.*

Larkin, P.J., and W.R. Scowcroft. 1981. Somaclonal variation — A novel source of variability from cell cultures for plant improvement. *Theoretical and Applied Genetics* 60: 197-214.

Lörz, H., P.J. Larkin, J. Thomson, and W.R. Scowcroft. 1983. Improved protoplast culture and agarose media. *Plant Cell, Tissue and Organ Culture* 2: 217-226.

Malaure, R.S., J.S. Al-Atabee, M.R. Davey, and J.B. Power. 1990. Somatic hybridization of ornamental Compositae. Page 215 in *Abstracts, Seventh International Congress on Plant Tissue and Cell Cultures, 24-29 June 1990, Amsterdam, The Netherlands.*

Nagata, T., and S. Ishii. 1979. A rapid method for isolation of mesophyll protoplasts. *Canadian Journal of Botany* 57: 1820-1823.

Ochatt, S.J., E.M. Patat-Ochatt, E.L. Rech, M.R. Davey, and J.B. Power. 1989. Somatic hybridization of sexually incompatible top-fruit tree rootstocks, wild pear (*Pyrus communis* var. *pyraster* L.) and Colt cherry (*Prunus avium* x *pseudocerasus*). *Theoretical and Applied Genetics* 78: 35-41.

Pehu, E., A. Karp, K. Moore, S. Steele, R. Dunckley, and M.G.K. Jones. 1989. Molecular, cytogenetic and morphological characterization of somatic hybrids of dihaploid *Solanum tuberosum* and diploid *Solanum brevidens*. *Theoretical and Applied Genetics* 78: 696-704.

Potrykus, I. 1988. Progress in plant protoplast research. Pages 1-5 in *Progress in Plant Protoplast Research* edited by K.J. Puite, J.J.M. Dons, H.J. Huizing, A.J. Kool, M. Koornnef, and S.A. Kreus. Kluwer Academic Publishers, Boston, Massachussetts, USA.

Ramulu, K.S., H.A. Verhoeven, and P. Dijkhuis. 1988. Mitotic dynamics of micronuclei induced by aminoprophos-methyl and prospects for chromosome-mediated gene transfer in plants. *Theoretical and Applied Genetics* 75: 575-584.

Roest, S., and L.J.W. Gilissen. 1989. Plant regeneration from protoplasts: A literature review. *Acta Botanica Neederlandica* 38: 1-23.

Sager, R., and D. Lane. 1972. Molecular basis of maternal inheritance. *Proceedings of the National Academy of Science, USA* 69: 2410-2413.

Shillito, R.D., J. Paszkowski, and I. Potrykus. 1983. Agarose plating and a bead culture technique enable and stimulate development of protoplast-derived colonies in a number of plant species. *Plant Cell Reports* 2: 244-247.

Webb, K.J. 1988. Recent developments in the regeneration of agronomically important crops from protoplasts. Pages 27-31 in *Progress in Plant Protoplast Research* edited by K.J. Puite, J.J.M. Dons, H.J. Huizing, A.J. Kool, M. Koornnef, and S.A. Kreus. Kluwer Academic Publishers, Boston, Massachussetts, USA.

Whitaker, R.J., and D.A. Evans. 1986. Organelle transfer for genetic transformation of plant protoplasts. Pages 172-191 in *Handbook of Plant Cell Culture* (Vol. 4) edited by D.A. Evans, W.R. Sharp, and P.V. Ammirato. Macmillan, New York, USA.

Wullems, G.J., F.A. Krens, and R.A. Schilperoort. 1986. Transformation of protoplasts via co-cultivation with *Agrobacterium* and via Ti plasmids. Pages 197-220 in *Handbook of Plant Cell Culture* (Vol. 4) edited by D.A. Evans, W.R. Sharp, and P.V. Ammirato. Macmillan, New York, USA.

3.4

Tissue culture of root and tuber crops at IITA[1]

S.Y.C. Ng

International Institute of Tropical Agriculture, PMB 5320, Ibadan, Nigeria

Abstract

At IITA, tissue culture methods, such as meristem, shoot-tip, and node-cutting cultures, are used routinely to eliminate virus infection from improved clones of cassava and yam and for micropropagating cassava, yam, sweet potato, and cocoyams. Virus-tested improved clones of cassava, yam, and sweet potato have been distributed to national programmes for evaluation and testing. The clonal germplasm is conserved by the in vitro reduced growth storage method; plantlets can be kept in the same tube for 1-2 years. An embryo culture technique for germinating mature embryos of *Manihot* species and immature embryos of cassava is being developed. Somatic embryos have been obtained in vitro from young cassava leaf material cultured in a medium containing picloram. Anther and unpollinated ovary culture of cassava, aimed at producing haploid plants, is under investigation. In vitro microtuber production of yams is an alternative that could aid international distribution of virus-tested clonal materials. The dormancy and uniform germination of microtubers need to be investigated.

Root and tuber crops (cassava, yam, sweet potato, and cocoyam) are major food sources in Africa. While the tubers are consumed as staples, the leaves (except for yam) are often used as a vegetable, providing protein, vitamins, and minerals. Root and tuber crops, vegetatively propagated through stem cuttings or tubers, usually have highly heterozygous seeds. Often there are problems of shy flowering, synchronization of flowering, and even non-flowering. Because these crops are vegetatively propagated, the global movement of their germplasm has been restricted by the danger of introducing diseases and pests to non-infected areas. Through its breeding efforts, the International Institute of Tropical Agriculture (IITA) has

---

1 Contribution No. IITA/91/CP/23 from the International Institute of Tropical Agriculture, PMB 5320, Ibadan, Nigeria.

developed improved clones of cassava, sweet potato, and yam, which are resistant to major diseases and pests. Although improved materials are sent to national programmes for evaluation in seed form, clonal materials could be sent only when meristem culture and virus indexing procedures had been developed.

The objectives of tissue culture of root and tuber crops are as follows: to disseminate improved clonal materials to national programmes and conserve clonal germplasm; to evaluate and develop tissue culture procedures to facilitate the selection/development of desirable clones; and to facilitate the transfer of technology to the national programmes. Cassava is given the highest priority, followed by yam, with some resources being devoted to sweet potato and cocoyam.

Research Activities and Applications

The tissue culture of root and tuber crops can be related to routine activities, to research activities, or to training and the transfer of technology.

Routine activities

Over the past few years, methodologies for disease elimination by meristem culture, micropropagation, and conservation of root and tuber crops have been developed (Ng and Hahn 1985; Ng 1988b). These are now used routinely at IITA. However, there may be room for further improvement.

Disease elimination

Procedures for the elimination of virus infections from cassava, yam, and sweet potato have been developed. A combination of heat treatment of the mother plants for 1 month, followed by meristem culture, is effective in eliminating African cassava mosaic virus from cassava and yam mosaic virus from white yam (Ng and Hahn 1985). Meristem culture alone is effective in eliminating the sweet potato virus complex from sweet potato. Culture media for meristem culture of cocoyam have also been developed. A total of 39 clones of cassava, 5 clones of white yam, and 38 clones of sweet potato have been tested and certified virus-free, and they are available for distribution.

Micropropagation and international distribution

Using single-node cuttings, procedures have been developed for the propagation of virus-tested clones of cassava, yam, and sweet potato. A two-step propagation method was developed for yam multiplication. Single-node cuttings are placed in a liquid culture medium for 1 month to induce multiple shoot formation, followed by subculturing the node cuttings in solid media for distribution.

Virus-tested clonal materials are micropropagated and distributed on request to national programmes, in plantlet form grown in sterile solid culture media for cassava and sweet potato, and as plantlets and microtubers for yam. Virus-tested cassava clones have been distributed to 43 countries in Africa and to the Centro Internacional de Agricultura Tropical (CIAT), sweet potato to over 50 countries throughout the world, and yam to 19 countries in Africa and to India and Fiji.

Germplasm conservation

A reduced growth storage method is used to conserve the germplasm collections of cassava and other *Manihot* species, yam, sweet potato, and cocoyam. This is accomplished by reducing the incubation temperature to 18-22°C for all these crops; for sweet potato, this treatment is combined with a reduced growth medium (addition of 3% mannitol to the medium). Under such conditions, the plantlets can be maintained for 1-2 years without subculturing. Over 1000 accessions of yam, 200 of cassava and other *Manihot* species, 1000 of sweet potato, and 100 of cocoyam are maintained in vitro using the reduced growth method.

Research activities

Research activities include embryo/ovule culture, anther/unpollinated ovary culture, somatic embryogenesis, and in vitro tuberization.

Embryo/ovule culture

The genetic improvement of cassava, aimed at increasing yield and conferring disease and pest resistance, involves the introduction of germplasm from other *Manihot* species and hybridization between cassava and its related wild species. The ability to germinate seeds from these wild species plays a vital role. In IITA's efforts to date, the germination rate has been low in some instances.

Embryo culture can be used to germinate isolated embryos from seeds that are difficult to germinate or that have low germination rates. Immature embryo culture is a refined technique that can rescue hybrids not obtainable by conventional methods. Procedures have been developed for the treatment of mature seeds, characterized by a hard seed coat, and for the culture of mature (Ng 1989) and immature embryos of cassava (Ng, unpubl.). Several culture medium formulations were used to evaluate the germination rate of both mature and immature embryos of cassava. Murashige and Skoog's (MS) medium and Gamborg's B5 medium were used with various modifications. Mature embryos germinated easily on all the media tested; however, the best result was obtained at 1/2 MS medium with 3% sucrose and agar (Table 1 *overleaf*). This medium was used to germinate embryos of several *Manihot* species. Immature embryos isolated from open pollinated fruits (3 weeks old) were induced to germinate on culture media containing coconut water (CW), and the best result obtained to date has been on 1/2 MS medium with 15% CW, 5 mg/L indoleacetic acid (IAA), 4% sucrose, and 0.7% agar (Table 2 *overleaf*).

Because of disease problems and physiological disorders in some water yam clones used in hybridization, the hybrid seeds did not grow to maturity. Recently, immature ovule culture of yam has been initiated.

Table 1 **Response of cassava embryos to different culture medium formulations (all media contained 3% sucrose)**

Media[a]	No growth (%)	Green leaf with root (%)	Root (%)	Plantlet (%)	Additional observation (%)
MS + agar	0	40	0	60	—
MS + 7% CW + agar	0	0	0	100	stunted
MS + 0.05% CH + agar	20	0	20	60	—
MS + 7% CW + gelrite	20	40	0	40	—
1/2 MS + agar	0	0	0	100	vigorous
1/2 MS + 7% CW + agar	0	40	0	60	—
1/2 MS + 0.05% CH + agar	0	40	0	60	—
1/2 MS + 7% CW + gelrite	0	0	0	100	long, thin

Note: a MS = Murashige and Skoog's medium; CW = coconut water; CH = casein hydrolysate; agar at 0.7%; gelrite at 0.2%.

Table 2 **Response of immature cassava embryos to different culture media (all media contained 4% sucrose and 0.7% agar)**

Media[a]	No growth (%)	Green cotyledon (%)	Green cotyledon with root (%)	Plantlet (%)
B5	60	30	0	10
B5 + CH + IAA	60	20	20	0
B5 + CW + IAA	60	10	30	0
MS	60	40	0	0
MS + CH + IAA	60	40	0	0
MS + CW + IAA	70	20	0	10
1/2MS	50	40	10	0
1/2MS + CH + IAA	80	10	10	0
1/2MS + CW + IAA	40	30	0	30

Note: a B5 = Gamborg's B5 medium; MS = Murashige and Skoog's medium; CH = casein hydrolysate at 500 mg/L; IAA = indoleacetic acid at 5 mg/L; CW = coconut water at 15%.

Anther/unpollinated ovary culture

Anther/unpollinated ovary culture for haploid plant production can be used to produce homozygous diploids, which is of value in genetic studies of cassava. Anther culture of a spontaneous tetraploid cassava has been investigated, and it has resulted in the production of protocorms and roots. Callus was obtained from the anthers cultured on B5 medium with sucrose, coconut water, and gelrite. Calluses were then transferred to MS medium with sucrose, kinetin, IAA, and gelrite, with or without adenine sulfate. A subsequent transfer to MS medium without a growth hormone resulted in the development of protocorm structures and roots (Ng, unpubl.).

Haploid plants have been obtained from unpollinated ovary culture of onion (Campion and Alloni 1990). Recently, unpollinated cassava ovaries have also been used to explore the possibility of obtaining haploid plants from such explants.

Somatic embryogenesis

The multiplication rate of virus-tested cassava materials for distribution by IITA could be increased if somatic embryogenesis is induced directly from somatic tissue, followed by the germination of somatic embryos into plantlets. This technique might also provide an opportunity for genetic engineering in cassava.

Although somatic embryo and plantlet formation has been obtained with Latin American cassava materials (Stamp and Henshaw 1982, 1987; Szabados et al. 1987), such work on African cassava materials has not yet been successful.

Young in vitro leaf material of eight IITA cassava clones was cultured on MS medium with sucrose and picloram (Ng 1989). It was found that there were varietal differences in somatic embryo formation but that, in general, a higher percentage of cultures producing somatic embryos was obtained in the medium containing 10 ppm picloram than in the one with 5 ppm (Table 3). Plant regeneration from somatic embryos was not obtained.

Table 3 **Somatic embryogenesis in eight cassava clones, as influenced by picloram concentration**

	Somatic embryogenesis (%)		
	Picloram		
Clones	**5 ppm**	**10 ppm**	**Clonal mean**
TMS 30001	0	0	0
TMS 30211	20	40	30
TMS 30395	40	60	50
TMS 30555	0	20	10
TMS 30572	20	40	30
TMS 4(2)1425	0	20	10
TMS 63397	60	40	50
TMS 91934	0	20	10
Concentration mean	18	30	

In vitro tuberization

In vitro microtuber formation can be used as a means of micropropagation in yam. It is also an effective method for the international distribution of virus-tested clonal germplasm. Tuberization in yam was promoted by increasing the sucrose concentration in the culture medium; 5% sucrose was optimal for white yam (Ng 1988a). Microtubers have been produced from over 30 accessions of white yam and water yam. Studies aimed at increasing the number of microtubers and aerial tuber formation are under way.

Training and transfer of technology

Training national scientists and technicians is one of IITA's major activities. Over 40 scientists from Africa, Asia, and Latin America have been trained by IITA in the tissue culture of root and tuber crops over the past 10 years. Technical assistance and information on setting up a tissue culture facility are also given to national programmes upon request.

Future Prospects

Routine tissue culture activities will continue in all the areas described. Further study on the culture of immature cassava embryos, aimed at rescuing some interspecific hybrids that may have incompatibility problems, is in progress. Studies on yam ovule culture will continue.

Research in plant regeneration from callus, anther, and unpollinated ovary cultures of cassava will be emphasized. Direct somatic embryogenesis, especially the germination of somatic embryos, is an important area of our research at IITA. *Agrobacterium*-mediated transformation in cassava and yam will be initiated, with a view to incorporating genes for improving protein quantity and quality as well as disease (fungal and virus) resistance.

Studies on the induction of aerial microtuber formation in vitro will continue to evolve standard procedures for micropropagation and international distribution of virus-tested yam germplasm. The control of dormancy and uniform germination of microtubers will need emphasis.

Discussion

FAUQUET: As regards somaclonal variation in cassava, CIAT has regenerated 15 cultivars from South America through somatic embryogenesis and has tested them in field trials with them. As far as I know, these cultivars were not different in shape, growth, or taste from the original cultivars.

References

Campion, B., and C. Alloni. 1990. Induction of haploid plants in onion (*Allium cepa* L.) by in vitro culture of unpollinated ovules. *Plant Cell, Tissue and Organ Culture* 20: 1-6.

Ng, S.Y.C. 1988a. In vitro tuberization in white yam (*Dioscorea rotundata* Poir.). *Plant Cell, Tissue and Organ Culture* 14: 121-128.

Ng, S.Y.C. 1988b. Meristem culture, multiplication and distribution: White yam (*Dioscorea rotundata*). Pages 46-48 in *Root, Tuber and Plantain Improvement Program 1988 Annual Report*. IITA, Ibadan, Nigeria.

Ng, S.Y.C. 1989. Embryo culture and somatic embryogenesis in cassava. Paper presented at the Fourth Triennial Symposium, International Society for Tropical Root Crops—Africa Branch, 4-8 December 1989, Kinshasa, Zaire.

Ng, S.Y.C., and S.K. Hahn.1985. Application of tissue culture to tuber crops at IITA. Pages 29-40 in *Proceedings, Inter-Centre Seminar on Biotechnology in International Agricultural Research, 23-27 April 1984, Philippines.* IRRI, Los Banos, the Philippines.

Stamp, J.A., and G.G. Henshaw. 1982. Somatic embryogenesis in cassava. *Zeitschrift für Pflanzenphysiologie* 105: 183-187.

Stamp, J.A., and G.G. Henshaw. 1987. Secondary somatic embryogenesis and plant regeneration in cassava. *Plant Cell, Tissue and Organ Culture* 10: 227-233.

Szabados, L., R. Hoyos, and W. Roca. 1987. In vitro somatic embryogenesis and plant regeneration of cassava. *Plant Cell Reports* 6: 248-251.

3.5

Biotechnological approaches to plantain and banana improvement at IITA[1]

D. Vuylsteke[2] and R. Swennen[3]

International Institute of Tropical Agriculture, PMB 5320, Ibadan, Nigeria[2]; Laboratory of Tropical Crop Husbandry, Katholieke Universiteit Leuven, Kardinaal Mercierlaan 92, 3001 Heverlee, Belgium[3]

Abstract

Plantains and bananas are an important staple food crop in developing countries. Genetic improvement programmes have been spurred by the rapid spread of black sigatoka disease, but the genus *Musa* is intractable in terms of conventional breeding strategies. The potential of biotechnology in banana and plantain improvement may therefore be considerable. Simple tissue culture techniques, such as shoot-tip culture and embryo culture, can overcome some of the obstacles impeding breeding progress. These techniques have contributed to IITA's success in the production of black sigatoka-resistant plantains. Assistance in the development of tissue culture laboratories has been provided to Nigerian national programmes, thereby strengthening their capability in biotechnology. Cell culture and RFLP mapping, both essential to the future implementation of molecular genetic techniques in *Musa* improvement, are being researched in collaboration with advanced laboratories.

The genus *Musa* provides one of the most important basic staple foods for large populations in the humid rainforests of lowland and upland Africa. Recent concern over declining yields of bananas and plantains as a result of the spread of black sigatoka, a virulent fungal leaf spot disease caused by *Mycosphaerella fijiensis*, has spurred genetic improvement programmes (Persley and De Langhe 1987; IITA 1988). Attention has focused on the collection, movement, and conservation of *Musa* germplasm, for which tissue culture is increasingly being used as an enabling technique (Vuylsteke 1989; Vuylsteke et al. 1990).

1 Contribution No. IITA/91/CP/25 from the International Institute of Tropical Agriculture, PMB 5320, Ibadan, Nigeria.

Efforts to breed *Musa* for disease resistance using conventional methods are fraught with obstacles (such as low fertility, triploidy, and lack of genetic variability) specific to the biology of the preferred parthenocarpic cultivars. This is illustrated by the absence, as yet, of a new man-bred banana cultivar that is commercially acceptable. A wide array of plant tissue culture and molecular genetic techniques may be used to overcome some of the problems limiting traditional breeding approaches (Krikorian and Cronauer 1984; Murfett and Clarke 1987; Persley and De Langhe 1987; Dale 1990). However, even the more advanced techniques (Table 1) need to be further developed before comprehensive benefits may accrue.

Table 1 Progress in the application of biotechnologies in plantains and bananas

Technique[a]	Present status[b]
Micropropagation	+++
Embryo rescue	++
Callus culture	++
Cell suspensions	+/++
In vitro selection	+
Anther culture	+
Microspore culture	na
Protoplast culture	+
Protoplast regeneration	na
Protoplast fusion	na
In vitro conservation	++
Cryopreservation	+
Somaclonal variation	++
RFLPs/RAPDs	+/++
Transgenic plants	na
Diagnostics	+

Note: a RFLP = restriction fragment length polymorphism; RAPD = random amplified polymorphic DNA.
b +++ = routine; ++ = widely used; + = just beginning; na = not available.
Source: Adapted from Sasson and Costarini 1990.

At the International Institute of Tropical Agriculture (IITA) existing in vitro culture methods, such as shoot-tip culture for micropropagation and germplasm conservation and exchange, and embryo culture to enhance the germination of botanical hybrid seed have been integrated with conventional multiplication and breeding activities. This work has been carried out in collaboration with the International Network for the Improvement of Banana and Plantain (INIBAP). Within 3 years of initiating a plantain breeding programme, IITA was able to produce hundreds of banana and plantain hybrids, some of which have already been selected as promising new varieties with resistance to black sigatoka (Swennen and Vuylsteke, in press). This feat can be partly attributed to the application of contemporary biotechnologies that complement orthodox breeding efforts. Although, as shown in Table 1, only conventional biotechnologies are currently applicable to *Musa*, these simple tissue-culture techniques have overcome significant obstacles that make *Musa* an intractable crop in terms of conventional genetic improvement.

Shoot-Tip Culture

Shoot-tip culture is a well-established, adequate, and relatively simple in vitro method for handling *Musa* germplasm (Jarret 1986; Vuylsteke 1989). It is routinely used in the rapid clonal multiplication of selected genotypes, the production of clean planting material, and the exchange and conservation of genetic resources. Its major advantages are that in vitro propagules are free of non-obscure pathogens and can be multiplied at rates that are several orders of magnitude higher than those obtained with conventional methods. This will be of great value for the multiplication and international dissemination of newly bred genotypes (Simmonds 1987), or to speed up the testing of selections by plant breeders.

Over 400 *Musa* accessions of seven genomic constitutions (AA, AAA, AAAA, AAB, AAAB, ABB, and BB) have been successfully propagated in vitro at IITA. This indicates that this standard methodology is applicable to a wide range of genotypes.

Micropropagation has been pivotal in the rapid deployment of IITA's plantain breeding programme by supplying 5800 plants of cultivars that possess workable levels of female fertility (Swennen and Vuylsteke, in press; Swennen et al., Paper 2.3, pp 69-74 this volume). Black sigatoka-resistant starchy alternatives to susceptible plantains have been identified among the ABB cooking bananas and five of these are being rapidly multiplied in vitro and distributed at a rate of 5000 plants annually to national programmes and farmers in Nigeria.

Shoot-tip culture also plays an essential role in the international movement of *Musa* germplasm, which is aimed at providing the genetic basis for the breeding programme. From 1985 to 1990, 320 new accessions of natural germplasm were introduced as in vitro cultures in 23 shipments, thereby quadrupling the number of accessions held in the IITA collection. These genetic resources were introduced in a joint effort with the Nigerian Plant Quarantine Service and INIBAP (Vuylsteke et al. 1990). At least 33 of these introductions have shown resistance to black sigatoka. Amongst these are several AA diploids which are useful sources of black sigatoka resistance in IITA's plantain breeding programme.

Over 400 accessions of natural germplasm and about 300 plantain hybrids produced by the breeding programme are maintained in vitro for germplasm storage and as duplicates of the field genebank.

Somaclonal Variation

With the increasing use of in vitro culture for plant production, the occurrence of somaclonal variation (that is, increased genetic variation among plants regenerated from tissue culture) has been found to be ubiquitous. Somaclonal variation is a potential hindrance to the in vitro propagation and conservation of germplasm, but it also has a potential benefit in terms of creating additional variability for crop improvement (Scowcroft 1984). The frequent use of in vitro culture techniques for the handling of *Musa* germplasm warrants investigation into the occurrence and nature of somaclonal variation in this genus.

In a study involving over 3500 micropropagated plants of seven representative plantain cultivars, large cultivar-dependent differences in the frequency of somaclonal variation were observed. Off-types constituted up to 69%, depending on the cultivar (Table 2). Different cultivars also differed in the characters affected by somaclonal variation. However, the spectrum of variation was rather narrow, which implies that some characters are not prone to change. The bulk of somaclonal variation mimicked natural inflorescence-type variation (Table 2), the occurrence of which was enhanced in vitro. Such inflorescence variation may be useful in breeding the sterile false horn plantains, as some of their variants show increased levels of female fertility (Vuylsteke and Swennen 1990).

Table 2 Somaclonal variation (SV) in micropropagated plantains (*Musa* spp., AAB group)

Cultivar	Plantain category	Number of plants screened	SV (%)	SV that mimics natural variability (%)
Bobby Tannap	medium french	391	0	—
Ntanga 2	giant french	393	1	50
Obino l'ewai	medium french	388	2	25
Agbagba	medium false horn	1555	6	56
Ubok Iba	medium horn	248	12	90
Big Ebanga	giant false horn	369	35	100
Bise Egome 2	medium french horn	181	69	99

Source: Vuylsteke and Swennen 1990.

Factors influencing the incidence of somaclonal variation in plantains are being investigated in order to establish guidelines for control of in vitro instability. In contrast to the choice of genotype (Table 2), length of time in culture did not appear to affect somaclonal variation frequency in plantain plants regenerated from shoot-tip culture (Vuylsteke et al. 1988; Vuylsteke and Swennen 1990). There was a tendency toward higher off-type frequencies with increasing proliferation rates of shoot-tips in culture, but the correlation was not significant because of the large genotypic effects.

Somaclonal variation arising among micropropagated plantains, and probably in *Musa* at large, should not be over-estimated as a source of novel variability for use in breeding schemes. Screening at the whole-plant level for somaclonal variants with disease resistance requires considerable space and labour. But if screening could be performed at the cellular level, with selection pressure applied in vitro, this technique could be useful.

Embryo Culture

Intractable fertilization barriers, such as female sterility and the polyploid genome, make genetic improvement of the preferred banana and plantain clones rather slow and technically difficult. In addition to the low seed set, hybrid plant production in the most common triploid clones is further complicated by low seed germination rates, due to embryo mortality or dormancy and disruption of embryo-endosperm relations.

The germination rate of botanical seed, obtained from plantain crosses, in soil is about 1%. Aseptic embryo culture techniques have been proposed to enhance seed germination rates, while the culture of immature embryos could rescue hybrid progeny before embryo abortion.

Initially, in vitro germination rates were also low (9%), but investigations into seed sterilization procedures and culture media composition, particularly the plant growth regulator addition, led to a fourfold increase in plantain embryo germination rates (Table 3). Over the past 2 years, an average of 725 plantain hybrid seeds have been handled monthly at IITA's tissue culture facilities. This involved the culturing of an average of 255 embryos per month, 20 of which germinated to produce plantain hybrid progeny.

Table 3 **Effect of culture protocol on germination rate of embryos from botanical seed of plantain (Bobby Tannap x Calcutta 4)**

Protocol	Number of embryos	Germination rate[a]
Seed sterilization[b]	280	0.09
Seed sterilization + plant growth regulators[c]	279	0.34

Note: a Means significantly different (P = 0.01).
b Seed sterilization in NaOCl (0.5%) for 15 min; medium devoid of plant growth regulators.
c Seed sterilization in $AgNO_3$ (1%) for 15 min; medium with NAA and BAP at 1 μmol/L.

Collaboration with National Programmes and Advanced Laboratories

In 1989, the Imo State Agricultural Development Project, Owerri, Nigeria, established a tissue culture laboratory with assistance from IITA. The laboratory is designed for the rapid in vitro multiplication of black sigatoka-resistant cooking bananas, but it can also be used in the future for resistant plantain varieties when they are released for distribution. This constitutes the first large-scale tissue culture operation for production of planting material in Nigeria for distribution to farmers. Through its assistance, IITA has functioned as a window of access to biotechnologies for a national agricultural system, thereby strengthening the latter's role as a governmental seed service for farmers.

Collaboration with advanced laboratories has been established in the areas of cell culture, restriction fragment length polymorphism (RFLP) and random amplified polymorphic DNA (RAPD) technologies, which are essential to the future implementation of molecular genetic techniques in the improvement of *Musa*.

At the Katholieke Universiteit Leuven, Belgium, somatic embryogenesis in cell suspensions (Dheda et al. 1991) is being researched in order to achieve controllable regeneration of plants from somatic cells, which is an important step in the production of transgenic plants and in the cryopreservation of germplasm (Panis et al. 1990). Also, plants regenerated from cell suspension cultures at the university, and from cryo-treated suspensions, are screened for clonal uniformity at IITA.

At the laboratories of the US Department of Agriculture (USDA) Regional Plant Introduction Station in Georgia, RFLP and RAPD technologies are being applied to plantains in order to estimate genetic diversity, in taxonomic studies, for clonal identification, and detect and study somaclonal variation (Gawel and Jarret, Paper 5.2, pp 231-235 this volume).

Conclusion

Bananas and plantains are important fruit crops and major sources of dietary starch in many African countries. Their importance has not been reflected in efforts to develop technologies aimed at sustaining and enhancing their production, despite increasing constraints to crop productivity. This is due partly to the intractable nature of *Musa* species with regard to conventional breeding strategies. The application of biotechnologies appears to offer great opportunities to overcome certain bottlenecks in conventional breeding of *Musa*.

Established tissue culture techniques are already contributing to banana/plantain improvement, particularly in rapid clonal propagation and the conservation and exchange of genetic resources. Embryo culture techniques are used by *Musa* breeding programmes to enhance seed germination rates and to rescue hybrid progeny. Advanced diagnostic techniques, such as monoclonal antibodies and DNA probes, for the identification and detection of banana bunchy-top virus (BBTV) and fusarium wilt are currently under development. These will be of considerable benefit to disease control and germplasm movement (Dale 1990). Molecular methods are becoming available for investigating genetic variability. RFLP mapping will assist breeding programmes by enabling early identification of promising new genotypes. RFLPs constitute a powerful instrument in the study of *Musa* phylogenesis and for detecting somaclonal variation in tissue cultures. They will also be useful to probe pathogen diversity.

Notwithstanding the recent success in conventional breeding at IITA, culminating in the production of black sigatoka-resistant plantains (Swennen and Vuylsteke, in press), recombinant DNA technology may be of great benefit in the improvement of *Musa* cultivars that are difficult to breed and in the incorporation of genes coding for characteristics not available in the *Musa* genepool. Concerning the latter, there is no known source of resistance to BBTV (Dale 1987). However, it may soon be possible to genetically engineer banana and plantain for BBTV resistance by introducing a viral coat protein gene. This requires a transformation and regeneration system. Plant regeneration from cell suspensions has recently been demonstrated in *Musa* cultivars (Novak et al. 1989; Panis et al. 1990; Dheda et al. 1991). In contrast, transformation systems for *Musa* have not yet been studied, but the technique of microprojectiles using the particle gun offers most promise (Dale 1990).

Discussion

GIBSON: There is evidence that micropropagated potatoes are more susceptible to viruses, as well as anecdotal evidence that micropropagated bananas are more susceptible to mosaic virus. Would you care to comment?

VUYLSTEKE: This may be true, but we have no data on it. Since 90% of all banana and plantain plants in IITA fields are currently micropropagated plants, we have occasionally observed increased damage by mosaic virus, but it has been rather erratic and is not a real problem. Following a long in vitro cycle and the likely elimination ofthis virus in such cultures, it could be that the plants were axenic, making them more vulnerable to subsequent infection in the field. In contrast, conventionally propagated plants are believed to be chronically infected, but seem to have fewer problems of mosaic virus damage.

KAHL: Does IITA have a collection of banana fungal pathogens, in particular *Mycosphaerella fijiensis* and *M. musicola*?

VUYLSTEKE: No, there is currently no collection of *M. fijiensis* at IITA. However, a plantain pathologist will soon start working at IITA and his research will include determining the genetic variability of the pathogen in West and Central Africa.

QUAYE: One of the major problems with plantain is that it is very susceptible to wind damage as the stem breaks easily. Do you plan to improve the mechanical strength of the varieties you are working with?

VUYLSTEKE: Only in the sense of incorporating dwarfism into plantains. Unfortunately, we have not yet seen any somaclonal variants exhibiting dwarfism, as has been the case elsewhere. However, this trait is a target for the plantain breeding programme at IITA and its attainment should be relatively easy, using conventional breeding methods.

References

Dale, J.L. 1987. Banana bunchy top: An economically important tropical plant virus disease. *Advances in Virus Research* 33: 301-325.

Dale, J.L. 1990. Banana and plantain. Pages 225-240 in *Agricultural Biotechnology: Opportunities for International Development* edited by G.J. Persley. CAB International, Wallingford, UK.

Dheda, D., F. Dumortier, B. Panis, D. Vuylsteke, and E. De Langhe. 1991. Plant regeneration in cell suspension cultures of the cooking banana cv. 'Bluggoe' (*Musa* spp., ABB group). *Fruits* 46: 125-135.

IITA. 1988. *Annual Report and Research Highlights 1987/88.* IITA, Ibadan, Nigeria.

Jarret, R.L. 1986. In vitro propagation and genetic conservation of bananas and plantains. Pages 15-33 in *Report, Third Meeting, IBPGR Advisory Committee on In Vitro Storage.* IBPGR, Rome, Italy.

Krikorian, A.D., and S.S. Cronauer. 1984. Aseptic culture techniques for banana and plantain improvement. *Economic Botany* 38: 322-331.

Murfett, J., and A. Clarke. 1987. Producing disease-resistant *Musa* cultivars by genetic engineering. Pages 87-94 in *Banana and Plantain Breeding Strategies. Proceedings, International Workshop 13-17 October 1986, Cairns, Australia* edited by G.J. Persley and E.A. De Langhe. ACIAR Proceedings No. 21. ACIAR, Canberra, Australia.

Novak, F.J., R. Afza, M. van Duren, M. Perea-Dallos, B.V. Conger, and X.L. Tang 1989. Somatic embryogenesis and plant regeneration in suspension cultures of dessert (AA and AAA) and cooking (ABB) bananas (*Musa* spp.). *Bio/technology* 7: 154-159.

Panis, B.J., L.A. Withers, and E.A.L. De Langhe. 1990. Cryopreservation of *Musa* suspension cultures and subsequent regeneration of plants. *Cryo-Letters* 11: 337-350.

Persley, G.J., and E.A. De Langhe (eds). 1987. *Banana and Plantain Breeding Strategies. Proceedings, International Workshop 13-17 October 1986, Cairns, Australia.* ACIAR Proceedings No. 21. ACIAR, Canberra, Australia.

Sasson, A., and V. Costarini (eds). 1990. *Plant Biotechnologies for Developing Countries. Proceedings, International Symposium, CTA/FAO, 26-30 June 1989, Luxembourg.* CTA/FAO, Ede, The Netherlands.

Scowcraft, W.R. 1984. *Genetic Variability in Tissue Culture: Impact on Germplasm Conservation and Utilization.* IBPGR, Rome, Italy.

Simmonds, N.W. 1987. Classification and breeding of bananas. Pages 69-73 in *Banana and Plantain Breeding Strategies. Proceedings, International Workshop 13-17 October 1986, Cairns, Australia* edited by G.J. Persley and E.A. De Langhe. ACIAR Proceedings No. 21. ACIAR, Canberra, Australia.

Swennen, R., and D. Vuylsteke. (in press). Breeding black sigatoka-resistant plantains with a wild banana. *Tropical Agriculture (Trinidad)*

Vuylsteke, D. 1989. *Shoot-Tip Culture for the Propagation, Conservation and Exchange of Musa Germplasm.* Practical Manuals for Handling Crop Germplasm In Vitro, 2. IBPGR, Rome, Italy.

Vuylsteke, D., and R. Swennen. 1990. Somaclonal variation in African plantains. *IITA Research* 1(1): 4-10.

Vuylsteke, D., R. Swennen, G.F. Wilson, and E. De Langhe.1988. Phenotypic variation among in vitro propagated plantain (*Musa* spp. cultivars AAB). *Scientia Horticulturae* 36: 79-88.

Vuylsteke, D., J. Schoofs, R. Swennen, G. Adejare, M. Ayodele and E. De Langhe. 1990. Shoot-tip culture and third-country quarantine to facilitate the introduction of new *Musa* germplasm into West Africa. *FAO/IBPGR Plant Genetic Resources Newsletter* 81/82: 5-11.

3.6

Status of plant regeneration of yam

C.E.A. Okezie and S.N.C. Okonkwo

Laboratory for Plant Cell and Tissue Culture and Biotechnology, Department of Botany, University of Nigeria, Nsukka, Nigeria

Abstract

Cultivated yams are a major carbohydrate staple in many tropical and subtropical countries, and many wild species are of medicinal importance. Current methods of improving yam are inadequate and thus plant scientists have begun to search for alternative regeneration materials, using biotechnology methods. These include the use of leafy vine cuttings and the culture of leafless nodal explants, leading to the production of uniform plantlets. Meristem culture is being used, especially for virus elimination. The culture of excised leaves, true seed, and zygotic embryos eliminates problems associated with the handling of large tubers. Somatic embryogenesis has been achieved in some species, and somaclonal variants arising in culture could form a basis for yam improvement. Tuber regeneration in culture clearly demonstrates the viability of non-tuber sources as useful alternatives for yam regeneration.

Yams (*Dioscorea* spp.) produce edible tubers, bulbils, or rhizomes of considerable economic importance (Ammirato 1984). They are a major carbohydrate staple in many tropical and subtropical countries (Onwueme 1978), and many wild species are of medicinal and pharmacological importance (Coursey 1967).

Large-scale yam propagation is restricted to areas of the world where agriculture is still largely in the hands of peasant farmers. Efforts at improvement have traditionally involved selection of the best-yielding plants and their propagation by setts from tubers. Persistent vegetative propagation has led to degeneration of the sexual reproduction process, which could otherwise have led to the formation of new genotypes, providing a basis for selection of high-yielding lines. The near absence of modern biotechnological approaches to yam propagation has resulted in a low presence for yam in international trade relative to other tuberous crops, such as potatoes. However, to help yam maintain its place as a major economic crop, scientists have embarked on a search for alternative methods of regeneration, other than through tubers, using modern biotechnological approaches.

Biotechnological Approaches to Yam Regeneration

Regeneration from leafy vine cuttings

Pioneering work was conducted by Correl et al. (1955) on wild yams (because of their medicinal and pharmacological importance) and by Njoku (1963) on edible yams. A cutting usually consisted of a single leafy node excised from the vine in such a way that 2-2.5 cm of the vine was left on either side of the node. The basal portion of the cutting was normally buried in a suitable substratum. Cuttings survived and rooted only when kept damp through periodic watering. Successful propagation of wild yams through vine nodal explants has also been reported by Blunden et al. (1966) on *D. spiculiflora*, and by Vasanthakumar (1981) on *D. floribunda* and *D. composita*. Okonkwo et al. (1986) reported efforts on the edible yams *D. alata*, *D. rotundata*, and *D. bulbifera*. Studies on *D. bulbifera* have revealed that there are three primordia in the leaf axil: primary, secondary, and bulbil. Of these, the bulbil primordium proliferates to form a mass of whitish callus tissue, from which roots are produced. Subsequently, part of the proliferation becomes organized as tubers, followed by shoot formation. The primary and secondary axillary buds remain quiescent and are never involved in the regenerative activities of the axillary complex. Age of the plant from which the cuttings were taken (Okonkwo et al. 1986), number of nodes per cutting (Akoroda and Okonmah 1982), and photoperiod (Okonkwo et al., unpubl.) affect the regenerative behaviour of the cuttings.

Culture of leafless nodal explants

In vitro culture of defoliated nodal explants on suitable agar nutrient media has been used in the regeneration of edible and medicinal yams. The regenerative behaviour of the explants in culture depends on the species (Ammirato 1984), the specific form (Lakshmi et al. 1976), concentrations of auxin and cytokinin in the medium (Uduebo 1971), and age of the parent plant (Mantell et al. 1978). It has been shown in edible yams that regeneration occurs from the existing bud and bulbil primordia, and multiple shoots could be produced from the nodal axillary complex (Okonkwo et al., unpubl.). There is often a gradual browning of the culture medium from the cut ends of the stem segments, probably because of exudation of phenolic compounds. Such browning may be minimized by subculturing, anti-oxidants in the medium, or the use of low light intensity (Ammirato 1984).

Meristem culture

The culture of excised meristem may be used for large-scale clonal propagation as well as for virus elimination (Mantell et al. 1980). Grewal et al. (1977) cultured 0.5 mm-long apical meristems of *D. deltoidea* and were able to regenerate large numbers of plants. Chaturvedi et al. (1977) compared apical meristems and single-node leaf cuttings and found that they had nearly the same rate of growth under one set of nutrient conditions. Plants which are produced clonally by nodal or meristem culture show a high degree of uniformity (Ammirato 1984).

Culture of excised leaves

Sinha and Chaturvedi (1977) initiated some cultures of *D. floribunda* from excised leaves. The leaf explants were devoid of the axillary buds but contained the pulvinus. It was demonstrated that regenerative proliferation and development of multiple shoots occurred from the pulvinus tissue. The plantlets which were obtained were successfully transplanted into soil.

Propagation from seed

Through recent understanding of the flowering behaviour of yams and the pattern of seed dormancy, it is now possible to raise many yam lines through seed (Sadik and Okereke 1975; Okezie et al. 1986). In addition to improving the handling and distribution of material for rapid propagation, this development offers the possibility that yam tubers now used for propagation will become available for consumption.

Zygotic embryo culture

The culture of excised embryos provides a level of experimental flexibility not easily attained through the use of seed or seed tubers in the propagation of yams. For example, in vitro embryo culture eliminates the constraints to seed germination (embryo growth and emergence of radicle) caused by the seed coat and endosperm (Okezie et al. 1984). With the efficient development of embryo culture, greatly increased rates of multiplication can be achieved. This technique could be exploited to generate uniform clonal material, both for initiating a viable breeding programme and for experimental work. It also provides a means for long-term storage of germplasm and rapid multiplication of improved cultivars in a disease- and insect-free form to meet the demands of national programmes anywhere in the world (IITA 1977; Henshaw 1979).

Somatic embryogenesis

The use of callus and cell cultures from various plant organs for the expression of totipotency through somatic embryogenesis holds promise for yam regeneration and propagation. Source materials vary from defoliated vine nodal explants (IITA 1974) to tuber pieces (IITA 1973). Grewal and Atal (1976) reported plantlet regeneration from callus cultures of *D. deltoidea*.

A common element in most of the reports is the use of 2,4-dichlorophenoxy acetic acid (2,4-D) as the auxin in the culture medium. It both initiated proliferation and maintained unorganized growth during subculture. Osifo (1988), however, used naphthalene acetic acid (NAA) in place of 2,4-D. Transfer of the material to another medium (secondary medium) without 2,4-D, with lower levels of 2,4-D, or with another auxin source coincided with the appearance of embryos or plants (Ammirato 1984). Somaclonal variations arising in culture (Larkin and Scowcroft 1981) may give rise to useful variants that could form a basis for improvement.

Conclusion

The present status of yam propagation is clearly inadequate to guarantee its place as a major economic crop in its current areas of cultivation, let alone worldwide. There is, however, great optimism that the application of modern technologies such as tissue and cell culture, meristem culture, and somatic hybridization, will provide viable alternatives to tubers as propagation material for yam regeneration. Tuber regeneration in culture (Uduebo 1971; Ammirato 1984; Ng 1988) clearly supports this. Persistent efforts by plant scientists using modern biotechnological approaches, adequate funding by governments in yam-producing countries, and support from international funding agencies for training and exchange programmes could help re-establish the economic importance of yams.

References

Akoroda, M.O., and L.U. Okonmah. 1982. Sett production and germplasm maintenance through vine cuttings in yams. *Tropical Agriculture (Trinidad)* 59: 311-314.

Ammirato, P.V. 1984. Yams. Pages 327-352 in *Handbook of Plant Cell Culture. Vol. 3: Crop Species* edited by P.V. Ammirato, D.A. Evans, and Y. Yamada. Macmillan, New York, USA.

Blunden, G., R. Hardman, and G. E. Trease. 1966. Some observations on the propagation of *Dioscorea belizensis* Lundell, and other steroid-yielding yams. *Planta Medica* 14: 84-89.

Chaturvedi, H.C., M. Sinha, and A.K. Sharma. 1977. Clonal propagation of *Dioscorea deltoidea* Wall, through in vitro culture of shoot apices and single-node leafy cuttings. Pages 500-505 in *Cultivation and Utilization of Medicinal and Aromatic Plants* edited by C.K. Atal and B.M. Kapur. Regional Research Laboratory, Jammu-Tawi, India.

Correl, D.S., B.G. Schubert, H.S. Gentry, and W.O. Hawley. 1955. The search for plant precursors of cortisone. *Economic Botany* 9: 307-375.

Coursey, D.G. 1967. *Yams.* Longman, London, UK.

Grewal, S., and C.K. Atal. 1976. Plantlet formation in the callus cultures of *Dioscorea deltoidea* Wall. *Indian Journal of Experimental Biology* 14: 352-353.

Grewal, S., S. Koul, U. Sachdeva, and C.K. Atal. 1977. Regeneration of plants of *Dioscorea deltoidea* Wall by apical meristem cultures. *Indian Journal of Experimental Biology* 15: 201-203.

Henshaw, G.G. 1979. Tissue culture and germplasm storage. *IAPTC Newsletter* 28: 2-6.

IITA. 1973. *Report, Root and Tuber Improvement Program.* IITA, Ibadan, Nigeria.

IITA. 1974. *Annual Report 1974.* IITA, Ibadan, Nigeria.

IITA. 1977. *Annual Report 1977.* IITA, Ibadan, Nigeria.

Lakshmi, S.G., R.K. Bammi, and G.S. Randhawa. 1976. Clonal propagation of *Dioscorea floribunda* by tissue culture. *Journal of Horticultural Science* 51: 551-554.

Larkin, P.G., and J.M. Scowcroft. 1981. Somaclonal variation — A novel source of variability from cell cultures for plant improvement. *Theoretical and Applied Genetics* 60: 197-214.

Mantell, S.H., S.Q. Haque, and A.P. Whitehall. 1978. Clonal multiplication of *Dioscorea alata* and *Dioscorea rotundata* Poir yams by tissue culture. *Journal of Horticultural Science* 53: 95-98.

Mantell, S.H., S.Q. Haque, and A.P. Whitehall. 1980. Apical meristem tip culture for eradication of flexous rod viruses in yams (*Dioscorea alata*). *Tropical Pest Management* 26: 170-179.

Ng, S.Y.C. 1988. In vitro tuberization in white yam (*Dioscorea rotundata* Poir). *Plant Cell, Tissue and Organ Culture* 14: 121-128.

Njoku, E. 1963. The propagation of yams (*Dioscorea* species) by vine cuttings. *Journal of the West African Science Association* 8: 29-32.

Okezie, C.E.A., F.I.O. Nwoke, and S.N.C. Okonkwo. 1984. In vitro culture of *Dioscorea rotundata* embryos. Pages 121-124 in *Tropical Root Crops: Proceedings, Second Triennial Symposium of the International Society of Tropical Root Crops—Africa Branch* edited by E.R. Terry, E.V. Doku, O.B. Arene and N.M. Mahungu. IDRC, Ottawa, Canada.

Okezie, C.E.A., F.I.O. Nwoke, and S.N.C. Okonkwo. 1986. Field studies on the growth pattern of *Dioscorea rotundata* Poir propagated by seed. *Tropical Agriculture (Trinidad)* 63: 22-24.

Okonkwo, S.N.C., F.I.O. Nwoke, and E. Njoku. 1986. Pattern of growth of whole plants and development of vine nodal cuttings of *Dioscorea bulbifera* L. *Proceedings of the Nigerian Academy of Science* 1: 30-38.

Onwueme, I.C. 1978. *The Tropical Tuber Crops: Yams, Cassava, Sweet Potato, and Cocoyam.* John Wiley, Chichester, UK.

Osifo, E.O. 1988. Somatic embryogenesis in *Dioscorea. Journal of Plant Physiology* 133: 378-380.

Sadik, S., and O.U. Okereke. 1975. A new approach to improvement of yam *Dioscorea rotundata. Nature (London)* 254: 134-135.

Sinha, M., and H. C. Chaturvedi. 1977. Rapid clonal propagation of *Dioscorea floribunda* by in vitro culture of excised leaves. *Current Science* 48: 176-178.

Uduebo, A. 1971. Effect of external supply of growth substances on axillary proliferation and development in *Dioscorea bulbifera. Annals of Botany* 35: 159-163.

Vasanthakumar, K. 1981. Studies on the propagation and growth of sapogenin-bearing yams, *Dioscorea floribunda* Mart and Gal, and *Dioscorea composita* items from single leaf stem cuttings. *Thesis Abstracts* 7: 69-70.

3.7

Status of in vitro regeneration of tropical plantation crops

E.B. Esan

Cocoa Research Institute of Nigeria, PMB 5244, Ibadan, Nigeria

Abstract

This paper summarizes the major achievements through the application of plant tissue culture, particularly rapid clonal propagation (micropropagation), in five economically important, tropical plantation crops — cocoa, oil palm, coconut, cashew, and tea. It also discusses the limitations and prospects of micropropagating tissues, organs, and plants of these woody perennials. Breakthroughs recorded for oil palm and coconut are highlighted.

The objectives of vegetative propagation are to regenerate, multiply, and conserve plants with desirable traits. Most plants have excellent natural ways of attaining these objectives, but they have not kept pace with the aspirations of man. Plant tissue culture techniques have remained more or less the exclusive preserve of the developed countries, although they have been adopted in a few developing world countries in Central and South America and Asia. In Africa, the use of these techniques has just begun. There is a need to apply plant tissue culture to all economic tree crops for the multiplication of superior plant individuals and their progenies, particularly progenies of interspecific crosses between inter-continental species, such as African *Elaeis guineensis* and South American *E. melanococca*.

Cocoa

The conventional vegetative propagation of cocoa (*Theobroma cacao* L.) by rooted cuttings was introduced in the late 1930s. Tissue culture in cocoa research was introduced by Evans (1951) and Archibald (1954), but the micropropagation of cocoa was not achieved until Esan (1977) induced somatic embryos, both directly and via callus, from embryo axes of mature seeds (beans) and cotyledons of immature embryos. This was achieved on Murashige-

Skoog (MS) (1962) medium, supplemented with naphthalene acetic acid (NAA), coconut milk, and casein hydrolysate, while sulphate and kinetin were optional additives.

When callus was produced as an intermediary to somatic embryogenesis, it came from cell's with characteristics of embryogenic-competent cells. From a single explant, 2-40 somatic embryos were produced; they were asynchronous and of various sizes, and attained various stages in embryogenesis before becoming dormant. They failed to grow into plants, a finding confirmed by Pence et al. (1979, 1980) and Novak et al. (1986). The induction of maturation processes and germination of somatic embryos by removing their cotyledons has been reported by Novak et al. (1986) and Adu-Ampomah et al. (1987, 1988).

Two approaches have dominated attempts to micropropagate cocoa: the stimulation of axillary shoot development; and the induction of adventitious shoots and somatic cell embryos. Some researchers have used explants from vegetative and reproductive adult tissues and organs, or seedling tissues such as nodal buds, cuttings, and shoot apices (Orchard et al. 1979; Passey and Jones 1983; Legrand et al. 1984; Litz 1986). Shoots and roots have been induced from callus derived from the plumule of a mature embryo axis when cultured on a modified MS medium containing NAA (50 mg/L), casein hydrolysate (500 mg/L), and supplementary $Ca(NO_3)_2$ (300 mg/L) (Esan 1982). Thompson et al. (1987) isolated protoplasts from cocoa leaf tissue and succeeded in maintaining them in culture. However, they failed to fuse or divide thereafter.

Attempts are still being made to induce germination and plantlet growth in vitro in cocoa somatic embryos which become dormant when obtained (Esan 1977; Novak et al. 1986). Scientists in the Biotechnology Unit of the Centro Agronomico Tropical de Investigacion y Ensenanze (CATIE) in Costa Rica have succeeded in producing composite cocoa plants by micrografting somatic embryos produced in vitro on to zygotic seedlings raised in vitro.

Recently, *Agrobacterium tumefaciens*, the crown gall bacterium, has been found associated with cocoa (Purdy and Dickstein 1989). Thus, the Ti plasmid of *A. tumefaciens* may perhaps be used as a vector for the introduction of selected DNA into cocoa.

To date no cocoa plants have been raised from cultures and planted in the field.

Cashew

Progress in cashew (*Anacardium occidentale* L.) improvement through conventional breeding methods has been hampered because the tree is an outbreeder and thus the plants are heterozygous. It takes 4-5 years for a cashew seedling to start bearing, and about 15 years to reach full production. Efforts to propagate it vegetatively by conventional methods have been laborious and usually unsuccessful (Rai 1970). Esan and Akinwale (in press) have reported that rooted marcots have produced flowers and mature-sized fruits within 9 months. However, rooting success was less than 25%.

A method for raising clones from existing premium cashew selections or progenies of their diallel crosses would be of great benefit. In vitro rapid clonal propagation of cashew has been done successfully by Philip (1984). Using Lin and Staba (1961) medium supplemented with indolacetic acid (IAA) and 6-furfurylaminopurine (KIN), at 0.5 mg/L each, multiple plantlets were induced in vitro directly from fragments of mature cotyledons. Jha (1988) reported similar achievements with immature embryos (5-10 mm long) when cultured to induce embryo callus as an intermediary to somatic embryos on MS medium supplemented with 2,4-dichlorophenoxy-acetic acid (2,4-D), NAA, KIN, and polyvidone.

Oil Palm

Tissue culture research on oil palm (*Elaeis guineensis* Jacq.) has been in progress for more than 20 years (Durand-Gasselin et al. 1990). The first successful regeneration of plantlets from leaf tissue was reported by Rabechault et al. (1972). Since then, several other researchers have successfully carried out in vitro vegetative propagation of oil palm and made improvements to the protocol. Most researchers have produced somatic cell embryos through a callus intermediary (Jones 1974; Rabechault and Martin 1976; Paranjothy and Othman 1982; Nwankwo and Krikorian 1983; Carley et al. 1986; Duval et al. 1988; Durand-Gasselin et al. 1990). Using this approach, direct somatic embryos were produced under conditions that do not favour rapid growth, and then proliferated in the absence of growth regulators. The age of trees from which explants were taken influenced the degree of success. Explants from younger plants produced more somatic embryos.

In general, explants were obtained from shoot apices of unopened young leaves of trees aged 3-25 years) (Duval et al. 1988), and cultured on MS medium augmented with an auxin (usually 2,4-D) or modifications of this. In about 60 days, primary callogenesis occurred, usually from secondary veins. The callus was allowed to multiply for about 6 months (secondary callus). Two types of calluses were obtained, fast and slow growing. The former produced few somatic embryos (Pannetier et al. 1981; Duval et al. 1988); the latter produced more somatic embryos as well as direct somatic embryos from even the primary callus (Duval et al. 1988). Thus, there are two defined stages in oil palm regeneration — callus initiation and development of plant somatic embryos — both of which occur within 60 days.

Over 50 000 in vitro oil palm plants are now being produced annually in some countries. In another year or two, it is anticipated that 7-8 million plantlets will be produced annually in Indonesia (Noiret et al. 1984/85). The main problem associated with in vitro oil palm regenerants so far is the high incidence of somatic variants, which are claimed to be of genetic origin (Durand-Gasselin et al. 1990). This is being tackled by several groups of researchers, the most prominent of whom are at the Unilever laboratories in Colworth, UK.

Tea

Tissue culture research on tea (*Camellia* spp.) began with the pioneering work of Forrest (1969). In the following 10 years, work was directed purely at studying biochemical processes associated with in vitro synthesis of phenols and alkaloids (Ogutuga and Northcote 1970; Koretskaya and Zaprometov 1975; Zagoskina and Zaprometov 1983).

Micropropagation studies have involved several species, some of which are cultivated purely as ornamentals, others for tea-seed oil and for beverage (*C. sinensis* L.O. Kuntze) (Wu et al. 1981; Carlisi and Torres 1986; Torres and Carlisi 1986; Yan et al. 1988; Samartin 1989; Vietez et al. 1989).

Several types of explants obtained from seeds, seedlings, and in a few cases from mature trees, have been cultured in vitro, including shoot tips, axillary buds, cotyledons, embryo axes, anthers, somatic embryos and callus produced in vitro, leaf, stem nodes, and roots (Mu-Quin and Ping 1983; Kim 1986; Esan 1989; Kato 1989; and Vietez et al. 1989). The most successful explants have been shoot tips, nodes, and cotyledons (Esan 1989). Regenerated structures have included plantlets, pseudobulbils, shoots, roots, cotyledons, and somatic embryos (Saha and Bhattacharya 1988; Yan et al. 1988; Kato 1989). The usual media have been MS or B5 supplemented with an auxin (especially 2,4-D) and cytokinin, particularly when anthers are cultured (Kim 1986; Chen and Liao 1988; Kutubidze et al. 1988). A mixture of glucose and fructose as a source of carbon has been reported to be better than sucrose alone (Haldeman and Thomas 1989). In general, there have been positive reports on the use of gibberellic acid, casein hydrolysate, and coconut water.

Genotypic differences among 10 clones have been observed with respect to rooting of shoots, shoot multiplication, and somatic embryo induction.

Rapid propagation methods, such as micropropagation, need to be developed to provide the large number of plants (6000-240 000/ha) required because of expanding tea plantations and higher recommended planting density (Etherington 1990). Iddagoda et al. (1988) claimed the production of 100 000 tea plants per micro-cutting per year in vitro. This would still need to be improved upon, popularized, and the plants assessed in the field to ensure their trueness to type. So far, few in vitro plants have survived transplanting to the field.

Coconut

Under unusual natural conditions, coconut (*Cocos nucifera* L.), a monoecious monocot and a perennial, will spontaneously reproduce vegetatively by branching and by bulbil formation on inflorescences (Davis 1969). These vegetative propagules (bulbils) have been rooted with some degree of success (Sudasrip et al. 1978). However, this phenomenon cannot be exploited for routine multiplication because of its sporadic occurrence.

Without micropropagation, selection and cross-breeding of the coconut palm would take several decades; thus, work on in vitro propagation of coconut was initiated more than 30 years ago. The earliest research investigations utilized the embryo axes of zygotic embryos as explants (Cutter and Wilson 1954; Abraham and Thomas 1962; De Guzman et al. 1971). More recently, most other parts of the coconut palm have been used (De Guzman et al. 1971; De Guzman et al. 1978; Gupta et al. 1984). Virtually every part of the coconut palm has been tried as a source of explants: stems, secondary and tertiary roots, leaves, inflorescences (mature and immature), anthers, rechillas, endosperm, seedling apices, bulbils, cotyledons, and haustoria (Branton and Blake 1983; Verdeil et al. 1989).

In vitro regeneration of coconut palm has been achieved through two-stage development, either after an initial (nodular) callogenesis followed by embryogenesis, or by independent development of shoots and then roots (Branton and Blake 1983; Raju et al. 1984; Buffard-Morel et al. 1988). The major problem with these in vitro plants, however, is their genetic instability (somaclonal variation), a condition usually induced by components in the culture medium (Cocking et al. 1981). Other, less important, problems are the quick decline, death, and decay of plantlets on transfer to field conditions (Iyer 1981) but this problem has been overcome recently by Assy-Bah et al. (1989). The high variability of expression of embryogenetic potentials, between and within selected trees, and the high frequency of embryogenesis neoformations that fail to produce complete embryos or plants (Verdeil et al. 1989) also cause problems. The most successful medium used for coconut regeneration so far consists of MS minerals, Morel and Westmore's vitamin complex, Fe-EDTA, 41 mg/L and sodium ascorbate 100 mg/L (Assy-Bah 1986; Assy-Bah et al. 1987).

Assy-Bah et al. (1989) outlined improvements to the technique for producing somatic embryos from leaf and inflorescence explants of coconut, and for ensuring successful field establishment (weaning). They eliminated the haustorium on the embryo and treated the embryos with auxin (NAA) to enhance simultaneous leaf and root system development. Alternatively, sucrose (60-90 g/L) adequately replaced the requirement for the auxin.

Conclusion

The choice of the five plant species discussed here was based on their economic significance in tropical countries. Micropropagation studies are also being carried out on other tropical crops such as date, coffee, citrus, rubber, banana, mango, and avocado. Currently these crops are being regenerated mainly under laboratory conditions (Sasson 1988).

Although breakthroughs have occurred with oil palm and coconut, the reproduced plants have shown high rates of genetic abberants. It has been suggested that the success recorded with oil palm and coconut micropropagation should be accepted and used with some caution, particularly as it is dependent on a synthetic auxin, 2,4-D. This is a potentially mild mutagen with persistent effects, which could account for the increase in genetic variations including these palms as well as in other plants (Benson 1990). As somaclones, the plants may even appear to be normal in anatomy, morphology, and physiology, but they may be biochemically deleterious to man through their products, a situation which may not be detectable by current quality monitoring systems. Thus there is a need for a proper awareness of biosafety in the use of some media constituents.

Efforts should be directed towards less use of "artificial" components and more of natural ones in media. A current aim in our laboratory is to devise a satisfactory nutrient medium for plant micropropagation from naturally occurring soil and plant components. This would be cheaper, safer, and less elitist as it would be easily degraded biologically and returned to the soil as harmless waste rather than potentially toxic and mutagenic waste.

Discussion

PERSLEY: How serious a threat do you think the biosynthesis of cocoa butter is to the conventional production of cocoa from field-grown trees?

ESAN: I do not believe the biosynthesis of cocoa butter in vitro to be a serious problem. There are three aspects of cocoa bean quality: butter, aroma, and flavour. The last two cannot be produced in vitro; they need to be produced from other sources, then incorporated into the butter fat. We are monitoring the economics and marketing of this biosynthetic to keep up with what is going on in advanced countries. We are also aware that shea butter and palm oil, after fermentation, are being considered as substitutes for cocoa butter.

HASEGAWA: From research conducted at Purdue University, it was not clear whether the efficiency of conversion of carbohydrate to cocoa triglycerides in vitro was high enough to make it a viable alternative economically.

ENE-OBONG: As advances in biotechnology appear in developed countries, developing countries may have problems with their natural raw materials. Apart from the crops already mentioned, there are many others, such as the proteinaceous sweetener (*Diocoreophyllum cumminsii*), which is thousands of times sweeter than sugar, and the gene for which is being introduced into microorganisms. What should we in Africa be doing with our own "new" plants? I believe that African scientists should start thinking about our own brand of biotechnology, to develop in parallel with global biotechnology.

WAITHAKA: The substitution of products from biotechnical innovations in developing countries may lead to deleterious effects. For example, the replacement of cane sugar by a synthetic sweetener (fructose) in the Philippines has resulted in high unemployment among those who used to work on the sugar plantations.

ADU-AMPOMAH: The micrografting of cocoa somatic embryos onto seedlings holds great promise and work in this field should be encouraged. Recently, a technician from the Ghanaian Cocoa Research Institute visited Dr Novak's laboratory at the International Atomic Energy Agency (IAEA) in Vienna; they were able to obtain 10 such plants, which are now in the greenhouse at Siebersdorf.

References

Abraham, A., and K.J. Thomas. 1962. A note on the in vitro culture on excised coconut embryos. *Indian Coconut Journal* 15: 84-87.

Adu-Ampomah, Y., F.J. Novak, R. Afza, and M. Van Durren. 1987. Embryoid and plant production from cultured cocoa explants. Pages 129-136 in *Proceedings, Tenth International Cocoa Research Conference, Santo Domingo.*

Adu-Ampomah, Y., F.J. Novak, R. Afaza, M. Van Durren, and M. Perea-Dallos. 1988. Initiation and growth of somatic embryos of cocoa (*T. cacao* L.). *Café, Cacao, The* 32: 187-200.

Archibald, J.F. 1954. Culture in vitro of cambial tissue of cacao. *Nature (London)* 173: 351-352.

Assy-Bah, B. 1986. Culture in vitro d'embryons zygotigues de cocotier. *Oleagineux* 41: 321-328.

Assy-Bah, B., T. Durand-Gasselin, and C. Pannetier. 1987. Use of zygotic embryo culture to collect germplasm of coconut. *Plant Genetic Resources Newsletter* 71: 4-10.

Assy-Bah, B., T. Durand-Gasselin, F. Engelmann, and C. Pannetier. 1989. The in vitro culture of coconut (*Cocos nucifera* L.) zygotic embryos. Revised and simplified method for obtaining coconut plantlets suitable for transfer to the field. *Oleagineux* 44: 515-523.

Benson, E.E. 1990. *Free Radical Damage in Stored Plant Germplasm.* IBPGR, Rome, Italy.

Branton, R.L., and J. Blake. 1983. A lovely clone of coconuts. *New Scientist* 98: 554-557.

Buffard-Morel, J., J.L. Verdeil, and C. Pannetier. 1988. Vegetative propagation of coconut (*Cocos nucifera* L.) through somatic embryogenesis. Poster presented at Eighth International Biotechnology Symposium, Paris.

Carley, R.H.V., C.H. Lee, L.H. Law, and C.Y. Wong. 1986. Abnormal flower development in oil palm clones. *Planter* 62: 233-240.

Carlisi, J.A., and K.C. Torres. 1986. In vitro shoot proliferation of *Camellia sinensis* "Purple Dawn". *HortScience* 21: 314.

Chen, Z.G., and H.H. Liao. 1988. Study on the induction of haploid plants from tea anther culture. *Journal of the Fujian Agricultural College* 17: 185-190.

Cocking, E.C., M.R. Davey, D. Pental, and J.B. Power. 1981. Aspects of plant genetic manipulation. *Nature (London)* 293: 265-270.

Cutter, V.M., and K.S. Wilson. 1954. Effect of coconut endosperm and other growth stimulants upon the development in vitro of embryos of *Cocos mucifera. Botanical Gazette* 115: 234-240.

Davis, T.A. 1969. Clonal propagation of the coconut. *World Crops* Sept/Oct: 253-255.

De Guzman, E.V., A.G. Del Rosario, and E.C. Eusebio. 1971. The growth and development of coconut makapuno in vitro III. Resumption of root growth in high sugar media. *The Philippine Agriculturist* 53: 566-579.

De Guzman, E.V., A.G. Del Rosario, and E.M. Ubalde. 1978. Proliferative growth and organogenesis in coconut embryo and tissue cultures. *Philippine Journal of Coconut Studies* 3: 1-10.

Durand-Gasselin, T., V. Le Geun, K. Konan, and Y. Duval. 1990. Oil palm (*Elaeis guineensis* Jacq.) plantation in Cote d'Ivoire, obtained through in vitro culture. *Oleagineux* 45: 1-11.

Duval, Y., T. Durand-Gasselin, K. Konan, and C. Pannetier. 1988. In vitro vegetative propagation of oil palm (*E. guineensis Jacq.*) I. Strategy and result. *Oleagineux* 43: 39-47.

Esan, E.B. 1977. Tissue culture studies on cacao (*Theobroma cacao* L.): A supplementation of current research. Pages 116-125 in *Proceedings, Fifth International Cacao Research Conference, 1975.* Cocoa Research Institute of Nigeria, Ibadan, Nigeria.

Esan, E.B. 1982. Shoot regeneration from callus derived from embryo axis cultures of *Theobroma cacao* in vitro. *Turrialba* 32: 359-364.

Esan, E.B. 1989. Progress in tea (*Camellia sinensis)* breeding in Nigeria. Pages 209-216 in *Progress in Tree Crop Research* (2nd edn). Cocoa Research Institute of Nigeria, Ibadan, Nigeria.

Esan, E.B., and S.A. Akinwale. (in press). Studies on the vegetative propagation of cashew (*Anacardium occidentale* L.) — Marcotting. *Nigerian Journal of Agricultural Technology*

Etherington, D.M. 1990. Economic analysis of a planting density experiment for tea in China. *Tropical Agriculture (Trinidad)* 62(3): 248-256.

Evans, H. 1951. Investigations on propagation of cacao. *Tropical Agriculture (Trinidad)* 28: 147-203.

Forrest, G.I. 1969. Studies on the polyphenol metabolism of tissue cultures derived from the tea plant (*Camellia sinensis* L.). *Biochemistry Journal* 113: 765-772.

Gupta, P.K., S.V. Kendurkar, V.M. Kulkarni, M.V. Shir-Gukka, and A.F. Mascarenhas. 1984. Somatic embryogenesis and plants from zygotic embryos of coconut (*Cocos nucifera* L.) in vitro. *Plant Cell Reports* 3: 222-225.

Haldeman, J.H., and R.L. Thomas. 1989. Multiple shoot production from single buds of tea (*Camellia sinensis*). *Camellia Journal* 44(2): 23-26.

Iddagoda, N., N.N. Kateva, and R.G. Butenko. 1988. In vitro clonal micro-propagation of tea (*Camellia sinensis*). I. Defining the conditions for culturing by means of mathematical design technique. *Indian Journal of Plant Physiology* 31: 1-10.

Iyer, R.D. 1981. Embryo and tissue culture for crop improvement especially of perennials, germplasm, concentration and exchange. Pages 229-230 in *Proceedings, International Symposium on Tissue Culture of Economically Important Plants* edited by A.N. Rao. COSTED/ANBS, National University, Singapore.

Jha, T.B. 1988. In vitro morphogenesis in cashew *Anacardium occidentale* L. *Indian Journal of Experimental Biology* 26: 505-507.

Jones, L.H. 1974. Propagation of clonal oil palms by tissue culture. *Oil Palm News* 17: 1-9.

Kato, M. 1989. Polyploids of camellia through culture of somatic embryos. *HortScience* 24: 1023-1025.

Kim, J.S. 1986. Studies on the propagation of Korean tea plants by tissue culture. *Journal of the Korean Forestry* Society 75: 25-31.

Koretskaya, T.F., and M.N. Zaprometov. 1975. Cultivation of tissue of the tea plant (*C. sinensis* L.) as a model for studying conditions of phenolic compound synthesis. *Fiziologiya Rastenii* 22: 282-288.

Kutubidze, V.V., S.K. Tavarkiladze, and N.A. Vecherina. 1988. Some characteristics of in vitro culture of tea axillary buds. *Subtropicheskie Kul'tur* 1: 80-84.

Legrand, B., C. Cilas, and E. Mississo. 1984. Comportment des tissue de *Theobroma cacao* L. var. Amelonado cultivaté in vitro. *Café, Cacao, The* 28: 245-250.

Lin, M.L., and E.I. Staba. 1961. Peppermint and spearmint tissue cultures. I. Callus formation and submerged culture. *Lioydior* 24: 139-145.

Litz, R.E. 1986. Tissue culture studies with *Theobroma cacao.* Pages 111-120 in *Proceedings, Cacao Biotechnology Symposium* edited by P.S. Dimick. Pennsylvania State University, University Park, USA.

Murashige, T., and F. Skoog. 1962. A revised medium for rapid growth and bioassays with tobacco tissue cultures. *Physiolologia Plantarum* 15: 473-497.

Mu-Quin, Y., and C. Ping. 1983. Studies on development of embryoids from the culture of cotyledons of *Thea sinensisi* L. *Scienta Silvae Sinicae* 19: 25-29.

Noiret, J.M., J.P. Gascon, and C. Pannetier. 1984. En france, les succes de la recherche 1.Production de palmier a huile par culture in vitro. *Cahier des Ingenieurs Agronomes (Paris)* 382: 57-64.

Novak, F.J., B. Donini, and G. Owusu. 1986. Somatic embryogenesis and in vitro plant development of cocoa (*Theobroma cacao*). Pages 443-449 in *Proceedings, International Symposium on Nuclear Techniques and In Vitro Culture for Plant Improvement.* IAEA, Vienna, Austria.

Nwankwo, B.A., and A.D. Krikorian. 1983. Morphogenetic potential of embryo and seedling derived callus of *Elaeis guineensis* Jacq. var. Pisifera Becc. *Annals of Botany* 51: 65-76.

Ogutuga, D.B.A., and D.H. Northcote. 1970. Caffeine formation in tea callus tissue. *Journal of Experimental Botany* 21: 258-273.

Orchard, J.E., H.A. Collin, and K. Hardwick. 1979. Culture of shoot apices of *Theobroma cacao*. *Physiologia Plantarum* 47: 207-210.

Pannetier, C., P. Arthuis, and D. Lievoux. 1981. Neoformation de jeunes plantes de *Elaeis guineensis* a partir de cals primaires obtenus sur fragment foliaires cultives in vitro. *Oleagineux* 36: 119-122.

Paranjothy, K., and R. Othman. 1982. In vitro propagation of oil palm. Pages 747-748 in *Proceedings, Fifth International Congress on Plant Tissue and Cell Culture* edited by A. Fujirawa. Japanese Association for Plant Tissue Culture, Tokyo.

Passey, A.J., and O.P. Jones. 1983. Shoot proliferation and rooting in vitro of *Theobroma cacao* L. type Amelonado. *Journal of Horticultural Science* 58: 589-592.

Pence, V.C., P.M. Hasegawa, and J. Janick. 1979. Asexual embryogenesis in *Theobroma cacao* L. *Journal of the American Society for Horticultural Science* 104: 145-148.

Pence, V.C., P.M. Hasegawa, and J. Janick. 1980. Initiation and development of asexual embryos of *Theobroma cacao* L. in vitro. *Zeitschrift für Pflanzenphysiologie*. 98(1): 1-14.

Philip, V.J. 1984. In vitro organogenesis and plantlet formation in cashew (*Anacardium occidentale* L.). *Annals of Botany* 54: 149-152.

Purdy, L.H., and E.R. Dickstein. 1989. *Theobroma cacao*, a host for *Agrobacterium tumefaciens*. *Plant Disease* 73: 638-639.

Rabechault, H., J.P. Martin, and S. Gas. 1972. Recherches sur la culture des tissue de palmier a huile (*Elaeis guinneensis* Jacq.) *Oleagineux* 27: 531-534.

Rabechault, H., and J.P. Martin. 1976. Multiplication vegetative du palmier a huile (*E. guinnensis* Jacq.) a l'aide de tissus foliair. *Comptes Rendus L'Academie des Sciences (Paris)* 238, Ser. D: 1735-1737.

Rai, B.G.M. 1970. Cashew propagation through vegetative propagation. *Intensive Agriculture* 7: 17-19.

Raju, C.E., P. Kumar Prakash, C. Min, and R.D. Iyer. 1984. Coconut plantlets from leaf tissues. *Journal of Plantation Crops* 12: 75-81.

Saha, D.K., and N.M. Bhattacharya. 1988. In vitro development of shoot apices from pollen callus in the anther cultures of tea (*Camellia sinensis* L. O. Kuntze). Pages 135-145 in *Proceedings, Seminar on Cell and Tissue Culture in Field Crop Improvement, 4-9 October 1987, Tsukuba, Japan*. Food and Fertilizer Technology Centre for the Asian and Pacific Region, Taipei, Taiwan.

Samartin, A. 1989. A comparative study of effects of nutrient media and cultural conditions on shoot multiplication of in vitro cultures of *Camellia japonica* explants. *Journal of Horticultural Science* 64: 73-79.

Sasson, A. 1988. *Biotechnologies and Development*. Unesco/CTA, Paris, France.

Sudasrip, H., H. Kaat, and A. Davis. 1978. Clonal propagation of the coconut via the bulbils. *Philippine Journal of Coconut Studies* 3(3): 5-14.

Torres, K.C., and J.A. Carlisi. 1986. Shoot and root organogenesis of *Camellia sasanqua*. *Plant Cell Reproduction* 5: 381-384.

Thompson, W., H.A. Collin, S. Isaac, and K. Hardwick. 1987. Isolation of protoplasts from cocoa *(Theobroma cacao L.)* leaves. *Café, Cacao, The* 31: 115-120.

Verdeil, J.L., J. Buffard-Morel, and C. Pannetier. 1989. Somatic embryogenesis of coconut from leaf and inflorescence tissue: research, result and prospects. *Oleagineux* 44: 403-411.

Vietez, A.M., J. Barciela, and A. Ballester. 1989. Propagation of *Camellia japonica* cv. Alba Plena by tissue culture. *Journal of Horticultural Science* 64: 177-182.

Wu, C.T., T. Huang, G.G. Chen, and S. Chen. 1981. A review on the tissue culture of tea plants and on the utilization of callus-derived plantlets. Pages 104-106 in *Proceedings, International Symposium on Tissue Culture of Economically Important Plants* edited by A.N. Rao. COSTED/ANBS, National University, Singapore.

Yan, M.Q., P. Chen, and Y.H. Wang. 1988. Rapid clonal propagation of seven species related to *Camellia chrysantha* of Guangxi. *Acta Biologiae Experimentalis Sinica* 21: 1-9.

Zagoskina, N.V., and M.N. Zaprometov. 1983. Effect of kinetin on formation of phenotypic compounds in the long passaged tea plant tissue culture. *Fiziologiya Biokhimii Kul'tur Rastenii* 15: 250-253.

3.8

Anther culture and its potential

E.E. Ene-Obong

Laboratory for Plant Cell and Tissue Culture and Biotechnology, Department of Botany, University of Nigeria, Nsukka, Nigeria

Abstract

Pollen-derived haploids are derived from a process that involves a switch from the normal development of pollen grains to a vegetative sporophytic pattern of development. The trigger is provided either by the sporophytic determinants carried over from the meiotic division preceding the formation of microspores or by totipotency. Although research on this technique is only just beginning in most African countries, marked advances have been made in other parts of the world. Attention is currently focused on methods for increasing the low embryogenic response of the pollen grains of most plants. Anther/pollen culture holds promise for the rapid production of pure lines from haploids or somaclonal variants, for use in direct crop improvement, or as a tool in other improvement strategies. It is also useful in classical genetic studies involving mutagenesis.

Anther culture refers to the production of haploid (monoploid) plants from the cultured male sex organ of flowering plants. The haploids are obtained either directly from aseptic cultures of whole (or parts of) anthers or isolated microspores (Nitsch and Norreel 1973), or indirectly through the induction of callus from pollen, which regenerates plants when subcultured. In fact, the process involves a switch from the normal development of pollen grains so that instead of male gametes, miniature plants are produced in culture. The change from the normal gametophytic development into a sporophytic pattern of development (George and Sherrington 1984) has long been thought the result of totipotency, the inherent ability of plant cells to regenerate the entire plant body under the right cultural conditions. However, an alternative model has been put forward recently (Bell 1970; Heberle-Bors 1985). This involves the so-called sporophytic determinants (polysomes with sporophytic mRNA, etc.), carried over from the meiotic division preceding the formation of pollen grains. Embryogenic pollen grains are rich in sporophytic determinants and the response in culture is associated with the percentage of them in the anther.

Two investigations led to the origin of anther culture. The first was the regeneration of haploid tissue from the pollen of *Ginkgo* species (Tulecke 1960). This was followed by the regeneration of haploid plants from cultured anthers of *Datura innoxia* (Guha and Maheshwari 1964, 1966). The anthers are normally cultured on a nutrient medium containing mineral salts and sugar, and supplemented with hormones (auxins and cytokinins) and amino acids, with or without complex organic addenda such as coconut milk, potato extract, and grape juice. The medium may be solidified with agar or used in liquid form, and the cultures are incubated under controlled illumination, temperature, and humidity (Sharp and Larsen 1979). Liquid cultures, which are best kept on a gyratory shaker, give better results, probably because of better access to nutrients and more efficient removal of inhibitors from the surroundings of the anther (Wernicke and Kohlenback 1976; Maheshwari et al. 1983).

Requirements for Enhancing Response

The most important requirements for the successful induction and enhancement of the response have been considered in detail in several reviews (Reinert and Bajaj 1977; Sunderland and Dunwell 1977; Vasil 1980; Maheshwari et al. 1983; Heberle-Bors 1985; Keller et al. 1986; Foroughi-Wehr and Wenzel 1989). The initial response is controlled by the frequency of occurrence of embryogenic pollen grains (small in size, with little starch) compared with the non-embryogenic pollen grains (large, rounded, and rich in starch). Embryogenic pollen grains are infertile but they are not necessarily aneuploids considering the regular haploid chromosomes seen in induced haploids. In some recent studies, several agents have been used to increase their frequency in anthers. These include auxins, antigibberellins, nitrogen starvation (Heberle-Bors 1985), and male gametocides (Schmid and Keller 1986).

It is becoming clear that there are two aspects to anther haploid regeneration. The first is the number of potential haploid plants, as determined by the number of embryogenic pollen grains in an anther. The second is the proportion of the embryogenic pollen grains that eventually develops into plantlets, without aborting. This second aspect is influenced by a wide range of factors, both genetic and environmental. Genotypic differences between donor plants determine the nature of culture medium used. The nutrient condition could be very simple, as in the case of *Datura innoxia,* where haploids could be induced in a 2% solution of sucrose alone, without the addition of mineral salts (Sunderland 1974; Vasil 1980). In other plants, such as cereals and members of the Cruciferae, the sucrose level is critical (Clapham 1977) while the level of iron or even the ionic nature of compounds in the medium affect development in many species (Maheshwari et al. 1983). In maize, Kuo et al. (1986) increased the induction frequency in anthers by up to 43% by using a medium supplemented with 500 mg/L casein hydrolysate, 0.5% active carbon, 2 mg/L 2,4-dichlorophenoxy-acetic acid (2,4-D) and 15% sucrose. More recalcitrant species, as in the Fabaceae, require more complex media.

Important environmental factors include asepsis, density and number of anthers per culture vessel, temperature, and light intensity and duration. Improved responses have also been obtained in some species through reduced atmospheric pressure, anaerobiosis, centrifugation, water saturated atmosphere (Maheshwari et al. 1983), low temperature (chilling) treatments of flower buds and anther explants (Kuo et al. 1986; Tsay and Widholm 1986; Zagorska and Pandeva 1986), gamma rays (Zagorska and Pandeva 1986), orientation of the anthers on the medium with regard to the loculi (Foroughi-Wehr and Wenzel 1989) and short exposures of cultures to high temperatures of 30-35°C (compared with normal temperatures of 25-30°C) (Keller and Stringam 1978; Maheshwari et al. 1983; Ouyang et al. 1983).

Uses and Potential

Pollen haploids are of special interest to the geneticist and plant breeder. New and high quality true breeding types (pure lines) can be obtained in a single generation through the selection of desirable haploids and their rescue by chromosome doubling (using colchicine or spontaneously in cultured cells). Compared with conventional breeding techniques, which often involve multiple hybridizations, laborious selections over several generations, and complicated chromosome manipulation techniques, this method is fast and easy, provided the desirable attribute was present in the original crop. There are also the added advantages of stability and uniformity of the new variety, two important breeding criteria when long-term yield and mechanical harvesting are required.

Pollen haploids are also used in hybrid variety breeding programmes (as equivalents of inbred lines) and in other breeding strategies. They have been incorporated in the improvement programmes of major crops, such as wheat, maize, rice, egg plant, rapeseed, rye, pepper, tobacco, barley, and potato (Maheshwari et al. 1983; Dattee 1988; Foroughi-Wehr and Wenzel 1989). The single set of chromosomes in haploids makes it easy to obtain induced mutants for genetic study. Plants have been regenerated from irradiated haploid protoplasts of tobacco, *Datura* species, and *Brassica* species (Vasil 1980). Also, several nitrate and auxotrophic mutants have been obtained through protoplast and cell cultures derived from pollen haploids of *D. innoxia*, *Hyoscyamus muticus*, and *Nicotiana* species (Maheshwari et al. 1983). Lastly, somaclonal variants, which are often produced in anther cultures (from the varying ploidy levels of callus cells as well as certain unseen gene changes), may include useful variants for improved yield and quality and for disease/pest resistance. Plant variants regenerated from sugar cane callus have shown improved resistance to mosaic virus disease (Dale 1975).

In conclusion, anther culture is presently in a transitory stage, moving towards the commercial exploitation of its potential in many aspects. There is an urgent need to encourage research on anther culture in African countries.

Discussion

ODEWALE: In view of the fact that anther culture leads to the production of homozygous plants (which may result in monoculture) and taking into consideration pathogens in perennial plantation crops, would you recommend anther culture to tree crop breeders?

ENE-OBONG: In all cases, whether plantation (tree) crops or other crops are involved, the problem of susceptibility to pathogens is very real. This is especially so for pathogens with high mutation rates. The problem with tree crops is the long life cycles involved and the possibility of any resistance breaking down with time. Where there is no other way (better source) of improvement, androgenesis could be used, but attempts should be made to obtain many lines that are similar phenotypically but possess certain genetic differences, thus maintaining a wide genetic base. In this way, at least some lines would possibly survive pathogen attack.

ROBERTSON: In 1977, a cultivar of tobacco (K110) was released in Zimbabwe to meet the need for wildfire resistance in tobacco. This cultivar was produced through anther culture, and currently (1990) it is still used by farmers, earning US $10 million each year. I wrote to *Science*, which had just published a review saying no anther-derived cultivar had reached the farmer, but they did not see fit to publish it. Unless the Chinese beat us to it, we think that we were the first to put anther culture into the hands of farmers. This work was done on our tobacco research station.

References

Bell, P.R. 1970. The archegoniate revolution. *Scientific Progress* 58: 27-45.

Clapham, D.H. 1977. Haploid induction in cereals. Pages 279-298 in *Plant Cell, Tissue and Organ Culture* edited by J. Reinert and Y.P.S. Bajaj. Springer-Verlag, Berlin, Germany.

Dale, P.J. 1975. Tissue culture in plant breeding. Pages 101-115 in *Welsh Plant Breeding Station Annual Report.* Welsh Plant Breeding Station, Aberystwyth, UK.

Dattee, Y. 1988. Impact of in vitro methods in plant improvements. Pages 1008-1014 in *Proceedings, Eighth International Biotechnology Symposium* edited by G. Durand, L. Bobichon and J. Florent. Société Française de Microbiologie, Paris, France.

Foroughi-Wehr, B., and G. Wenzel. 1989. Androgenetic haploid production. *IAPTC Newsletter* 58: 11-18.

George, E.F., and P.D. Sherrington. 1984. *Plant Propagation by Tissue Culture: Handbook and Directory of Commercial Laboratories.* Exegetics Limited, Basingstoke, UK.

Guha, S., and S.C. Maheshwari. 1964. In vitro production of embryos from anthers of *Datura. Nature (London)* 204: 497.

Guha, S., and S.C. Maheshwari. 1966. Cell division and differentiation of embryos in the pollen grains of *Datura* in vitro. *Nature (London)* 212: 97-98.

Heberle-Bors, E. 1985. In vitro haploid formation from pollen: A critical review. *Theoretical and Applied Genetics* 71: 361-374.

Keller, W.A., P.G. Arnison, and B.J. Cardy. 1986. Haploids from gametophytic cells: Recent developments and future prospects. Pages 223-241 in *Plant Tissue and Cell Culture, Proceedings, Sixth International Congress on Plant Tissue and Cell Culture, IAPTC, 3-8 August 1986, University of Minnesota.* Alan R. Liss, New York, USA.

Keller, W.A., and G.R. Stringam. 1978. Production and utilization of microspore-derived haploid plants. Pages 113-122 in *Frontiers of Plant Science* edited by T.A. Thorpe. University of Calgary Press, Calgary, Canada.

Kuo, C.S., W. Lu, and Y.L. Kui. 1986. Progress on anther culture of maize. Page 125 in *Abstracts, Sixth International Congress on Plant Tissue and Cell Culture, IAPTC, 3-8 August 1986, University of Minnesota.*

Maheshwari, S.C., A. Rashid, and A.K. Tyagi. 1983. Anther/pollen culture for production of haploids and their utility. *IAPTC Newsletter* 41: 2-9.

Nitsch, C., and B. Norreel. 1973. Effet d'un choc thermique sur le pouvoir embryogene du pollen de *Datura innoxia* cultivé dans l'anthere in isolé de l'anthere. *Comptes Rendus L'Academie des Sciences (Paris)* Ser. D 276: 303-306.

Ouyang, J.W., S.M. Zhou, and S.E. Jia. 1983. The response of anther culture to culture temperature in *Triticum aestivum. Theoretical and Applied Genetics* 66: 101-109.

Reinert, J., and Y.P.S. Bajaj. 1977. Anther culture: Haploid production and its significance. Pages 251-267 in *Applied and Fundamental Aspects of Plant Cell, Tissue and Organ Culture* edited by J. Reinert and Y.P.S. Bajaj. Springer-Verlag, Berlin, Germany.

Schmid, J., and E.R. Keller. 1986. Improved androgenetic response in wheat (*Triticum aestivum*) as a result of gametocide application to anther donor plants. Page 122 in *Abstracts, Sixth International Congress on Plant Tissue and Cell Culture, IAPTC, 3-8 August 1986, University of Minnesota.*

Sharp, W.R., and P.O. Larsen. 1979. Plant cell and tissue culture: Current applications and potential. Pages 115-120 in *Plant Cell and Tissue Culture: Principles and Application* edited by W.R. Sharp, P.O. Larsen, E.F. Paddock and V. Raghavan. Ohio State University Press, Columbus, Ohio, USA.

Sunderland, N. 1974. Anther culture as a means of haploid induction. Pages 91-122 in *Haploids in Higher Plants: Advances and Potential* edited by K.J. Kasha. University of Guelph Press, Ontario, Canada.

Sunderland, N., and J.M. Dunwell. 1977. Anther and pollen cultures. Pages 223-265 in *Plant Tissue and Cell Culture* edited by H.E. Street. Blackwell Scientific Publications, Oxford, UK.

Tsay, H.S., and J.M. Widholm. 1986. Factors affecting haploid plant regeneration from maize anther culture. Page 125 in *Abstracts, Sixth International Congress on Plant Tissue and Cell Culture, IAPTC, 3-8 August 1986, University of Minnesota.*

Tulecke, W. 1960. Arginine-requiring strains of tissue obtained from *Ginkgo* pollen. *Plant Physiology* 35: 19-24.

Vasil, I.K. 1980. Androgenetic haploids. Pages 195-233 in *Perspectives in Plant Cell and Tissue Culture: International Review of Cytology* (Supplement 11A) edited by I.K. Vasil. Academic Press, New York, USA.

Wernicke, W., and H.W. Kohlenback. 1976. Investigations on liquid culture medium as a means of anther culture in *Nicotiana*. *Zeitschrift für Pflanzenphysiologie* 79: 189-198.

Zagorska, N.A., and R.S. Pandeva. 1986. In vitro androgenesis in *Lycopersicon, Medicago* and *Capsicum*. Page 123 in *Abstracts, Sixth International Congress on Plant Tissue and Cell Culture, IAPTC, 3-8 August 1986, University of Minnesota.*

3.9

Tissue culture in disease elimination and micropropagation[1]

S.Y.C. Ng, G. Thottapilly and H.W. Rossel

International Institute of Tropical Agriculture, PMB 5320, Ibadan, Nigeria

Abstract

Meristem-tip culture is commonly used to eliminate viruses from plants. The fact that some viruses are more difficult to eliminate than others has led to the use of thermotherapy and chemotherapy, along with meristem-tip culture, to increase the efficacy of disease elimination. Tissue culture material is considered to be the most suitable form for international exchange of vegetative plant material and is accepted by quarantine authorities. Micropropagation is used in both research institutions and commercial companies to produce large quantities of high quality planting materials. It is applied to vegetatively propagated crops, those crop species that are difficult to multiply, and new hybrids. It is achieved through the induction of axillary shoot formation, direct and indirect adventitious shoot formation, somatic embryogenesis, and storage organ formation. This paper describes tissue culture techniques used for both disease elimination and micropropagation and discusses the advantages and disadvantages of different techniques.

Control of plant bacterial and fungal diseases may be obtained through the application of chemicals. However, it has not been possible to control virus diseases through the use of chemicals. Chemicals that can affect virus multiplication usually exert high phytotoxicity on the host plant. In addition, they are usually very expensive and once the treatment stops, the virus concentration may rapidly build up again.

Since Morel and Martin (1952) first used meristem-tip culture to produce virus-free dahlias, this technique has been applied widely to many vegetatively propagated crops to

---

1 Contribution No. IITA/91/CP/26 from the International Institute of Tropical Agriculture, PMB 5320, Ibadan, Nigeria.

obtain virus-free plants from originally diseased materials. Regenerated plants obtained from meristem-tip culture can be declared free from viruses only through a sound understanding of disease etiology and the availability of reliable virus-indexing methods, particularly as some viruses may be latent. Meristem-tip culture has greatly facilitated the international distribution of vegetative materials because shipment in this form circumvents most quarantine restrictions. Through this method, virus-free clones have been regenerated from a wide range of economically important crops (Quak 1977; Walkey 1980).

The successful rapid multiplication of orchids by shoot meristem culture, demonstrated by Morel (1965), increased the interest in the application of tissue culture techniques as an alternative means of asexual propagation. The size of the propagule in culture is so minute that the in vitro asexual propagation technique has been referred to as "micropropagation". The use of micropropagation for asexual propagation is the most advanced area of plant tissue culture. It also currently has the most practical application.

Micropropagation is especially valuable for propagation of new hybrids, virus-free plants, species that are difficult to propagate, those that are vegetatively propagated, and those that have a low multiplication ratio (de Fossard 1976; Hussey 1978). There have been several reviews on species successfully propagated through tissue culture (Conger 1981; Hu and Wang 1983; George and Sherrington 1984; Sagawa and Kunisaki 1990).

In the early 1980s there were over 200 commercial micropropagation laboratories worldwide (George and Sherrington 1984). A study by Murashige (1990) estimated that there are about 1000 species that can be propagated by tissue culture, with about 300 commercial laboratories worldwide dealing in micropropagation. An increasing number of universities, government institutions, research venture companies, and botanical gardens are also actively involved in the application of tissue culture technologies.

Tissue Culture and Virus Elimination

Meristem-tip culture is now commonly used for disease elimination in crop plants. One of the important features of meristem-tip culture is that the regenerated plants usually retain the genetic integrity of the parent plants. This is probably due to the more uniformly diploid nature of the meristematic cells (Murashige 1974). Other tissue culture methodologies, such as callus culture and in vitro micrografting, have also been used for virus elimination.

Meristem-tip culture

A procedure for obtaining virus-free plants using meristem-tip culture is presented in Figure 1. Obviously, it is imperative to know what kind of virus(es) are present in the crops concerned in order to make an assessment of the disease status of regenerated plants. The relative ease with which viruses are eliminated by meristem-tip culture alone depends on the type of virus and the host-virus combination.

Usually, meristem tips, about 0.5-1 mm long and consisting of the meristematic dome and two leaf primordia, are excised from surface-disinfected apical or axillary buds. Generally, the percentage of virus-free plants obtained is inversely proportional to the size of the tips cultured. The excised tips are then placed on a culture medium under aseptic conditions. Culture medium formulations, such as that of Murashige and Skoog (1962),

Figure 1 A procedure for obtaining virus-free plants using meristem-tip culture

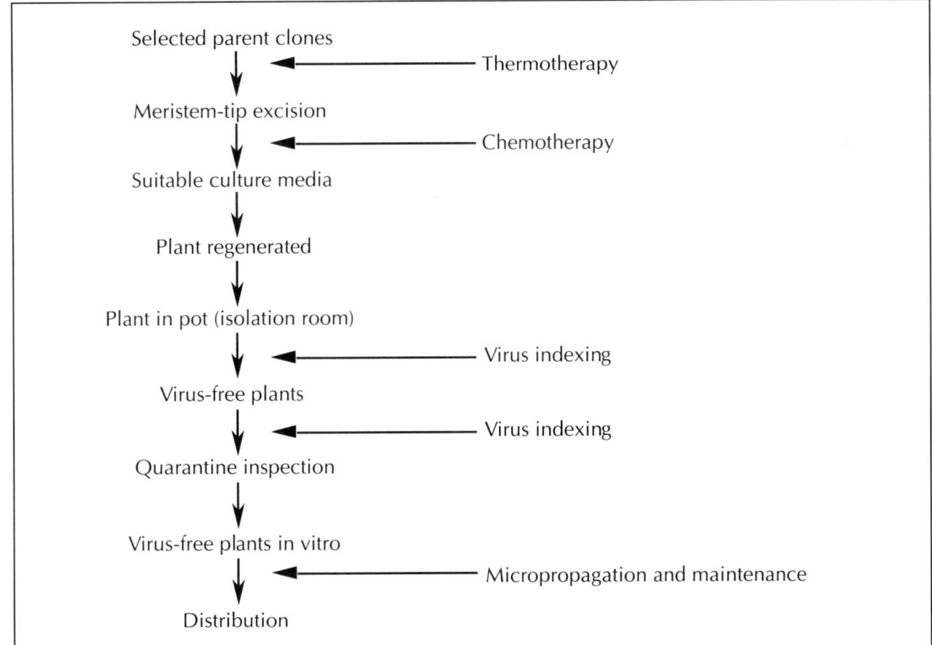

have been commonly used with slight modifications, depending on the crop species. The presence of growth substances, such as auxin, cytokinin, and gibberellic acid, are important in the regeneration of meristem tips. Quak (1977) and Frison and Ng (1981) have given details on surface disinfection, explant removal, medium composition, and culture conditions.

Meristem-tip culture after heat treatment

Heat treatment (thermotherapy) at temperatures ranging from approximately 33°C to 40°C (constantly or alternately) of the parent plant materials for a certain length of time, followed by meristem-tip culture, has been reported to increase the number of meristem tips that may be regenerated to plantlets and the percentage of virus-free plants obtained from such meristem tips (for chrysanthemum, see Hakaart and Quak 1964; for potato, see Stace-Smith and Mellor 1968 and MacDonald 1973; for sweet potato, see Over de Linden and Elliott 1972; and for cassava, Kartha and Gamborg 1975 and Ng and Hahn 1985). Heat treatment prior to the excision of explants should be considered where it has proved difficult to eliminate viruses by meristem-tip culture alone.

Meristem-tip culture and chemotherapy

Chemotherapy has not been generally considered as a means of eradicating viruses from infected plants. However, a synthetic riboside, known as ribavirin, has been shown to increase the frequency of production of virus-free progeny; it helps to eliminate viruses that are unresponsive to current techniques. Ribavirin is a guanosine analogue which exhibits broad spectrum antiviral activity against both RNA and DNA animal viruses (Sidwell et al. 1972); it has also been shown to have some activity against virus replication in whole plants (Lerch 1977).

Chemicals reported to inactivate or inhibit plant viruses have been reviewed (Kartha 1986; Long and Cassells 1986). They include purine and pyrimidine analogues, amino acids, plant growth regulators, antibiotics, and other substances. Culturing meristem tips or other explants in the presence of a suitable chemical, such as ribavirin, it may be possible to maintain inhibitory conditions long enough to eliminate viruses successfully (Cassells and Long 1982; Klein and Livingston 1982; Hansen and Lane 1985; Wambugu et al. 1985).

There are limitations to consider. Cultures treated with ribavirin take longer to differentiate and grow than heat-treated cultures (Klein and Livingston 1982; Ng, unpubl.). Ribavirin at high concentration can also cause severe phytotoxicity to the explants (Wambugu et al. 1985). In addition, the treatment is expensive.

Other tissue culture methods

In a number of plant species where meristem-tip culture continues to be difficult, shoot-tip grafting has been studied and found to be effective in virus elimination (Navarro et al. 1975; Huang and Millikan 1980; Navarro 1984). Kartha (1986) provides a list of viruses that were successfully eliminated through micrografting.

Virus elimination through callus cultures has also been reported. Virus-free plants were obtained from the callus culture of plants infected with tobacco mosaic virus (TMV) (Hansen and Hildebrandt 1966; Murakishi and Carlson 1976). Virus-free plants were also obtained from mesophyll protoplast cultures of tobacco leaves infected with potato virus X (PVX) (Shepard 1975). The application of this technique is limited to those crops that are able to regenerate plants from callus cultures. The regenerated virus-free plants so obtained may be genetically altered, with advantages and disadvantages.

Virus indexing of regenerated plants

Plants regenerated from meristem-tip cultures are not automatically free from viruses and must be indexed for freedom from virus infection. It may be the most time-consuming stage in a disease elimination scheme.

Sensitive and reliable methods of virus detection are crucial in obtaining disease-free germplasm. There are several important factors to be taken into account in developing sensitive and reliable indexing methods, including:

- Proper identification of pathogens, especially viruses and their possible strains.
- Evaluation of various methods for detection of viruses and their strains. The method(s) finally adopted should be easy to handle but sensitive and reliable.

The regenerated plants are transplanted from tubes into sterile soil in pots and are kept in an isolation room for further growth and monitoring for disease expression. Virus indexing methods vary depending on the type of virus(es) that are involved, and on the availability of facilities. Ideally, regenerated plants should be tested for viruses by a variety of methods and these tests should be repeated several times. Virus indexing may take up to 1 year.

Some methods used for the indexing of plant materials are:

- Monitoring for possible occurrence of symptoms over a long period of time
- Sap inoculation of test plants
- Grafting on to suitable indicator plants
- Inspection of materials under the electron microscope for possible presence of virus particles by the simple "dip method"
- Immunosorbent electron microscopy (ISEM); antiserum is either added to the electron microscope carrier films to absorb more virus particles specifically, so that sensitivity is increased, or the virus particles are decorated with antibody in order to detect and identify the virus under investigation.
- Enzyme-linked immunosorbent assay (ELISA)

In addition to the highly sensitive serological techniques, such as ELISA and ISEM, the so-called "dot-blot" immuno assay may be applied. This method seems to be even more sensitive than ISEM.

Recent developments in biotechnology permit detection of viruses by means of monoclonal antibodies (Thomas et al. 1986) as well as by the so-called nucleic acid hybridization technique (Robinson et al. 1984). These modern techniques provide highly efficient tools for the diagnosis of viruses and viroids. The ultimate benefit resulting from the use of such techniques is a more rapid, efficient, and safer transfer of vegetative materials.

IITA applies meristem-tip culture techniques to eliminate viruses from its mandated root and tuber crops (Ng and Hahn 1985). The regenerated plants are tested extensively to ensure absence of all known viruses. Virus-tested, certified plants are then multiplied in vitro and prepared for international distribution. The plantlets are grown in transparent containers with a sterile culture medium that does not contain charcoal. Charcoal is omitted in order to facilitate the detection of any fungal or bacterial contamination.

Tissue Culture and Micropropagation

Micropropagation can be obtained through enhanced precocious axillary shoot formation, by production of adventitious shoots and somatic embryogenesis either directly on organ explants or indirectly from callus, or by storage organ formation (Figure 2 *overleaf*).

Among these possible methods, somatic embryogenesis is potentially the most rapid for cloning plants in vitro (Murashige 1978). Artificial seeds (encapsulated somatic embryos) may be produced by this method and delivered to growers/farmers for sowing. However, with its high multiplication rate, there is a problem with phenotypically altered plants resulting from genetic changes.

According to Murashige (1974), micropropagation can be divided into three stages: establishing the culture in vitro; propagating the materials and rooting; and transplanting and establishment in soil.

Figure 2 Micropropagation methods

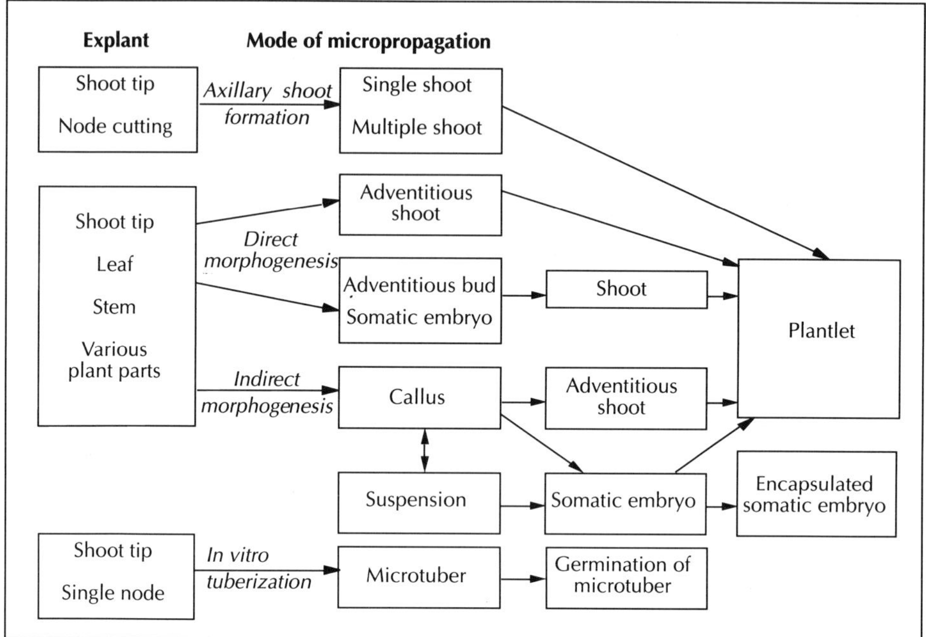

Axillary shoot formation

The production of plants from axillary shoots has proved to be the most generally applicable and reliable method of in vitro propagation. Shoot tips became popular as explants for micropropagation only after the discovery by Morel (1964) of protocorm formation from cymbidium orchid shoot tip, which suggested that shoot-tip culture could be used for efficient and rapid clonal propagation of orchids.

Two methods are usually used, shoot-tip culture and single-node culture. The shoot tip referred to here is usually taken from the tender tip of the growing shoot (about 2 cm long), and node cuttings are from either terminal or axillary buds with the stem segment attached. These two types of explants are preferred over meristem culture in micropropagation when viral elimination is not part of the objective. Use of larger explants is desirable as they are easier to dissect, and they have much higher survival and growth rates than smaller explants. On the other hand, the larger the explant the more difficult it may be to decontaminate in some plant species. Shoot-tip and single-node cultures are often started indirectly from the shoots obtained from meristem-tip cultures so that the plants are indexed for virus.

In most herbaceous plants, shoot-tip explants may be derived from apical or lateral buds of an intact plant. Shoot tips from trees or other woody perennials may be difficult to decontaminate and usually have problems with blackening. Shoot tips and lateral buds are more responsive to the culture medium when taken from juvenile shoots.

After surface disinfection, the shoot tips, lateral buds, or node cuttings are transferred to solid or liquid media. Clusters of shoots, single unbranched shoots, or complete plants may be obtained from these type of cultures. The shoot clusters are then subcultured for further induction of multiple shoots and growth of the shoots. The shoots are then separated and transferred to rooting media and subseqently transplanted to soil.

Unbranched shoots bearing several discrete and separated nodes obtained from shoot-tip or node-cutting cultures may be placed on a fresh medium in an horizontal position, or each shoot may be cut into several single nodes and subcultured. The sprouted axillary buds can be grown into unbranched shoots for subculturing or they can be grown to form roots and be transplanted into soil.

The multiplication rate through this method varies with genotype. In general, if an optimal concentration of cytokinin is incorporated in the culture medium (in the range of 0.5-10 mg/L) and optimal culture conditions are maintained, a 4-10 times multiplication rate can be achieved on a regular 4-8 week micropropagation cycle (Mantell et al. 1985). Monthly multiplication rates of 4.2 and 4 have been reported for sweet potato and cassava, respectively (Ng 1986), and a multiplication rate of 6 was obtained for yams when cultured in liquid medium (Ng, unpubl.). It has been estimated that about 65 000 yam plantlets may be obtained from a single node in 6 months (Mantell et al. 1978). The culture stocks can be maintained in vitro and may be subcultured at regular intervals (Ng 1989).

Direct adventitious shoot and somatic embryo formation

The ability of plant species to develop secondary meristems and adventitious shoots has been exploited in many micropropagation schemes. Adventitious shoots arise directly from the tissues of the explant. The induction of direct shoot regeneration depends on the plant part used, and on the plant species. In some species, adventitious shoots arise in vitro on pieces of tissue such as leaves, stems, flower petal, roots, bulbs, scales, and rhizomes. Successful methods using this type of micropropagation have been developed for many ornamental species (Mantell et al. 1985). In plantain/banana, shoot tips were induced to form adventitious buds which have been finally regenerated into whole plants. Thus a multiplication potential of over 10^6 meristem tips/explant/year can be achieved (Vuylsteke and De Langhe 1985).

Somatic embryos can also be induced directly from explant tissue without going through a callus stage. They may arise directly from groups of cells or individual cells within the original explant. The somatic embryos can then be separated from the explant and transferred to a culture medium where the embryos germinate and grow into complete plants. Large numbers of embryoids have been produced at the tips of leaves of the orchid *Malaxis paludosa*; leaves of *arabusta* coffee trees were induced to form embryos directly when cultured on MS medium containing high levels of cytokinin (Dublin 1981); and young cassava leaves were induced to form embryos directly with the use of auxin in the culture medium (Stamp and Henshaw 1982; Ng 1989). Somatic embryos are also frequently formed on the nucellus tissue of cultured ovules as found in citrus (George and Sherrington 1984).

Indirect adventitious shoot and somatic embryo formation

Somatic cells in culture may undergo organogenesis to produce shoots and roots, or they may produce somatic embryos which will then germinate into complete plants when transferred to an appropriate culture medium. The usual procedure involves establishing an actively dividing callus from a suitable explant such as a leaf, stem or root segment, a piece of storage tissue, a shoot tip, a seed embryo or an immature inflorescence. Callus is obtained when an explant is placed in a culture medium containing relatively high levels of auxin, with or without cytokinin. Callus can be multiplied and a suspension culture can be obtained by subculturing calluses in liquid media of the same compositions. When auxin is reduced or omitted from the medium, shoot buds or embryos may form in the cultures. This method of micropropagation has provided high multiplication ratios; however, the drawback of using it is its low capacity to regenerate plants and, more importantly, the genetic instability of the regenerants (Barbier and Dulieu 1983). Despite the problems mentioned, there are some reports indicating the ability to establish stable regenerative callus cultures. These stable callus cultures have been described for genera such as *Lilium* (Sheridan 1968), *Freesia* (Davies 1972), *Chrysanthemum* (Earle and Langhans 1974), *Lycopersicon* (De Langhe and De Bruijne 1976), and *Hemerocallus* (Krikorian et al. 1981).

Artificial seeds, consisting of somatic embryos enclosed in a protective coating, have been proposed as a low-cost, high-volume propagation system (Redenbaugh et. al. 1986). The aim is to produce "clonal seeds" similar to true seeds. Redenbaugh (1990) has summarized and discussed the possibility of using an artificial seed delivery system in tropical crops. More research needs to be carried out in the following areas: production of somatic embryos; attainment of high quality, conversion frequency of somatic embryos; embryo viability; encapsulation methods; identification of conversion environment; field testing to determine the efficacy of the propagation system; and somaclonal variation.

Storage organ formation

Many ornamental and crop species are normally propagated and stored in the form of storage organs. It has been reported that these storage organs can be produced in vitro and may provide a convenient means of microprogation. Examples are amaryllis hyacinth, lily, onion, and narcissus, which produce bulbils; gladiolus, which produces cormlets; orchids, which produce protocorms; and potato and yams, which produce microtubers. Some of the storage organs produced in vitro can be planted directly into soil; others have a period of dormancy. Methods of obtaining such storage organs vary according to the type of tissue being cultured (for example, for gladiolus see Ziv et al. 1970 and Ziv and Halevy 1970; for hyacinths see Hussey 1975; for potato see Hussey and Stacey 1981, Wang and Hu 1982, Hussey and Stacey 1984, and Abbot and Bulcher 1986; and for yams see Uduebo 1971, Ammirato 1984, Forysth and Van Staden 1984, and Ng 1988).

Microtubers have been used in the propagation and delivery of potato cultivars (Wang and Hu 1982) and in the international distribution of potato germplasm (Estrada et al. 1986) and white yam germplasm (Ng 1988).

Conclusion

Tissue culture techniques have been used to eliminate disease from a wide range of crop species. However, the most commonly used method is meristem-tip culture, or a combination of thermotherapy and meristem-tip culture. The need to develop virus detection methods for each virus cannot be over-emphasized.

Micropropagation has proved successful for a wide range of crop species. From the point of view of genetic stability of the regenerants, axillary shoot formation, direct adventitious shoot formation, and somatic embryogenesis are preferred. However, if a stable callus culture can be maintained, indirect adventitious shoot formation and somatic embryogenesis can provide higher multiplication rates. Although the encapsulated somatic embryo system may serve as an alternative plant delivery system, several problems still require investigation.

Discussion

OKORO: Those of us working in the national agricultural research and extension systems are concerned with the implications and application of biotechnology on the improvement of the overall productivity of the small farmers. In view of the various systems and crops in Africa, national priorities will vary. We should, as a matter of policy, address problems of crop production through the synthesis of knowledge available via gene manipulation and disease elimination, especially in respect of crops such as rice, which has suffered from gall midge in rice-producing areas of Asia and Africa.

ASIEDU: Cassava germplasm as well as wild *Manihot* species are introduced from Latin America through our sister institution, Centro Internacional de Agricultura Tropical (CIAT) in Colombia. The pathologists there make sure that as far as can be detected, the materials are free from seed transmissible diseases that cannot be treated, such as viruses. The seeds are then treated against fungal and bacterial infection and mailed to the International Institute of Tropical Agriculture (IITA) through the Nigerian quarantine authority in Ibadan. After they have been cleared by the quarantine officials, they are brought to IITA where pathologists again check for diseases. When the materials are confirmed as being disease free, they are germinated in a protected screenhouse and kept under observation by the pathologists for at least 3 weeks before being transplanted to the field. If at any stage in the process, or even after transplanting, any outbreak of a new disease or insect pest is observed, the introductions are destroyed.

References

Abbott, A.J., and A.R. Belcher. 1986. Potato tuber formation in vitro. Pages 113-122 in *Plant Tissue Culture and Its Agricultural Applications* edited by L.A. Withers and P.G. Alderson. Butterworths, London, UK.

Ammirato, P.V. 1984. Yams. Pages 327-354 in *Handbook of Plant Cell Culture (Vol. 3) Crop Species* edited by P.V. Ammirato, D.A. Evans, W.R. Sharp, and Y. Yamada. Macmillan, New York, USA.

Barbier, M., and H. Dulieu. 1983. Early occurrence of genetic variants in protoplast cultures. *Plant Science Letter* 29: 201-206.

Cassells, A.C., and R.D. Long 1982. The elimination of potato viruses X, Y, S and M in meristem and explant cultures of potato in the presence of Virazole. *Potato Research* 25: 165-173.

Conger, B.V. (ed.) 1981. *Cloning Agricultural Plants via In Vitro Techniques.* CRC Press, Boca Raton, Florida, USA.

Davies, D.R. 1972. Speeding up the commercial propagation of freesias. *Grower* 77: 711.

de Fossard, R.A. 1976. *Tissue Culture for Plant Propagators*. University of New England Press, Armidale, Australia.

De Langhe, E., and E. De Bruijne 1976. Continuous propagation of tomato plants by means of callus cultures. *Scientia Horticulturae* 4: 221-227.

Dublin, P. 1981. Direct somatic embryogenesis on fragments of *Arabusta* coffee tree leaves. *Café, Cacao, The* 25: 237-242.

Earle, E.D., and R.W. Langhans. 1974. Propagation of *Chrysanthemum* in vitro. II. Production, growth and flowering of plantlets from tissue culture. *Journal of the American Society for Horticultural Science* 99: 352-358.

Estrada, R., P. Tover, and J.H. Dodds. 1986. Induction of in vitro tubers in a broad range of potato genotypes. *Plant Cell, Tissue and Organ Culture* 7(1): 3-10.

Forsyth, C., and J. Van Staden. 1984. Tuberization of *Dioscorea bulbifera* stem nodes in culture. *Journal of Plant Physiology* 115: 79-83.

Frison, E.A., and S.Y. Ng. 1981. Elimination of sweet potato virus disease agents by meristem-tip culture. *Tropical Pest Management* 27: 452-454.

George, E.F., and P.D. Sherrington. 1984. *Plant Propagation by Tissue Culture: Handbook and Directory of Commercial Laboratories*. Exegetics Limited, Basingstoke, UK.

Hakkaart, F.A., and F. Quak. 1964. Effect of heat treatment of young plants on freeing chrysanthemums from virus B by means of meristem culture. *Netherlands Journal of Plant Pathology* 70: 154-157.

Hansen, A.J., and A.C. Hildebrandt. 1966. The distribution of tobacco mosaic virus in plant callus cultures. *Virology* 28: 15-21.

Hansen, A.J., and W.D. Lane. 1985. Elimination of apple chlorotic leafspot virus from apple shoot cultures by ribavirin. *Plant Disease* 69: 134-135.

Hu, C.Y., and P.J. Wang. 1983. Meristem, shoot tip and bud cultures. Pages 177-227 in *Handbook of Plant Cell Culture* (Vol. 1) edited by D.A. Evans, W.R. Sharp, P.V. Ammirato and Y. Yamada. Macmillan, New York, USA.

Huang, S.C., and D.F. Millikan. 1980. In vitro micrografting of apple shoot tips. *HortScience* 15: 741-742.

Hussey, G. 1975. Propagation of hyacinths by tissue culture. *Scientia Horticulturae* 3: 21-28.

Hussey, G. 1978. The application of tissue culture to the vegetative propagation of plants. *Science Progress* 65: 185-208.

Hussey, G., and N.J. Stacey. 1981. In vitro propagation of potato (*Solanum tuberosum* L.). *Annals of Botany* 48: 787-796.

Hussey, G., and N.J. Stacey. 1984. Factors affecting the formation of in vitro tubers of potato (*Solanum tuberosum* L.). *Annals of Botany* 53: 565-578.

Kartha, K.K. 1986. Production and indexing of disease free plants. Pages 219-238 in *Plant Tissue Culture and Its Agricultural Applications* edited by L.A. Withers and P.G. Alderson. Butterworths, London, UK.

Kartha, K.K., and O.L. Gamborg. 1975. Elimination of cassava mosaic disease by meristem culture. *Phytopathology* 65: 826-828.

Klein, R.E., and C.H. Livingston. 1982. Eradication of potato virus X from potato by Ribavirin treatment of cultured potato shoot tips. *American Potato Journal* 59: 359-365.

Krikorian, A.D., S.A. Staicu, and R.P. Kann. 1981. Karyotype analysis of a daylily clone reared from aseptically cultured tissues. *Annals of Botany* 47: 121-131.

Lerch, B. 1977. Inhibition of biosynthesis of potato virus X by Ribavirin. *Phytopathologische Zeitschrift* 89: 44-49.

Long, R.D., and A.C. Cassells. 1986. Elimination of viruses from tissue cultures in the presence of antivirus chemicals. Pages 239-248 in *Plant Tissue Culture and Its Agricultural Applications* edited by L.A. Withers and P.G. Alderson. Butterworths, London, UK.

MacDonald, D.M. 1973. Heat treatment and meristem culture as a means of freeing potato varieties from viruses X and S. *Potato Research* 16: 263-269.

Mantell, S.H., S.Q. Haque, and A.P. Whitehall. 1978. Clonal multiplication of *Dioscorea alata* L. and *D. rotundata* Poir. yams by tissue culture. *Journal of Horticultural Science* 53: 95-98.
Mantell, S.H., J.A. Mathews, and R.A. McKee. 1985. *Principles of Plant Biotechnology — An Introduction to Genetic Engineering in Plants*. Blackwell Scientific Publications, Oxford, UK.
Morel, G. 1964. A new means of clonal propagation in orchids. *American Orchid Society Bulletin* 33: 473-478.
Morel, G.M. 1965. Clonal propagation of orchid by meristem culture. *Cymbidium Society News* 20: 3-11.
Morel, G.M., and Q.C. Martin. 1952. Guerison de dahlia attients d'une maladie a virus. *Comptes Rendues Academie de Sciences (Paris)* 235: 1324-1325.
Murakishi, H.H., and P.S. Carlson. 1976. Regeneration of virus-free plants from dark-green islands of tobacco mosaic virus-infected tobacco leaves. *Phytopathology* 66: 931-932.
Murashige, T. 1974. Plant propagation through tissue culture. *Annual Review of Plant Physiology* 25: 135-166.
Murashige, T. 1978. The impact of plant tissue culture on agriculture. Pages 15-26 in *Frontiers of Plant Tissue Culture* edited by T.A. Thorpe. Calgary University Press, Calgary, Canada.
Murashige, T. 1990. Plant propagation by tissue culture. A practice with unrealized potential. Pages 3-9 in *Handbook of Plant Cell Culture* (Vol. 5) edited by P.V. Ammirato, D.A. Evans, W.R. Sharp, and Y.P.S. Bajaj. McGraw-Hill, New York, USA.
Murashige, T., and F. Skoog. 1962. A revised medium for rapid growth and bioassays with tobacco tissue culture. *Physiologia Plantarum* 15: 473-497.
Navarro, L. 1984. Citrus tissue culture. Pages 113-154 in *Micropropagation of Selected Root Crops, Palms, Citrus and Ornamental Species*. FAO Plant Production and Protection Paper 59. FAO, Rome, Italy.
Navarro, L., C.N. Roistacher, and T. Murashige. 1975. Improvement of shoot-tip grafting in vitro for virus free citrus. *Journal of the American Society for Horticultural Sciences* 100: 471-479.
Ng, S.Y.C. 1986. Rapid propagation techniques for sweet potato, yams and cocoyams. Pages 117-118 in *Proceedings, Global Workshop on Root and Tuber Crops Propagation*. CIAT, Cali, Colombia.
Ng, S.Y.C. 1988. In vitro tuberization in white yam (*Dioscorea rotundata* Poir). *Plant Cell, Tissue and Organ Culture* 14: 121-128.
Ng, S.Y.C. 1989. In vitro technique in pathogen elimination and germplasm conservation. Pages 141-158 in *Root Crops and Low Input Agriculture* edited by M.N. Alvarez and S.K. Hahn. IITA, Ibadan, Nigeria.
Ng, S.Y.C., and S.K. Hahn. 1985. Application of tissue culture to tuber crops at IITA. Pages 24-40 in *Proceedings, Inter-Center Seminar on Biotechnology in International Agricultural Research*. IRRI, Los Baños, Philippines.
Over de Linden, A.J., and R.F. Elliott. 1972. Virus infection in *Ipomoea batatas* and a method for its elimination. *New Zealand Journal of Agricultural Research* 14: 720-724.
Quak, F. 1977. Meristem culture and virus-free plants. Pages 598-615 in *Applied and Fundamental Aspects of Plant Cell, Tissue and Organ Culture* edited by J. Reinert, and Y.P.S. Bajaj. Springer-Verlag, Berlin, Germany.
Redenbaugh, K. 1990. Application of artificial seed to tropical crops. *HortScience* 25: 251-255.
Redenbaugh, K., B. Passch, J. Nichol, M. Kossler, P. Viss, and K. Walker. 1986. Somatic seeds encapsulation of asexual plant embryos. *Bio/technology* 4: 797-801.
Robinson, D.J., B.D. Harrison, J.C. Sequeira, and G.H. Duncan.1984. Detection of strains of African cassava mosaic virus by nucleic acid hydridization and some effects of temperature on their multiplication. *Annals of Applied Biology* 105: 483-493.
Sagawa, Y., and J.T. Kunisaki. 1990. Micropropagation of floricultural crops. Pages 25-56 in *Handbook of Plant Cell Culture* (Vol. 5) edited by P.V. Ammirato, D.A. Evans, W.R. Sharp, and Y.P.S. Bajaj. McGraw-Hill, New York, USA.

Shepard, J.F. 1975. Regeneration of plants from protoplasts of potato virus X-infected tobacco leaves. *Virology* 66: 492-501.

Sheridan, W.F. 1968. Tissue culture of monocot *Lilium. Planta* 82: 189-192.

Sidwell, R.W., J.H. Huffman, G.P. Khare, L.B. Allen, J.T. Witkowski, and R.K. Robins. 1972. Broad-spectrum antiviral activity of Virazole: 1,ßD-Ribofuranosyl-1,2,4-triazole-3-carboxamide. *Science* 177: 705-706.

Stace-Smith, R., and F.C. Mellor. 1968. Eradication of potato virus X and S by thermotherapy and axillary bud culture. *Phytopathology* 53: 199-203.

Stamp, J., and G. Henshaw. 1982. Somatic embryogenesis in cassava. *Zeitschrift für Pflanzenphysiologie* 105:183-187.

Thomas, J.E., P.R. Massalski, and B.D. Harrison. 1986. Production of monoclonal antibodies to African cassava mosaic virus and differences in their reactivities with other whitefly-transmitted geminiviruses. *Journal of General Virology* 67: 2739-2748.

Uduebo, A.E. 1971. Effect of external supply of growth substances on axillary proliferation and development in *Dioscorea bulbifera.* Annals of Botany 35: 159-163.

Vuylsteke, D., and E. De Langhe. 1985. Feasibility of in vitro propagation of banana and plantains. *Tropical Agriculture (Trinidad)* 62(4): 323-327.

Walkey, D.G.A. 1980. Production of virus-free plants by tissue culture. Pages 109-117 in *Tissue Culture Methods for Plant Pathologist* edited by D.S. Ingram and J.P. Helgeson. Blackwell Scientific Publications, Oxford, UK.

Wambugu, F.M., G.A. Secor, and N.C. Gudmestad.1985. Eradication of potato virus Y and S from potato by chemotherapy and cultured axillary bud tips. *American Potato Journal* 62: 667-672.

Wang, P.J., and C.Y. Hu, 1982. In vitro mass tuberization and virus-free seed potato production in Taiwan. *American Potato Journal* 59: 33-37.

Ziv, M., A.H. Halevy, and R. Shilo. 1970. Organ and plantlets regeneration of *Gladiolus* through tissue culture. *Annals of Botany* 34: 671-676.

Ziv, M., and A.H. Halevy. 1970. Organs and plantlets regeneration of *Gladiolus* through tissue culture. *Proceedings International Horticultural Congress* 1: 211-212.

3.10

Micropropagation techniques and the production of pathogen-free plants

K. Waithaka

Department of Crop Science, University of Nairobi, Kenya

Abstract

The main applications of in vitro plant tissue culture in crop production are in micropropagation and plant sanitation. Tissue culture is currently being applied to a large variety of ornamental plants on a commercial basis. It has been applied effectively in the clonal propagation of many plant species, including ornamental and crop plants such as ferns, orchids, gerbera, carnation, chrysanthemum, strawberry, apple and potato. The use of meristem and shoot-tip cultures for the recovery and establishment of pathogen-free plants has also become common practice in the production of virus-free stock of vegetatively propagated plants in many commercial nurseries in the developed world.

In vitro plant tissue culture is an important tool for the study of various basic problems in experimental biology and agriculture. This paper discusses two of the main applications of tissue culture: plant propagation; and the recovery of pathogen-free plants.

Tissue Culture for Plant Propagation

Commercial laboratories have adopted tissue culture techniques for rapid clonal propagation of ornamentals (Hackett and Anderson 1967; Earle and Langhans 1974; Murashige 1974a; Okioga et al. 1989), vegetables (Hasegawa et al. 1973), and fruit (Cheema and Sharma 1983).

Vegetative propagation through tissue culture has played a major role in the mass production of propagating material (Murashige 1974a). It is fast and requires less space than that required for conventional methods of preparing cuttings. The application of tissue culture to woody plants still remains difficult, but successes have been reported for some species, such as *Terminalia brownii* and *T. kilimandcharica* (Mbaratha 1985), *Ulmus americana* (Durzan and Lopushanski 1975), and *Acacia koa* (Skolmen and Mapes 1976).

The propagation of plants by tissue culture involves three stages:

1. The establishment of an aseptic culture and development of an explant by cell division.

2. A series of subculturings to achieve rapid multiplication of the propagules. This multiplication may be through somatic embryogenesis, adventitious organogenesis, or axillary bud development. While adventitious organogenesis may result in faster multiplication of the propagule, the occurrence of genetically aberrant plants is not uncommon (Waithaka 1988). Axillary bud multiplication may be slow but genetically aberrant propagules are rare.

3. Rooting of established plantlets and their hardening to impart some tolerance to moisture stress and pathogens. This stage also involves the conversion of the plants from a heterotrophic mode of nutrition to the autotrophic state. Rooting can be promoted by excluding growth hormones from the medium or supplementing the medium with auxins such as indolebutyric acid (IBA) or naphthalene acetic acid (NAA). Hardening occurs because of a change in the physical environment (for example, increasing the temperature and light intensity compared with that in stage 2 and gradually exposing plantlets to an environment similar to that expected in the field).

The regeneration of plantlets from an explant at any stage of culture differs from plant to plant. These differences are influenced by the culture medium, culture environment, and source and stage of physiological development of the explant.

Culture medium

The culture medium should be appropriate in terms of both chemical composition and physical qualities during all the three stages. The ingredients of a basic medium consist of inorganic mineral salts, a carbon and energy source (sugar), vitamins, organic supplements, and growth regulators where needed. Concentration of these ingredients in a basic medium will depend on the type of plant being cultured and the stage of culture development (Street 1977; Hartmann and Kester 1983). The medium for stage 2 is considered most critical, since it is intended for multiplication of the propagule and is, therefore, enriched with substances that enhance organogenesis. Root development is promoted in stage 3 and may involve either supplementing the medium with auxins or using a medium without any growth regulators.

Various basic media have been developed for different types of cultures and stages of culture development (Murashige and Skoog 1962; Linsmear and Skoog 1965; Gamborg et al. 1976; Lloyd and McCown 1980). Auxins promote cell division and root initiation, while cytokinins promote cell division and shoot initiation. For some plants, the success or failure of in vitro culture depends on whether the medium is in liquid or solid form. For instance, the culture of many bromeliads can be started only in liquid medium, while others require starting in liquid medium and then transfer to a solid one.

The pH (usually between 5.5 and 6.2) of a nutrient medium is a critical factor, influencing the regenerative capacity of explants in culture (Gamborg et al. 1976).

Culture environment

Light, temperature, and relative humidity are the critical enviromental factors influencing in vitro regeneration of plantlets from an explant. Light has been reported to be essential in shoot and root differentiation (Gautheret 1966). The general practice in plant tissue culture is to maintain the cultures under constant light intensity, photoperiod, and temperature. Maintenance of high relative humidity is essential for 2-3 weeks to protect the propagules from desiccation when in vitro regenerated plants are introduced into soil.

Source and age of explant

The regeneration capacity of explants in vitro is influenced by the physiological age of the plant part used. Murashige (1974a) reported that tobacco stem explants which were nearer the apical region produced more adventitious roots and shoots more rapidly than those from the basal region . Somatic embryogenesis in coffee was more easily induced in younger leaf explants than in older ones (Sondahl and Sharp 1977). The source of explant for in vitro propagation differs widely in plants, for example, leaves (coffee, begonia, African violet), stems (rose, carnation, *Alstroemeria* spp., chrysanthemum), roots (raspberry), and nucellar tissue (citrus).

Tissue Culture for the Recovery of Pathogen-Free Plants

It is advisable to assume that any cultivar that has been propagated by conventional asexual cloning for an extended period is infected with one or more pathogens. Many carnation and potato cultivars are reportedly infected with several viruses (Hollings 1965). The pathogens reduce crop yield by depressing the growth performance.

The recovery of pathogen-free clones through the use of cell and tissue culture techniques is based on the premise that pathogen concentration is not uniform throughout the infected plant. To obtain pathogen-free plants, the following methods can be applied: shoot-tip culture; meristem culture; in vitro shoot-tip grafting; or regeneration of somatic embryos or adventitious shoots from callus cultures or specialized tissue such as the nucellus.

Shoot-tip culture

This has been the method widely used for the recovery of pathogen-free plants (Baker and Phillips 1962). Explants for shoot-tip culture range in length from 0.1 mm to 2 mm. The explant may be all or part of an apical or lateral growing point of a stem or a stem section of several nodes (Murashige 1977).

This explant size, although convenient for propagation, may not be free of viruses and other systemic pathogens. Usually, shoot-tip culture produces one rooted plant per culture, but at times the explant may be accompanied by proliferation of adventitious buds from callus on the base of the propagule (Abbort and Whitely 1976). The method is applicable to many herbaceous and some woody plants.

Meristem culture

The explant of a true meristem culture should be under 0.1 mm long. Only the meristematic dome and a few subtending leaf primordia are included. Complete plant formation has been achieved from such meristem explants for tobacco, carrot, coleus, and nasturtium (Smith and Murashige 1970). Meristem culture has been successful with carnation (Os 1964), chrysanthemum (Hollings and Stone 1970), and strawberry (McGrew 1980).

The size of the meristematic dome determines the ability of explants to survive on a nutrient medium, even at optimal growth conditions, and the recovery of pathogen-free plants. The smaller the explant, the more difficult the procedure and the lower the survival rate. In strawberry, the optimal size for both plant regeneration and pathogen elimination was found to be 0.5-0.9 mm (McGrew 1980). The time required to establish a plant of transplantable size from a true meristem is much longer (several months) than from a shoot-tip culture (1-2 months) (Murashige 1974b).

Shoot-tip grafting

This method has been used mainly for producing virus-, viroid- or mycoplasma-free woody species that do not regenerate shoots or roots easily through shoot-tip culture, such as citrus (Navarro 1981), apple (Huang 1980), and *Prunus* species (Negueroles and Jones 1979).

In practice, the shoot-tip is grafted onto a seedling rootstock grown in vitro. The rootstock must be free of any infection, which requires that the pathogen is not transmitted through the seed that develops into a seedling rootstock. Seeds are germinated aseptically; grafting is also done aseptically when the seedling is at least 2-3 weeks old.

Regeneration of shoots from callus and specialized tissues

Plants which are obtained through organogenesis or somatic or zygotic embryogenesis are frequently free of pathogens that might have systemically infected the mother plant. Adventitious embryos in nucellar cultures of citrus appear to escape all the known citrus viruses and viroids (Navarro and Juarez 1977). Virus-free plants have been regenerated from callus cultures of several horticultural crops, including geranium (Abo El-Nil and Hildebrandt 1971), gladiolus (Simonsen and Hildebrandt 1971), and potato (Wang and Huang 1975).

However, this method is limited by the resumption of the juvenile state, which may persist for substantial periods (Weathers and Calavan 1959; Kochba et al. 1972). In addition, plants regenerated by this method show a significant incidence of genetic aberrations. Consequently, shoot-tip grafting is more practical for the recovery of pathogen-free plants in woody species.

Heat treatment

Growing a mother plant at elevated temperatures (35-40°C) and high relative humidity (80-90%) reduces the incidence of viruses at the apical meristems (Hollings and Stone

1970; MacDonald 1973). High temperatures partially or completely inactivate most, if not all, viruses without damaging the host plant (Baker and Phillips 1962; Berg and Bustamante 1974). However, this method is not effective in reducing the concentration of some pathogens, such as citrus exocortis viroid and the citrus stubborn spiroplasma (Bitters et al. 1972). Nevertheless, shoot-tip and meristem cultures are more certain of producing pathogen-free plants when mother plants have been exposed to elevated temperatures for a few days before obtaining explants from them.

References

Abbort, A.J., and E. Whitely. 1976. Cultures of *Malus* tissues in vitro. I. Multiplication of apple plants from isolated shoot apices. *Scientia Horticulturae* 4: 183-189.

Abo El-Nil, M.M., and A.C. Hildebrandt. 1971. Differentiation of virus-symptomless geranium plants from anther callus. *Plant Disease Reporter* 55: 1017-1020.

Baker, L.A., and D.J. Phillips. 1962. Obtaining pathogen-free stock by shoot-tip culture. *Phytopathology* 52: 1242-1244.

Berg, L.A., and M. Bustamante. 1974. Heat treatment and meristem culture for the production of virus-free bananas. *Phytopathology* 64: 320-322.

Bitters, W.P., T. Murashige, S. Raugan, and E. Nauer. 1972. Investigations on establishing virus-free citrus plants through tissue culture. Pages 267-271 in *Proceedings, Fifth Conference of International Organization of Citrus Virologists (IOCV)* edited by W.C. Price. IOCV, Riverside, California, USA.

Cheema, G.S., and D.P. Sharma. 1983. In vitro propagation of apple rootstock EMLA 25. *Acta Horticulturae* 131: 75-88.

Durzan, D.J., and S.M. Lopushanski. 1975. Propagation of American elm via cell suspension cultures. *Canadian Journal of Forest Research* 5: 273-277.

Earle, E.D., and R.W. Langhans. 1974. Propagation of chrysanthemum in vitro I. Multiple plantlets from shoot tips and the establishment of tissue cultures. *Journal of the American Society for Horticultural Science* 99: 128-131.

Gamborg, O.L., T. Murashige, T.A. Thorpe, and I.K. Vasil. 1976. Plant tissue culture media. *In Vitro* 12: 473-478.

Gautheret, R.J. 1966. Factors affecting differentiation of plant tissue growth in vitro. Pages 55-95 in *Cell Differentiation and Morphogenesis*. North-Holland, Amsterdam, The Netherlands.

Hackett, W.P., and J.M. Anderson. 1967. Aseptic multiplication of differentiated carnation shoot tissue derived from shoot apices. *Journal of the American Society for Horticultural Science* 90: 365-369.

Hartmann, H.T., and D.E. Kester. 1983. *Plant Propagation: Principles and Practices* (4th edn). Prentice-Hall, Englewood Cliffs, New Jersey, USA.

Hasegawa, P.M., T. Murashige, and F.H. Takatori. 1973. Propagation of asparagus through shoot apex culture II. Light and temperature requirements, transplantability of plants and cytohistological characteristics. *Journal of the American Society for Horticultural Science* 98: 143-148.

Hollings, M. 1965. Disease control through virus-free stock. *Annual Review of Phytopathology* 3: 367-396.

Hollings, M., and O.M. Stone. 1970. Attempts to eliminate chrysanthemum stunt from chrysanthemum by meristem-tip culture after heat treatment. *Annals of Applied Biology* 65: 311-315.

Huang, S.C. 1980. In vitro micrografting of apple shoot tips. *Horticultural Science* 15: 741-743.

Kochba, J., P. Spiegel-Roy and H. Safran. 1972. Adventive plants from ovules and nucelli in Citrus. *Planta* 106: 237-245.

Linsmaier, E.M., and F. Skoog. 1965. Organic growth factor requirements of tobacco tissue cultures. *Physiologia Plantarum* 18: 100-127.

Lloyd, G., and B. McCown. 1980. Commercially feasible micropropagation of mountain laurel, *Kalmia latifolia* by use of shoot tip culture. *International Plant Propagator Society Combined Proceedings of Annual Meetings* 30: 421-427.

MacDonald, D.M. 1973. Heat treatment and meristem culture as a means of freeing potato varieties from viruses X and S. *Potato Research* 16: 263-269.

Mbaratha, J.M. 1985. Conventional and non-conventional propagation of tree species: Development of appropriate methods for arid land tree propagation. MS thesis, University of Nairobi, Kenya.

McGrew, J.R. 1980. Meristem culture for production of virus free strawberries. Pages 80-85 in *Proceedings, Conference on Nursery Production of Fruit Plants through Tissue Culture: Applications and Feasibility* edited by R.H. Zimmerman. ARR-NE-II, USDA.

Murashige, T. 1974a. Plant propagation through tissue culture. *Annual Review of Plant Physiology* 25: 135-165.

Murashige, T. 1974b. Plant cell and organ culture methods in the establishment of pathogen-free stock. Second Annual A.W. Dimock Lecture, Cornell University, Ithaca, New York, USA.

Murashige, T. 1977. Manipulation of organ initiation in plant tissue cultivars. *Botanical Bulletin of Academia Sinica* 18: 1-24.

Murashige, T., and F. Skoog. 1962. A revised medium for rapid growth and bioassays with tobacco tissue cultures. *Physiologia Plantarum* 15: 473-497.

Navarro, L. 1981. Citrus shoot-tip grafting in vitro (STG) and its applications: A review. *International Society of Citriculture, Proceedings* 1: 452-456.

Navarro, L., and J. Juarez. 1977. Tissue culture techniques used in Spain to recover virus-free citrus plants. *Acta Horticulturae* 78: 425-434.

Negueroles, J., and O.O. Jones. 1979. Production of in vitro rootstock-scion combinations of *Prunus* cultivars. *Journal of Horticultural Science* 54: 279-281.

Okioga, D.M., L. Muriithi, W.G.M. Ottaro, and S.P. Gichuru. 1989. Clonal propagation of pyrethrum. In *Proceedings, Third Conference on International Plant Biotechnology Network, Nairobi, Kenya.*

Os, H. van. 1964. Production of virus-free carnations by means of meristem culture. *Netherlands Journal of Plant Pathology* 70: 18-26.

Simonsen, J. and A.C. Hildebrandt. 1971. In vitro growth and differentiation of *Gladiolus* plants from callus cultures. *Canadian Journal of Botany* 49: 1819.

Skolmen, R.G., and M.O. Marpes. 1976. *Acacia koa* (Gray) plantlets from somatic callus tissue. *Journal of Heredity* 67: 114-115.

Smith, R.H., and T. Murashige. 1970. In vitro development of isolated shoot apical meristems of angiosperm. *American Journal of Botany* 57: 652-668.

Sondahl, M.R., and W.R. Sharp. 1977. High frequency induction of somatic embryos in cultured leaf explants of *Coffea arabica. Zeitschrift für Pflanzenphysiologie* 81: 395-408.

Street, H.E. (ed). 1977. *Plant Tissue and Cell Culture.* Blackwell Scientific Publications, Oxford, UK.

Waithaka, K. 1988. Application of plant tissue and cell culture in horticultural production. *Acta Horticulturae* 218: 131-139.

Wang, P.J., and L.C. Huang. 1975. Callus cultures from potato tissue and the exclusion of potato virus X from plants regenerated from stem tips. *Canadian Journal of Botany* 53: 2565-2567.

Weathers, L.G., and E.C. Calavan. 1959. Nucellar embryony — A means of freeing citrus clones of viruses. Pages 197-202 in *Citrus Virus Diseases* edited by J.M. Wallace. University of California (Davis), Berkeley, California, USA.

4.1

Genotype identification and gene isolation

R. Rao[1] and S. Grandillo[2]

Department of Agronomy and Plant Genetics, University of Naples, Via Universita 100, 80055 Portici, Italy[1]; Research Centre for Vegetable Breeding, National Research Council, Via Universita 133, 80055 Portici, Italy[2]

Abstract

The use of morphological, protein and DNA markers has been reported for genotype identification, as well as for genetic linkage map construction. Several approaches towards isolating genes whose final products or mRNA are known have been successfully applied. In this paper, examples are given for the application of expression vectors, DNA-RNA hybridization, differential screening, synthetic oligonucleotide probes, and heterologous probes. However, for the majority of plant genes the final product is not known and other strategies must be taken into consideration. While transposon tagging has already allowed the isolation of genes in maize and snapdragon, only a few examples are reported for the use of mutant complementation. With the recent availability of restriction fragment length polymorphism (RFLP) maps for several important crops, map-based gene cloning is now taken into consideration.

One of the most common activities pursued by geneticists and breeders since the beginning of genetic research has been monitoring, inducing, and mapping single gene markers in higher plants. Linkage maps of tomato, maize, or pea provide evidence of the large amount of work performed in this area of research.

Genotype Identification

The study of genetic variation for the identification of genotypes to be used in higher plant genetics has previously been performed on genes affecting morphological traits, such as leaf morphology or dwarfism. Because of the very limited utilization of these markers in several plant breeding approaches, new protein and DNA markers have been developed, which can be used at the whole plant, tissue, or cellular levels. Proteins (such as vicilin from *Vicia faba*) can be easily separated using mono- or bidimensional gel electrophoresis, enabling the presence or absence of alleles of a specific locus to be detected (Figure 1).

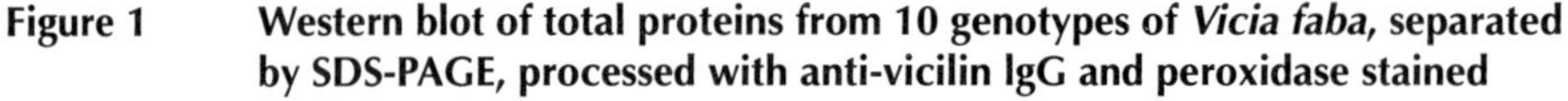

Figure 1 Western blot of total proteins from 10 genotypes of *Vicia faba*, separated by SDS-PAGE, processed with anti-vicilin IgG and peroxidase stained

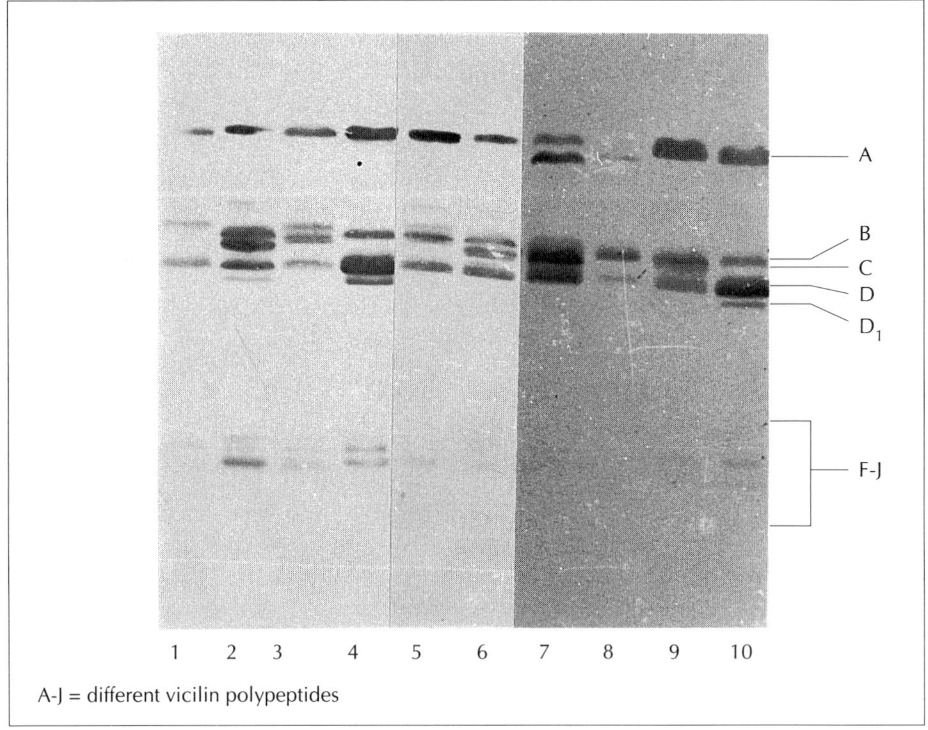

A-J = different vicilin polypeptides

The existence of qualitative genetic variation in the structure of seed proteins has allowed species identification within genera such as *Pisum* (Przybylska et al. 1984; Grillo and Rao 1985) and *Vigna* (Rao and Del Vaglio, unpubl.), predictability of interspecific crosses (Sullivan and Freytag 1986), and inheritance and linkage studies in many important crops. Several loci, both simple and complex, have been identified and mapped using seed proteins in different plant species (Soave and Salamini 1984; Casey et al. 1986).

Other widely used protein markers in plant breeding and applied genetics are isoenzymes which, after separation on gel electrophoresis, can be stained for specific enzyme activity, thus aiding recognition of allelic bands. Isoenzymes are useful markers for the identification of important genetic materials such as hybrids, lines with introgression of genes from wild species, and anther culture-derived haploid plants (Tanksley 1983). They have also been successfully used for the production of linkage maps in different species, such as tomato and pea (Weeden and Marx 1984; Tanksley and Bernatzky 1987).

Advances in molecular biology have provided new methodologies which extend the list of useful genetic markers. Cloned DNA sequences can be used to probe specific regions of eukaryotic genomes which have been digested with restriction endonucleases, for detection of variation in the length of DNA fragments (restriction fragment length polymorphism, RFLP). RFLPs have been used for constructing detailed genetic linkage maps, fingerprinting for identification of individuals, and studying genome organization (Landry and Michelmore 1987; Tanksley et al. 1989).

Gene Isolation

Isolation of single genes became feasible after the discovery that purified messenger RNA (mRNA) could be copied to produce sequences of complementary DNA (cDNA). The fragments of cDNA obtained can be transferred into suitable vectors, cloned, sequenced, possibly modified in vitro, and utilized.

In plants, this cloning technique is complicated by the complexity of the genome which has many repeated sequences (Flavell et al. 1974). Thus, the first genes cloned in plants are members of gene families producing high levels of mRNA in specific tissue, like zeins in maize. Different approaches for molecular cloning of genes are summarized in Figure 2.

Figure 2 Strategies for the identification and characterization of single- and low-copy number genes

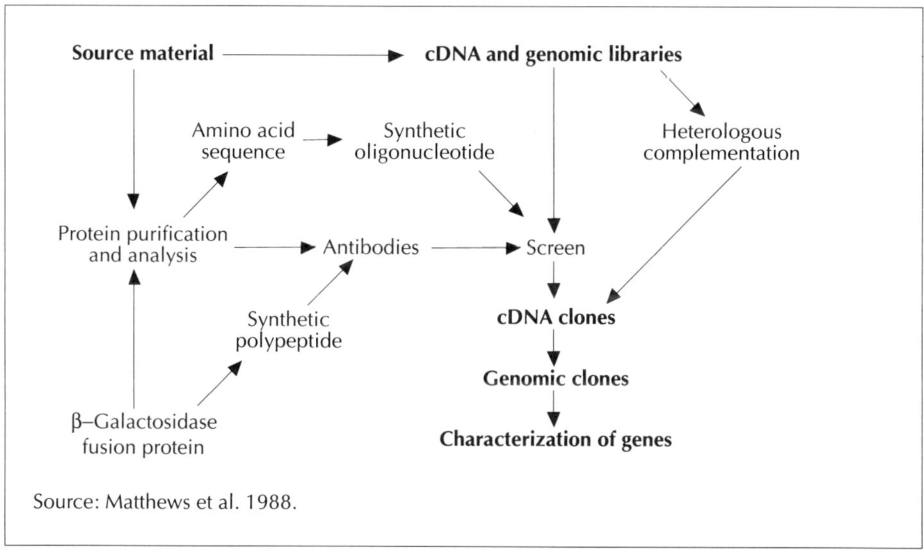

Source: Matthews et al. 1988.

Gene cloning requires information about the final gene product or its mRNA. When the protein coded by the gene of interest has been purified and characterized it is possible to screen a cDNA library, constructed from mRNA (poly A(+)) extracted from the tissue where the gene is expressed, and isolate clones containing the gene using different strategies.

One possibility is to use expression vectors such as *λgt11* or *λZap*, in which the cDNA is inserted within a bacterial gene and under the control of a promoter that, once in

Escherichia coli, allows production of the hybrid protein. The recombinant plaque can be detected by reaction with specific antibody. The procedure has been used to isolate genes coding for seed proteins of maize and rice (Di Fonzo et al. 1988; Masumura et al. 1989).

Another possibility is to screen the cDNA library with RNA extracted from the tissue in which the gene is expressed, to produce DNA-RNA hybrids. After hybridization the RNAs are eluted, translated in vitro, and the product analysed with immuno techniques using antibodies produced against the protein coded by the gene of interest. The cDNA to isolate is that which hydridized to the mRNA that, when translated in vitro, produces a protein identical to that observed in vivo. This procedure allowed the molecular cloning of the *Wx* and *Sh1* genes from maize (coding for UDPG-glucose-starch transferase and saccarose-synthatase, respectively) as well as some maize zeins genes (Marocco and Salamini 1987). A variation of this strategy is the use of differential hybridization of mRNAs extracted from protoplasts (or a particular tissue, or at a specific developmental stage) where the gene is expressed and mRNA extracted from cells where the gene is silent, in order to eliminate from further consideration those mRNAs present in the tissue where the gene is not expressed. A similar procedure is performed by differential screening, where poly (A+) RNAs extracted from tissue, both where the gene is expressed and where it is silent, are used as templates to produce radiolabelled cDNA probes that are hybridized separately to duplicate copies of the same cDNA library. Clones that hybridize to both probes correspond to genes that are expressed in both tissues, whereas clones that hybridize to only one probe correspond to mRNAs that are expressed in only one tissue. Enrichment of cDNA clones corresponding to poorly abundant RNAs can be obtained by constructing a subtracted cDNA library in which poly (A+) RNA extracted from the tissue where the gene is expressed is used as a template to make cDNA, which is then hybridized to poly (A+) RNA extracted from the tissue in which the gene is not expressed. The cDNAs which do not form hybrids are purified by hydroxyapatite chromatography, made double stranded and then used to construct the cDNA library for use in the isolation of the corresponding genomic clones. These strategies have proved particularly useful for the molecular cloning of several genes involved in development or regulation (Bartels et al. 1988; Mundy and Chua 1988; Bartels et al. 1990; Evans et al. 1990; Kiss et al. 1990; Reddy et al. 1990; Sommer et al. 1990).

In some cases, a gene isolated from a species can be used to isolate the analogous gene in other species. The gene encoding chalcone synthase (CHS), the key enzyme of flavonoid biosynthesis, which was isolated originally from parsley (Kreuzaler et al. 1983), has been successfully used to obtain the CHS gene from *Anthirrinum majus* (Sommer and Saedler 1986), *Petunia hybrida* (Reif et al. 1985), *Phaseolus vulgaris* (Ryder et al. 1987), and *Arabidopsis thaliana* (Feinbaum and Ausubel 1988).

Despite these achievements, it must be emphasized that the successful use of heterologous probes is limited because the sequence of some genes is not sufficiently conserved. Moreover, it is advisable to use heterologous probes only for the isolation of cDNA clones in order to avoid the risk of isolating silent genes.

A different strategy, particularly useful when attempting isolation of genes which are poorly expressed, is to use synthetic oligonucleotide probes. This approach needs a total or partial amino acid sequence of the gene product (5 to 15 consecutive aminoacids) used for chemical synthesis of the oligonucleotide, which will then be the probe for cDNA or genomic libraries and isolation of the gene of interest (Anderson et al. 1986; Aeschbacher et al. 1989; Palme et al. 1989; Claes et al. 1990). However, the encoded product of most plant genes, especially those involved in developmental processes, plant growth and morphology, and photosynthesis is not known and the only available information is the phenotype they affect. In such cases, different strategies of gene cloning should be performed.

Some plants, such as maize, contain transposable segments (TS) of DNA with the unique ability of moving from place to place in the genome through a process of excision and reintegration. Their insertion or excision from a specific locus causes mutations that may affect the phenotype. The direct correlation between a mutant phenotype and a specific DNA sequence allows the use of these elements to clone genes for unknown products. The basic approach to a transposon tagging experiment is to cross a plant harbouring an active transposable element to another genetically distinct and homozygous recessive for the gene to tag, and to screen the progeny for the presence of a specific mutation. A genomic library created from the mutant is then screened with TS to isolate clones containing, besides the TS, a partial or complete sequence of the gene of interest (Chandlee 1990). This approach is relatively straightforward for the isolation of genes, as well as promoters, when the TS have been cloned and characterized at the molecular level. TS are active in heterologous systems such as the *Ac* elements of maize, which transpose in tobacco, potato, and tomato (Baker et al. 1987; Knapp et al. 1988; Yoder et al. 1988) and tagging can be performed with the T-DNA of the Ti plasmid of *Agrobacterium tumefaciens* (Koncz et al. 1989), thus increasing the number of species to which this strategy can be applied. Transposon tagging allowed the isolation of several genes, such as *viviparous-1* in maize, a gene controlling multiple developmental responses associated with maturation of seeds (McCarty et al. 1989), genes involved in the anthocyanin pathway in maize and *Antirrhinum majus*, the regulatory locus *opaque-2* which regulates zein seed storage proteins, and genes involved in the thylakoid membrane (Chandlee 1990).

Gene isolation through mutant complementation is possible for genes that are expressed at the morphological level or that can be selected in vitro (Lorenzetti and Salamini 1989). A defective mutant is transformed with cDNA clones randomly selected from an expression library. The DNA that restores the function contains the gene of interest, which can then be isolated. This technique requires an expression vector with promoters suitable for plant cells, a large number of protoplasts prepared from the mutant plant (which is not easy to obtain from all plants species), and a highly efficient transformation technique.

The recent availability of RFLP maps for important crops provides an alternative and independent method for gene isolation based on map position (map-based gene cloning or reverse genetics). This approach has been used to clone genes involved in hereditary deseases in humans (Orkin 1986). Together with other molecular techniques, such as the polymerase chain reaction (Peterhans et al. 1990), they represent a powerful means for the isolation of genes underlying quantitative traits or characteristics difficult to score (Tanksley et al. 1989).

References

Aeschbacher, R.A., M.W. Saul, and I. Potrykus. 1989. Isolation of a putative transcription factor from *Arabidopsis t.* by a novel strategy. Page 74 in *Proceedings, Fifteenth Annual Symposium on Molecular Communication in Higher Plants, EMBO, 18-21 September 1989, Heidelberg, Germany.*

Anderson, M.A., E.C. Cornisch, S.L. Mau, E.G. Williams, R.M. Hoggart, A. Atkinson, I. Bonig, B. Grego, R.J. Simpson, P. Roche, J. Haley, J. Penschow, H. Niall, G. Tregear, J. Coghlan, R. Crawford, and A.E. Clarke. 1986. Cloning of cDNA for a stylar glycoprotein associated with expression of self-incompatibility in *Nicotiana alata. Nature (London)* 321: 38-44.

Baker, B., G. Coupland, N. Fedoroff, P. Starlinger, and J. Schell. 1987. Phenotypic assay for excision of the maize controlling element *Ac* in tobacco. *EMBO Journal* 6: 1547-1555.

Bartels, D., M. Singh, and F. Salamini. 1988. Onset of desiccation tolerance during development of the barley embryo. *Planta* 175: 485-492.

Bartels, D., K. Schneider, G. Terstappen, D. Piatkowski, and F. Salamini. 1990. Molecular cloning of abscisic acid-modulated genes which are induced during desiccation of the resurrection plant *Craterostigma plantagineum. Planta* 181: 27-34.

Casey, R., C. Domoney, and N. Ellis. 1986. Legume storage proteins and their genes. Pages 1-95 in *Oxford Survey of Plant Molecular and Cell Biology* (Vol. 3). Oxford University Press, Oxford, UK.

Chandlee, J.M. 1990. The utility of transposable elements as tools for the isolation of plant genes. *Physiologia Plantarum* 79: 105-115.

Claes, B., R. Dekeyser, R. Villarroel, M. Van den Bulcke, G. Bauw, M. Van Montagu, and A. Caplan. 1990. Characterization of a rice gene showing organ-specific expression in response to salt stress and drought. *The Plant Cell* 2: 19-27.

Di Fonzo, N., H. Hartings, M. Brembilla, M. Motto, C. Soave, E. Navarro, J. Palav, W. Rohde, and F. Salamini. 1988. The B-32 protein from maize endosperm, an albumin-regulated by the O2 locus: Nucleic acid (cDNA) and amino acid sequences. *Molecular and General Genetics* 212: 481-487.

Evans, I.M., L.N. Gatehouse, J.A. Gatehouse, N.J. Robinson, and R.D. Croy. 1990. A gene from pea (*Pisum sativum* L.) with homology to metallothionin genes. *FEBS Letters* 262 (1): 29-32.

Feinbaum, R.L., and F.M. Ausubel. 1988. Transcriptional regulation of the *Arabidopsis thaliana* chalcone synthase gene. *Molecular and Cellular Biology* 8: 1985-1992.

Flavell, R.B., M.D. Bennett, J.B. Smith, and D.B. Smith. 1974. Genome size and the proportion of repeated sequence DNA in plants. *Biochemical Genetics* 12: 257-269.

Grillo, S., and R. Rao. 1985. Polypeptide composition of major albumins in different pea species. Pages 45-57 in *Proceedings, Eucarpia Meeting on Pea Breeding, 10-13 June 1985, Sorrento, Italy.*

Kiss, G.B., E. Vincze, Z. Vegh, G. Toth, and J. Soos. 1990. Identification and cDNA cloning of a new nodule-specific gene, Nms-25 (nodulin-25) of *Medicago sativa. Plant Molecular Biology* 14: 467-475.

Knapp, S., G. Coupland, U. Uhrig, P. Starlinger, and F. Salamini. 1988. Transposition of the maize transposable element *Ac* in *Solanum tuberosum. Molecular and General Genetics* 213:285-290.

Koncz, C., N. Martini, R. Mayerhofer, Z. Koncz-Kalman, H. Korber, G.P. Redei, and J. Schell. 1989. High-frequency T-DNA-mediated gene tagging in plants. *Proceedings of the National Academy of Science, USA* 86: 8467-8471.

Kreuzaler, F., H. Ragg, E. Hautz, D.N. Kuhn, and K. Hahlbrock. 1983. UV induction of chalcone synthase mRNA in cell suspension cultures of *Petroselium hortense. PANS* 80: 2591-2593.

Landry, B.S., and R. Michelmore. 1987. Methods and applications of restriction fragment length polymorphism analysis to plants. Pages 25-44 in *Tailoring Genes for Crop Improvement* edited by G. Bruening, J. Harada, T. Kosuge and A. Hollaender. Basic Life Sciences Series (Vol. 41). Plenum Press, New York, USA.

Lorenzetti, F., and F. Salamini. 1989. Biotecnologie e innovazione in agricoltura. *Rivista di Agronomia* 23: 337-371.

Marocco, A., and F. Salamini. 1987. Strategie di clonaggio dei geni delle piante. *Agricoltura e Ricerca* 77: 75-84.

Masumura, T., D. Shibata, T. Hibino, T. Kato, K. Kawabe, G. Takeba, K. Tanaka, and S. Fujii. 1989. cDNA cloning of an mRNA encoding a sulfur-rich 10 kDa prolamin polypeptide in rice seeds. *Plant Molecular Biology* 12: 123-130.

Matthews, B.F., E.M. Reardon, F.J. Turano, and B.J. Wilson. 1988. Amino acid biosynthesis in plants: Approaching an undertanding at the molecular level. *Plant Molecular Biology Reporter* 6 (3): 137-154.

McCarty, D.R., C.B. Carson, P.S. Stinard, and D.S. Robertson. 1989. Molecular analysis of *Viviparous*.1: An abscisic acid-insensitive mutant of maize. *The Plant Cell* 1: 423-532.

Mundy, J., and N-H. Chua. 1988. Abscisic acid and water-stress induce the expression of a novel rice gene. *EMBO Journal* 7: 2279-2286.

Orkin, S.H. 1986. Reverse genetics and human disease. *Cell* 47: 845-850.

Palme, K., N. Campos, T. Diefenthal, J. Feldwisch, C. Garbers, T. Hesse, K. Nitschke, S. Schwonke, and J. Schell. 1989. Molecular cloning and structural analysis of genes from *Zea mays* (L.) coding for auxin-binding proteins and members of the guanine-nucleotide binding regulatory protein family. Pages 48-49 in *Proceedings, Fifteenth Annual Symposium on Molecular Communication in Higher Plants, EMBO, 18-21 September 1989, Heidelberg, Germany*.

Peterhans, A., H. Schlpmann, C. Basse, and J. Paszowski. 1990. Intrachromosomal recombination in plants. *EMBO Journal* 9: 3437-3445.

Przybylska, J., Z. Zimniak-Przybylska, E. Kozubek, and S. Blixt. 1984. Comparative study of seed proteins in the genus *Pisum*. VIII. Further investigation on variation in electrophoretic albumin patterns. *Genetica Polonica* 25 (2): 139-147.

Reddy, A.S.N., P.K. Jena, S.K. Mukherjee, and B.W. Poovaiah. 1990. Molecular cloning of cDNAs for auxin-induced mRNAs and developmental expression of the auxin-inducible genes. *Plant Molecular Biology* 14: 643-653.

Reif, H.J., U. Niesbach, B. Deumling, and H. Saedler. 1985. Cloning and analysis of two genes for chalcone synthase from *Petunia hybrida*. *Molecular General Genetics* 199: 208-215.

Ryder, T.B., S.A. Hedrick, J.N. Bell, X. Liang, S.D. Clouse, and C.J. Lamb. 1987. Organization and differential activation of a gene family encoding the plant defense enzyme chalcone synthase in *Phaseolus vulgaris*. *Molecular and General Genetics* 210: 219-233.

Soave, C., and F. Salamini. 1984. Organization and regulation of zein genes in maize endosperm. *Physiological Transactions of the Royal Society of London* B 304: 341-347.

Sommer, H., and H. Saedler. 1986. Structure of the chalcone synthase gene of *Antirrhinum majus*. *Molecular and General Genetics* 202: 429-434.

Sommer, H., J-P. Beltran, P. Huijser, H. Pape, W-E. Lonnig, H. Saedler, and Z. Schwarz-Sommer. 1990. *Deficiens*, a homeotic gene involved in the control of flower morphogenesis in *Antirrhinum majus*: The protein shows homology to transcription factors. *EMBO Journal* 9: 605-613.

Sullivan, J.G., and G. Freytag. 1986. Predicting interspecific compatibilities in beans (*Phaseolus*) by seed protein electrophoresis. *Euphytica* 35: 201-209.

Tanksley, S.D. 1983. Molecular markers in plant breeding. *Plant Molecular Biology Reporter* 1: 3-8.

Tanksley, S.D., and R. Bernatzky. 1987. Molecular markers for the nuclear genome of tomato. Pages 37-44 in *Tomato Biotechnology* edited by D.J. Nevins and R.A. Jones. Alan R. Liss, New York, USA.

Tanksley, S.D., N.D. Young, A.H. Paterson, and M.W. Bonierbale. 1989. RFLP mapping in plant breeding: New tools for an old science. *Biotechnology* 7: 257-264.

Weeden, N.F., and G.A. Marx. 1984. Chromosomal locations of twelve izozyme loci in *Pisum sativum*. *Journal of Heredity* 75: 365-370.

Yoder, J.I., J. Palys, K. Alpert, and M. Lassner. 1988. *Ac* transposition in transgenic tomato plants. *Molecular and General Genetics* 213: 291-296.

4.2

Agrobacterium tumefaciens-mediated gene transfer[1]

E. Filippone[2] and R. Penza[3]

Research Centre for Vegetable Breeding, National Research Council, Via Universita 133, Portici, Italy[1]; Department of Agronomy and Plant Genetics, University of Naples, Via Universita 100, Portici, Italy[3]

Abstract

Single gene transfer in plants has already yielded exciting results as well as considerable expectation in some areas of research and development. Co-culture of tissues with *Agrobacterium tumefaciens* seems to be one of the most promising systems to transfer genes into the plant cell. A number of dicotyledonous crops have been genetically transformed, showing new traits amenable to further field selection. Nevertheless, there are still some limitations, such as the host range, the virulence of bacteria, and the genotype-specific response of the plant to the infection.

Many papers have been published on plant genetic transformation (Gasser and Fraley 1989). On the basis of results achieved to date, *Agrobacterium tumefaciens* is now considered the best natural plant genetic engineering system (Lurquin 1987).

Plant Cell Infection by *Agrobacterium tumefaciens*

Agrobacterium tumefaciens is a soil-borne gram-negative bacterium, causing the plant disease, "crown gall". Cells isolated from crown galls grown in vitro without any external source of hormones show tumorous or neoplastic behaviour, even in the complete absence

1 Contribution No. 74 from Research Centre for Vegetable Breeding, National Research Council, Via Universita 133, Portici, Italy.

of bacteria. It is well established that during infection *Agrobacterium* transfers part of a large plasmid, the pTi or "tumor inducing plasmid", in the nuclear genome of the infected cell (Binns and Bock 1989). The pTi fragment transferred (T-DNA) is flanked by two regions, "left border" and "right border". These borders constitute the signal that is recognized by the transfer system of *Agrobacterium*. The T-DNA transferred contains genes, known as "oncogenes", for the synthesis of plant hormones such as cytokinin and auxins; it also contains genes for opins synthesis, that are used by the bacteria for feeding (Zambryski et al. 1989).

Co-cultivation of *Agrobacterium* with plant tissues has been seen as a powerful system for transferring genes into plants, after having cloned those in the T-DNA. To have a successful and repeatable genetic transformation, it is necessary to overcome two major problems: the limited host range of *Agrobacterium* and the presence of the oncogenes carried on the T-DNA and transferred into cells.

With reference to the former problem, the range of plant species that have been submitted to co-cultivation is very broad (Binns 1990). For a long time, it was thought that the host range of *Agrobacterium* species was restricted to dicotyledonous plants (Usami et al. 1987), although Hooykaas-Van Slogteren et al. (1984) have demonstrated that T-DNA transfer and expression of genes does occur in at least a few monocots. However, Potrykus (1990) has pointed out that no clear proof of genetic transformation after co-cultivation has been reported in cereals to date.

With reference to the oncogenes that are present on T-DNA, these have been replaced by marker genes that confer on the bacteria resistance to antibiotics, such as neomycin phospho transferase II, using genetic engineering techniques. Two strategies have been adopted to obtain a vector suitable for the transfer of genes into plants: the use of the co-integrative pTi (Fraley et al. 1985) and the binary vector (Hoekhema et al. 1983). The latter strategy is used mainly for genetic transformation in plants. It involves a plasmid that contains a T-DNA fragment carrying at least the left border sequence, while the genes involved in the transfer are already present on the pTi, referred to as helper plasmid, deprived of the oncogene region. This system allows easy manipulation of the binary plasmid for the whole cloning process in *Escherichia coli* (Zyprian and Kado 1990). Some *Agrobacterium* strains and their helper plasmids used in genetic transformation programmes are listed in Table 1.

Plant Cell and Tissue Co-Cultivation

Co-cultivation of tissues with *Agrobacterium* has been performed on various kinds of tissues: roots, stems, leaves, meristems, epicotyls, cotyledons, pollen, mature and immature embryos, cells, and protoplasts (Klee and Rogers 1989). The method is widely applied to plants in which regeneration from already formed tissues is an easy process, such as tobacco, potato, tomato and petunia. For those species in which regeneration is not an easy task, co-cultivation with meristematic tissues could overcome the problem (Penza et al. 1991), although it has been demonstrated that the efficiency of genetic transformation is very low in these tissues. Moreover, shoots coming from buds differentiated from transformed meristems generally show a chimeric pattern. To reduce chimerism, it is also necessary to induce in vitro lateral branching, and to select, among the newly formed shoots, for the solid transformed ones.

Table 1 **Some *Agrobacterium tumefaciens* strains and their helper plasmids used in genetic transformation programmes**

***Agrobacterium* strain**	**Helper plasmid[a]**
A208	pGV3850::pA41003
	pTIT37ASE::pMON200
	pTi201
	pGV2298
	pGV2441
	pGV2775
	pGV3304
	pGV3850
A281	pTiBo542
A479	pAL4404
Ach5	pTiAch5
EHA101	pEHA101
	pAL1050
EHA105	pEHA105
LBA1050	pAl1050
LBA4301	pRAL4404
LBA4401	pTiAch5
LBA4404	pAL4404
LBA4434	pAL1050
LBA4434	pAL4404
GV3101	pGV3850::pCAT4
GV3101	pGV3850::pLGV1103
GV3111SE	pTiB6S3::pMON120
T37	pTiT37

Note: a :: = the co-integrative vector.

The co-culture period is generally 2 days. At the end of this period, explants are transferred into a medium containing antibiotics to kill bacteria and to select transformed cells from untransformed ones. It is necessary to set up the appropriate level of antibiotic to select transformed cells without affecting the plant regeneration pathway, and at the same time to prevent the possibility of escape of untransformed plants. Afterwards, it is necessary to check the stable integration of the newly inserted gene in the nuclear genome, using dot or Southern blots.

Gene Transfer Efficiency using *Agrobacterium tumefaciens*

Many papers have been published on the physical and chemical methods for enhancing the *Agrobacterium* infection rate. *Agrobacterium* is unable to penetrate intact tissues, and thus scalpel blades, forceps, or even carborundum powder have been adopted to injure explants.

Chemical substances, such as acetosyringone (Mathews et al. 1990) or glucose or xylose (Cangelosi et al. 1990), can activate genes involved in *Agrobacterium* virulence. These compounds are produced in nature by wounded plant cells and they act as signal molecules for inducing infection (Bolten et al. 1986). Many monocots do not produce these substances and this might explain why these plants are inefficiently transformed by *Agrobacterium* (Potrykus 1990). Other techniques have been adopted to increase the efficiency of transformation as well as plant regeneration after co-cultivation, such as a feeder-layer made by a cell culture from a healthy plant, or reduction of *Agrobacterium* density in co-culture medium. A list of some crops genetically transformed by *Agrobacterium* after co-culture is presented in Table 2.

Table 2 Gene transfer in some crops via co-cultivation of plant tissues with *Agrobacterium tumefaciens*

Species	*Agrobacterium* strain and binary vector[a]	Gene(s) transferred[b]
Brassica oleracea	EHA101; nr	*NptII + aux 2*
Chrysanthemum spp.	nr; nr	*NptII + Gus*
Cucumis melo	GV3111; pMON200	*Dhfr + Gus*
Cucumis sativus	C58; pGA482	*NptII*
Gossypium hirsutum	GV3850; nr	*NptII*
Mangifera indica	C58; pGV3850	*NptII*
Medicago sativa	nr; nr	*NptII + sod*
	nr; nr	*Bar + adh*
Nicotiana tabacum	C58;pGA482	*NptII*
	LBA4404; pBI121	*Gus + prot. inhib.*
Phaseolus vulgaris	nr; pBI121	*NptII + Gus*
	pBI121	*NptII + Gus + fl*
Pisum sativum	GV2260; p35GUS-INT	*NptII + Gus*
	GV3850; nr	*Hpt*
Solanum tuberosum	nr; nr	*Hpt + Gus + cpv*
Vigna unguiculata	GV2260; p35GUS-INT	*NptII + Gus*

Note: a nr = not reported.
b *adh* = alcohol dehydrogenase; *aux 2* = auxin synthesis from *Agrobacterium rhizogenes*; *Bar* = phosphinotricin acetyltransferase; cpv = BNYV-virus coat protein; *Dhfr* = dihidrofolate reductase; *fl* = firefly luciferase; *Gus* = glucuronidase; *Hpt* = hygromycin phosphotransferase; *NptII* = neomycin phospho-transferase II; *prot. inhib.*= protease inhibitor from potato; sod = superoxide dismutase.
Source: IAPTC 1990.

An important factor affecting transformation efficiency is the *Agrobacterium* strain (Melchers and Hooykaas 1987). In fact, it is well known that some strains, such as A281, EHA101, and EHA105, are more virulent than others, such as LBA4404 (Filippone and Lurquin 1989; Filippone 1990). Another important factor is the influence of the plant genotype on the transformation rate (Kohler et al. 1987).

Discussion

HASEGAWA: From what cowpea tissue was the DNA isolated for use in the dot-blot shown in your slide?

FILIPPONE: DNA was isolated from calli growing in vitro in the presence of kanamycin and the absence of hormones. These calli were grown on epicotyl of cowpea submitted to agroinfection with the hypervirulent and oncogenic strain A281.

MAKAMBILA: Are there no other types of bacteria, apart from *Agrobacterium tumefaciens*, that could be used for gene transfer? What is the role of plasmid in the bacterial cells in this case for *Agrobacterium*?

FILIPPONE: There are some other biotic vectors to transfer DNA, or even RNA, in plant cells, but the knowledge we have on *Agrobacterium* genetics and physiology today is detailed enough to use it solely. Moreover, the development of the binary vector strategy has allowed us to clone genes in plasmid small enough to be manipulated in vitro easily; in fact, despite the dimension of pTi (a 200 kbp), binary vectors range between 6 and 20 kbp, the role of the binary plasmid is to transfer an "artificial" T-DNA, carrying the cloned gene, in the plant cell; the transfer is carried out by genes acting in trans and already present on pTi.

FAUQUET: Is there a correlation between the virulence of an *Agrobacterium* strain, in terms of tumour formation onto a particular host, and the rate of integration of foreign DNA into the same host?

FILIPPONE: On the basis of the experience my colleagues and I have in the co-cultivation of leaf discs of egg plant and potato and the embryo co-culture of chickpea, pea, and cowpea, we have shown a strict dependence of the *Agrobacterium* genotype (strain) with the transformation rate. The best results we had were using the hypervirulent strains, A281 and EHAIOI, while few or no transformants were collected using LBA 4404 strain.

References

Binns, A.N. 1990. *Agrobacterium*-mediated gene delivery and the biology of host range limitation. *Physiologia Plantarum* 79: 135-139.

Binns, A.N., and A. Bock. 1989. Cell biology of *Agrobacterium* infection and transformation of plants. *Annual Review of Microbiology* 42: 1-35.

Bolten, G.W., E.W. Nester, and M.P. Gordon. 1986. Plant phenolic compounds induce expression of *Agrobacterium tumefaciens* loci needed for virulence. *Science* 232: 981-985.

Cangelosi, G.A., R.G. Ankenbauer, and E.W. Nester. 1990. Sugars induce the *Agrobacterium* virulence genes through a periplasmic binding protein and a transmembrane signal protein. *Proceedings of the National Academy of Science, USA* 87: 6708-6712.

Filippone, E. 1990. Genetic transformation of pea (*Pisum sativum* L.) and cowpea (*Vigna unguiculata* L.) by co-cultivation of tissues with *Agrobacterium tumefaciens* carrying binary vectors. Pages 175-181 in *Cowpea Genetic Resources* edited by N.Q. Ng and L.M. Monti. IITA, Ibadan, Nigeria.

Filippone, E., and P.F. Lurquin. 1989. Stable transformation of egg plant (*Solanum melongena* L.) of co-cultivation of tissues with *Agrobacterium tumefaciens* carrying a binary plasmid vector. *Plant Cell Reports* 8: 370-373.

Fraley, R.T. S.G. Rogers, R.B. Horsch, D.A. Eichholtz, C.L. Flick, N.L. Hoffmann, and P.R. Saunders. 1985. The SEB system: A new disarmed Ti plasmid vector system for plant transformation. *Bio/technology* 3: 629-635.

Gasser, C.S., and R.T. Fraley. 1989. Genetically engineering plants for crop improvement. *Science* 244: 1293-1299.

Hoekhema, A., P.R. Hirsch, P.J.J Hooykaas, and R.A. Schilperoort. 1983. A binary plant vector strategy based on separation of vir-and T-region of the *Agrobacterium tumefaciens* Ti plasmid. *Nature (London)* 303: 179-180.

Hooykaas-Van Slogteren, G.M.S., P.J.J. Hooykaas, and R.A. Schilperoort. 1984. Expression of Ti plasmid genes in monocotyledonous plants infected with *Agrobacterium tumefaciens. Nature (London)* 311: 763-764.

IAPTC. 1990. *Abstracts, Seventh International Congress on Plant Tissue and Cell Cultures, 24-29 June 1990, Amsterdam, the Netherlands.*

Klee, H.J., and S.G. Rogers. 1989. Plant gene vectors and transformation: Plant transformation systems based on the use of *Agrobacterium tumefaciens*. Pages 2-25 in *Cell Culture and Somatic Cell Genetics. 6. Molecular Biology of Plant Nuclear Genes* edited by J. Schell and I.K. Vasil. Academic Press, San Diego, USA.

Kohler, F., C. Golz, S. Eapen, and O. Schieder. 1987. Influence of plant cultivar and plasmid DNA on transformation rates in tobacco and moth bean. *Plant Science* 53: 87-91.

Lurquin, P.F. 1987. Foreign gene expression in plant cells. *Progress in Nucleic Acid Research* 34: 143-188.

Mathews, H., N. Bharathan, R.E. Litz, K.R. Narayanan, P.S. Rao, and C.R. Bhatia. 1990. The promotion of *Agrobacterium*-mediated transformation in *Atropa belladonna* L. by acetosyringone. *Journal of Plant Physiology* 136: 404-409.

Melchers, L.S., and P.J.J. Hooykaas. 1987. Virulence of *Agrobacterium. Oxford Survey of Plant Molecular Cell Biology* 4: 167-220.

Penza, R., P.F. Lurquin, and E. Filippone. 1991. Gene transfer by co-cultivation of mature embryos with *Agrobacterium tumefaciens*: Application to cowpea (*Vigna unguiculata* Walp.). *Journal of Plant Physiology* 138: 39-43.

Potrykus, I. 1990. Gene transfer to cereals: An assessment. *Bio/technology* 8: 535-542.

Usami,S., S. Morikawa, I. Takebe, and Y. Machida. 1987. Absence in monocotyledonous plants of the diffusible plant factors inducing T-DNA circularization and vir gene expression in *Agrobacterium. Molecular and General Genetics* 209: 221-226.

Zambryski, P., J. Tempe, and J. Schell. 1989. Transfer and function of T-DNA genes from *Agrobacterium* Ti-and Ri plasmids in plants. *Cell* 56: 193-201.

Zyprian, E., and C.I. Kado. 1990. *Agrobacterium*-mediated plant transformation by novel mini-T vectors in conjunction with a high-copy vir region helper plasmid. *Plant Molecular Biology* 15: 245-256.

4.3

Physical methods for gene transfer in plants

N.K. Singh and J.J. Shaw

Department of Botany and Microbiology, Alabama Agricultural Experiment Station, Auburn University, Auburn, Alabama 36849, USA

Abstract

The use of gene transfer technology to produce transgenic plants promises to provide both qualitative and quantitative improvements in crop production. *Agrobacterium tumefaciens*- and viral-mediated gene transfer technologies are restricted to certain plant species and cell types. To overcome such limitations, physical methods for gene transfer have been developed. These are based on the transformation of protoplasts or intact cells, from which whole plants must be regenerated. Such methods include DNA uptake by the protoplast (for example, electroporation or chemically mediated uptake) or the mechanical introduction of DNA into the cell (for example, microinjection or use of high velocity microprojectiles). These methods require varying degrees of skill and equipment.

Recombinant DNA techniques have been successfully applied to study the regulation of gene expression and transfer of genes into agronomically important crops. However, progress in the transformation of a wide variety of plants has been restricted by the limitations of available gene transfer and regeneration systems. The best available method for gene transformation is based upon the ability of the bacterium *Agrobacterium tumefaciens* to transfer and integrate the T-DNA region of its Ti plasmid into the recipient genome (Bevan and Chilton 1982; Rogers et al. 1986). *Agrobacterium* is a natural genetic engineer and its Ti plasmid has been extensively studied and used in gene transfer, but the actual process by which T-DNA traverses the bacterial and plant cell walls and becomes integrated into the host genome is not understood. Also, the limited host range of *Agrobacterium* has prevented its widespread use for the transformation of important crops. The Ti plasmid has been used to deliver DNA sequences to the cells of intact plant organs by the process of agroinfection (Grimsley and Bisaro 1987). The functional virus is reconstituted inside the recipient cell, where it replicates and spreads systemically in the host, causing amplification of the viral gene products, including desired recombinant nucleic acid sequences. In addition to agroinfection, direct use of both DNA and RNA viruses as vectors for plant transformation has been reported in recent years (Ward et al. 1988; Dawson et al. 1989; Gronenborn and Matzeit 1989). Their routine use will require a better understanding and control of the processes involved. In general, agroinfection or viral strategies produce cytoplasmic or nuclear replicons, but they do not produce stable transgenic plants.

Integrative transformation, such as T-DNA insertion, is often desirable, since it leads to stable transformants which can be studied and used further through classical plant breeding approaches. An essential feature in this regard is the stability that integration confers upon the heritability of the introduced trait(s). While the integrative transformation of plants has been attempted for a number of years, it has only recently become possible to combine molecular and breeding techniques to verify such events. Indeed, the physical presence of the introduced sequences in the genome of the recipient is routinely demonstrated by Southern hybridization and can be followed in succeeding generations.

A number of physical methods have been developed to produce stable transgenic plants. Unlike the *Agrobacterium*-mediated methods, physical methods are not limited by host range or tissue types and can theoretically be applied to all plants and recipient cells. Strategies to transfer genes directly into cells must be coupled with the ability of the introduced genetic material to integrate into the recipient genome, and the ability to regenerate whole plants. As the cell wall offers a formidable barrier for DNA uptake, protoplasts have typically been targeted for such transformations.

Once the DNA has been introduced into cells, its expression depends upon several factors, such as regulation of transcription, translation, and replication of the introduced DNA. In addition, it is of concern whether or not this DNA will be integrated into the host genome.

Protoplast Transformation

Protoplasts are more amenable to the uptake of DNA because the cell wall barrier is removed. While advances have been made in the quality of enzymes used in the production of protoplasts, a largely empirical approach is still required. Generally, protoplast transformation has a number of advantages, provided access to good quality and viable protoplasts is assured: it is relatively simple to perform; a very high frequency of transformants can be obtained; there is a theoretically unlimited range of hosts; both stable and transient transformants can be obtained; and the transformation frequencies are reproducible. Genetically transformed cells can be selected when the protoplasts are transformed with suitable markers, and the frequency of transformation can be greatly increased by chemical or physical factors that promote DNA uptake.

DNA uptake mediated by chemicals

Polyethylene glycol (PEG), poly-L-ornithine, or calcium phosphate have each been used in the incubation mixture of protoplasts and plasmid DNA to increase the frequency of transformation (Davey et al. 1980; Draper et al. 1982; Krens et al. 1982) and a wide variety of plant species have been successfully transformed by protoplast transformation augmented by these chemicals. Significant improvements have been made in increasing transformation frequency by optimizing transformation parameters, such as component concentrations and buffer formulations, duration of treatment, use of eukaryotic carrier DNA and bivalent metal ions, and the form and concentration of transforming DNA. With the isolation of totipotent cell lines of cereals, transient and stable transformation has even been achieved with several important grain crop species (Lorz et al. 1985; Potrykus et al. 1985; Rhodes et al. 1988).

Electroporation

A short pulse of high voltage electricity produces transient pores in the plasma membrane, facilitating uptake of macromolecules into the protoplast. This process of DNA uptake is known as electroporation and requires specialized equipment by which field strength and pulse durations can be manipulated. Two alternative protocols have been used for protoplast transformation: a short high voltage pulse, about 1500 V/cm for 10 seconds (Shillito et al. 1985); and a low voltage, long duration pulse, about 350 V/cm for 54 seconds (Fromm et al. 1985). High voltage pulses combined with PEG have been used to improve transformation frequency (Shillito et al. 1985). Stable transformants have been obtained from the electroporation of protoplasts (Fromm et al. 1986; Toriyama et al. 1988). Although general guidelines have been reported, these will vary and optimal electroporation conditions must be determined empirically for different types of protoplasts.

Microinjection

Solutions containing DNA can be directly delivered to individual nuclei by microinjection using fine capillary tubes. However, the protoplasts need to be immobilized by embedding in low melting agarose or by application of slight suction from holding pipettes. Microinjection has been successfully used to obtain stable transformation of tobacco, alfalfa, and rape protoplasts (Crossway et al. 1986; Reich et al. 1986; Neuhaus et al. 1987) and transformation frequencies up to 30% have been reported. Nonetheless, this procedure is very tedious and requires specialized instrumentation and recovery systems for regeneration of protoplasts. Microinjection of plant cells with intact walls has also been reported. Microinjection of *Brassica napus* embryoids derived from microspores (Neuhaus et al. 1987) resulted in an 80% transformation rate and about 50% of such events represented stable transformants.

Liposome fusion

Direct delivery of DNA into protoplasts may be affected by the presence of nucleases in the medium. The use of DNA encapsulated into liposomes, composed of lecithin and cholesterol, for stable transformation of tobacco protoplasts has been documented (Deshayes et al. 1985). This synthetic membrane encapsulation process provides a protective cover for the DNA, preventing nuclease attack. However, the frequency of transformation mediated by liposome is considerably lower than that obtained either by electroporation or by chemical agents. Delivery of DNA to pea at the time of pollination has yielded segregating transformants (Ahokas 1987). In this case, liposome-associated DNA was assumed to be taken up by the pollen at the time of germination, but this has not been verified in other laboratories.

Intact Cell Wall Transformation

Agronomically important crop plants often cannot be regenerated from protoplasts. In such situations, it is desirable to introduce genes into regeneration competent cells or other tissue with morphogenic potential. To accomplish such transformation, DNA must be introduced across the cell wall barrier.

Transformation of intact cells offers several advantages over the use of protoplasts: regeneration capability and viability of the transformed cells are not lost; special protoplasting skills are not required; and transformation can be independent of tissue types. Thus, intact morphogenic tissues in whole plants/seedlings can be targeted, facilitating the task of regenerating new plants. Several new approaches have been attempted to obtain reproducible transformation of intact plant cells.

Biolistics

The recent development of high velocity microprojectiles to deliver DNA into intact plant cells offers enhanced potential for transformation. Because the DNA is shot into the plant cell, it is analogous to ballistics in a non-biological system and thus is referred to as biolistics. This system of transformation is also referred to as the use of a gene gun or particle gun. In general, the technique uses small (0.3-0.5 micron) high density metal (usually tungsten or gold) particles coated with DNA. These microprojectiles are accelerated by the use of a macroprojectile (a large bullet-like object of metal or plastic). This process is performed more efficiently under a slight vacuum. The macroprojectiles can be accelerated to a very high velocity by a variety of means, including rapidly decompressed air or an explosion. When the macroprojectile (carrying the DNA-coated microballistic particles) is accelerated towards the target tissue, it is stopped by a retaining plate, but pores in the plate allow the continued forward movement of the microprojectiles. The microballistic particles penetrate cell walls and may even traverse entire cells, and the DNA coating may be unloaded as the particles pass through nuclei or organelles.

The first demonstration of the use of biolistics in plant transformation was shown in onion epidermal cells (Klein et al. 1987). Since then, a large number of economically important plants have been successfully and stably transformed, such as soybean (McCabe et al. 1988), tobacco (Klein et al. 1988), maize (Fromm et al. 1990; Gordon-Kamm et al. 1990; Spencer et al. 1990), papaya (Fitch et al. 1990), and poplar (McCown et al. 1991).

The transformation of plant cells by biolistics appears to be very promising. This process is independent of tissue type and is simple and rapid. The morphogenic tissues, pollen grains, as well as embryos and meristematic tissues, can all be easily targeted for transformation without the loss in ability to produce regenerated plants. It has also been possible to transform mitochondria and chloroplasts using the biolistic approach (Daniell et al. 1990; Svab et al. 1990). One of the major limitations of biolistics, however, is the need for specialized instrumentation. At present, the best available commercial biolistic system is manufactured by DuPont (USA), and it is expensive. However, several homemade versions of "gene guns" have been developed inexpensively in individual laboratories (McCabe et al. 1988; Morikawa et al. 1989; Oard et al. 1990). While biolistics has broad applications, the transformation frequency and reproducibility of transformation frequency by this approach is still considerably lower than that of protoplast transformation methods. There is a need for refinement and optimization of this process. The use of biolistics has not become a routine transformation method, but with its widespread adoption, improvements in transformation frequency can be expected.

Macroinjection

Hamilton-type microsyringe needles have been used to deliver DNA into wound sites of ovules, embryos, and immature inflorescences. Transgenic progeny of rye were obtained by injecting DNA containing the aminoglycoside phosphotransferase (*aph*) gene under the control of the nopaline synthase promoter into the immature inflorescences (De la Pena et al. 1987). The frequency of transformation was extremely low (two plants out of 3023 seeds). Yet, the transformed rye progeny displayed *aph* activity which confers antibiotic resistance, and the *aph* DNA was detected by Southern hybridization. Although this procedure is fairly inexpensive and simple to perform, reports from other laboratories need to be gathered before this can be judged useful in routine applications.

Pollen tube pathway

Immediately after pollination, pollen tubes have been used as a conduit to deliver *Npt*II DNA under the control of the CaMV 35S promoter into fertilized eggs in rice (Luo and Wu 1988). Transformed progeny with *Npt*II activity have been obtained and the presence of the corresponding DNA has been demonstrated by Southern hybridization. This appears to have potential as an inexpensive and rapid gene transfer technique in a wide variety of plant species but, again, these results have yet to be repeated in other laboratories.

Vortex mixing

Attempts have been made to deliver DNA into cell-suspension cultures of citrus and sweet corn by vortexing DNA and silica fibres together with the cells. Transient expression of beta-glucuronidase activity has been reported in a small number of cells (Kaeppler et al. 1990). There is no evidence of integrative transformation and, as with macroinjection and the pollen tube protocol, this method requires more study before its general usefulness will be known. If integrative transformation can be demonstrated by this method, it will probably be performed inexpensively and routinely on a wide range of plants.

Seed imbibition

Dry seed embryos have been shown to take up plasmid DNA during germination. Such imbibition has produced seedlings of cereals and grain legumes showing transient expression of the *Npt*II gene (Topfer et al. 1989). Again, it is unclear if integrative transformation can be routinely obtained by this method.

Other potential transformation methods

Several other innovative approaches have been attempted to produce transgenic plants, including the use of a laser microbeam (Weber et al. 1988), pollen transformation (Ohta 1986), and electrophoresis of intact cells (Ahokas 1989). Such methods have potential, but at present they are not supported by sufficient data to show transformation beyond a reasonable doubt.

Conclusion

While transformation of intact cells offers great hope for the improvement of important plants, some of these procedures require expensive instrumentation and considerable skill. Also, the transformation frequencies are not generally reproducible. Many of these procedures are not likely to be routinely practised in developing countries because of lack of instrumentation or other resources. However, some of these approaches (such as vortex mixing and the pollen tube procedure) offer hope for routine transformation experiments without the need for sophisticated and expensive laboratory facilities, provided these processes can be optimized to produce stable transformants.

References

Ahokas, H. 1987. Transfection by DNA-associated liposomes evidenced at pea pollination. *Heriditas* 106: 129-138.

Ahokas, H. 1989. Transfection of germinating barley seed electrophoretically with exogenous DNA. *Theoretical and Applied Genetics* 77: 469-472.

Bevan, M., and M.D. Chilton. 1982. T-DNA of *Agrobacterium* Ti and Ri plasmids. *Annual Review of Genetics* 16: 357-384..

Crossway, A., J.V. Oakes, J.M. Irvine, B. Ward, V.C. Knauf, and L.K. Shewmaker. 1986. Integration of foreign DNA following microinjection of tobacco mesophyll protoplasts. *Molecular and General Genetics* 202: 179-185.

Daniell H., J. Vivekananda, B.L. Nielsen, G.N. Ye, K.K. Tewari, and J.C. Sanford. 1990. Transient foreign gene expression in chloroplasts of cultured tobacco cells after biolistic delivery of chloroplast vectors. *Proceedings of the National Academy of Science, USA* 87: 88-92.

Davey, M.R., E.C. Cocking, E.C. Freeman, J. Pearce, and N. Tudor. 1980. Transformation petunia protoplasts by isolated *Agrobacterium* plasmid. *Plant Science Letter* 18: 307-313.

Dawson, W.O., D.J. Lewendowski, M.E. Hilf, P. Bubrick, A.J. Raffo, J.J. Shaw, G.W. Grantham, and P.R. Desjardins. 1989. A tobacco mosaic virus-hybrid expresses and looses an added gene. *Virology* 172: 285-292.

De la Pena, A., J. Lorz, and J. Schell. 1987. Transgenic plants obtained by injecting DNA into young floral tillers. *Nature (London)* 325: 274-276.

Deshayes, A., L. Herrera-Estrella, and M. Caboche. 1985. Liposome-mediated transformation of tobacco mesophyll protoplasts by an *Escherichia coli* plasmid. *EMBO Journal* 4: 2731-2737.

Draper, J., M.R. Davey, J. Freeman, E.C. Cocking, and B.J. Cox. 1982. Ti plasmid homologous sequences present in tissues from *Agrobacterium* plasmid-transformed *Petunia* protoplasts. *Plant Cell Physiology* 23: 451-458.

Fitch, M.M.M., R.M. Manshardt, D. Gonsalves, J.L. Slightom, and J.C. Sanford. 1990. Stable transformation of papaya via microprojectile bombardment. *Plant Cell Reports* 9: 189-194.

Fromm, M.E., L.P. Taylor, and V. Walbot. 1985. Expression of genes transferred into monocot and dicot plant cells by electroporation. *Proceedings of the National Academy of Science, USA* 82: 5824-5828.

Fromm, M.E., L.P. Taylor, and V. Walbot. 1986. Stable transformation of maize after gene transfer by electroporation. *Nature (London)* 319: 791-793.

Fromm, M.E., F. Morrish, C. Armstrong, R. Williams, J. Thomas, and T.M. Klein. 1990. Inheritance and expression of chimeric genes in the progeny of transgenic maize plants. *Biotechnology* 8: 833-839.

Gordon-Kamm, W.J., T.M. Spencer, M.L. Mangano, T.R. Adams, R.J. Daines, W.G. Start, J.V. O'Brien, S.A. Chambers, W.R. Adams Jr., N.G. Willetts, T.B. Rice, C.J. Mackey, R.W. Krueger, A.P. Kausch, and P.G. Lemaux. 1990. Transformation of maize cells and regeneration of fertile transgenic plants. *The Plant Cell* 2: 603-618.

Grimsley, N.H., and D. Bisaro. 1987. Agroinfection. Pages 87-107 in *Plant Gene Research: Plant DNA Infectious Agents* edited by T. Hohn and J. Schell. Springer-Verlag, New York, USA.

Gronenborn, B., and V. Matzeit. 1989. Plant gene vectors and genetic transformation: Plant viruses as vectors. Pages 69-100 in *Cell Culture and Somatic Cell Genetics of Plants* edited by J. Schell and I.K. Vasil. Academic Press, San Diego, California, USA.

Kaeppler, J.F., W. Gu, D.A. Somers, H.W. Rines, and A.F. Cockburn. 1990. Silicon carbide fiber-mediated DNA delivery into plant cells. *Plant Cell Reports* 9: 415-418.

Klein, T.M., E.D. Wolf, R. Wu, and J.C. Sanford. 1987. High-velocity microprojectiles for delivering nucleic acids into living cells. *Nature (London)* 327: 70-73.

Klein, T.M., E.C. Harper, Z. Svab, J.C. Sanford, M.E. Fromm, and P. Maliga. 1988. Stable genetic transformation of intact *Nicotiana* cells by the particle bombardment process. *Proceedings of the National Academy of Science, USA* 85: 8502-8505.

Krens, F.A., L. Molendijk, G.J. Wullems, and R.A. Schilperoort. 1982. In vitro transformation of plant protoplasts with Ti plasmid DNA. *Nature (London)* 296: 72-74.

Lorz, J., B. Baker, and J. Schell. 1985. Gene transfer to cereal cells mediated by protoplast transformation. *Molecular and General Genetics* 199: 178-182.

Luo, Z., and R. Wu. 1988. A simple method for the transformation of rice via the pollen-tube pathway. *Plant Molecular Biology Reporter* 6: 165-174.

McCabe D.E., W.F. Swain, B.J. Martinell, and P. Christou. 1988. Stable transformation of soybean (*Glycine max*) by particle acceleration. *Biotechnology* 6: 923-926.

McCown B.H., D.E. McCabe, D.R. Russell, D.J. Robinson, K.A. Barton, and K.F. Raffa. 1991. Stable transformation of *Populus* and incorporation of pest resistance by electric discharge particle acceleration. *Plant Cell Reports* 9: 590-594.

Morikawa, H., A. Iida, and Y. Yamada. 1989. Transient expression of foreign genes in plant cells and tissues obtained by a simple biolistic device (particle gun). *Applied Microbiology and Biotechnology* 31: 320-322.

Neuhaus, G., G. Spangenberg, O. Mittelsten Scheid, and H-G. Schweiger. 1987. Transgenic rapeseed plants obtained by the microinjection of DNA into microspore-derived embryoids. *Theoretical and Applied Genetics* 75: 30-36.

Oard, J.H., D.F. Paige, J.A. Simmonds, and T.M. Gradziel. 1990. Transient gene expression in maize, rice, and wheat cells using an air gun apparatus. *Plant Physiology* 92: 334-339.

Ohta, U. 1986. High efficiency genetic transformation of maize by a mixture of pollen and exogenous DNA. *Proceedings of the National Academy of Science, USA* 83: 715-719.

Potrykus, I., M.W. Saul, J. Petruska, J. Paszkowski, and R.D. Shillito. 1985. Direct gene transfer to cells of a graminaceous monocot. *Molecular and General Genetics* 199: 183-188.

Reich, T.J., V.N. Iver, and B.L. Miki. 1986. Efficient transformation of alfalfa protoplasts by the intranuclear microinjection of Ti plasmids. *Biotechnology* 4: 1001-1004.
Rhodes, C.A., D.A. Pierce, I.J. Mettler, D. Mascarenhas, and J.J. Detmer. 1988. Genetically transformed maize plants from protoplasts. *Science* 240: 204-207.
Rogers, S.G., R.B. Horsch, and R.T. Fraley. 1986. Gene transfer in plants: Production of transformed plants using Ti plasmid vectors. *Methods in Enzymology* 118: 627-641.
Shillito, R.D., M.W. Saul, J. Paszkowski, M. Muller, and I. Potrykus. 1985. High efficiency direct gene transfer to plants. *Biotechnology* 3: 1099-1103.
Spencer, T.M., W.J. Gordon-Kamm, R.J. Daines, W.G. Start, and P.G. Lemaux. 1990. Bialaphos selection of stable transformants from maize cell culture. *Theoretical and Applied Genetics* 79: 625-631.
Svab Z., P. Hajdukiewicz, and P. Maliga. 1990. Stable transformation of plastids in higher plants. *Proceedings of the National Academy of Science, USA* 87: 8526-8530.
Topfer, R., B. Gronenborn, J. Schell, and H.H. Steinbib. 1989. Uptake and transient expression of chimeric genes in seed-derived embryos. *The Plant Cell* 1: 133-139.
Toriyama, K., Y. Arimoto, H. Uchimiya, and K. Hinata. 1988. Transgenic rice plants after direct gene transfer into protoplasts. *Biotechnology* 6: 1072-1074.
Ward, A., P. Etessami, and J. Stanley. 1988. Expression of a bacterial gene in plants mediated by infectious geminivirus DNA. *EMBO Journal* 7: 1583-1587.
Weber, G., S. Monajembashi, K.O. Greulich, and J. Wolfrum. 1988. Injection of DNA into plant cells with a UV laser microbeam. *Naturwissenschaften* 75: 35-36.

Acknowledgements

The authors wish to thank Dr P. Lemke for his valuable comments on the draft manuscript of this paper.

4.4

Plant viruses as gene vectors

P.G. Markham

Department of Virus Research, The John Innes Centre for Plant Science Research, Colney Lane, Norwich NR4 7UH, UK

Abstract

Plant viruses usually replicate to a high copy number in plant cells and, although they have a limited coding capacity, they all encode one or more enzymes to initiate and control autonomous replication, independent of plant chromosomes. The genome is usually a messenger sense RNA, but several groups of plant viruses have genomes of DNA. DNA viruses usually have a nuclear phase whereas RNA viruses are normally cytoplasmic. Not all viruses are deleterious to the growth of plants and viral genes may provide essential components for designing gene vectors. Plant viruses have already been used as gene vectors to transfer foreign genes into plants. The use of reporter genes has proved useful in establishing the technology, but their true potential will be realized only when desirable genes have been isolated and transferred to the crop of choice. The technique is useful in studying gene expression and in elucidating molecular aspects of viral diseases.

Some plant viruses cause serious crop losses. Control strategies are numerous and may involve, for instance, agronomic changes to avoid exposure of the crop to the virus, selection of different plant genomic backgrounds to resist or ameliorate the replication of the virus, pest management to control a vector or, more likely, several approaches in an integrated control programme. However, not all plant viruses are harmful to their hosts; some, such as the "cryptic" viruses, induce no outward signs of their presence, others induce dramatic colour changes in their hosts but otherwise produce no detrimental effects. These latter plants are often intentionally propagated as ornamentals. The widespread application of molecular biology to virus research has contributed enormously to our understanding of how plant viruses function and interact with their hosts. This knowledge, together with the biological understanding of how these agents interact with their environment, has led not only to ideas for plant virus control but to a more general application of viruses as gene vectors. For example, one of these new approaches for viral control necessitates transferring and integrating, into the chromosomal DNA of a plant, a whole viral genome or part of the virus nucleic acid sequence to protect against another or more severe virus (Baulcombe 1989; Wilson 1989). This biotechnology is just one of the more recent strategies available to the agricultural producer.

Biotechnology in agriculture is aimed not only at the protection and improvement of crop species but also at harnessing the energy conversion systems of plants to produce useful by-products. Inducing plants to produce the desired product usually involves the transfer of a "foreign" gene from another organism or species, be it another species of plant, prokaryote, or animal, and such plants are "transformed" to express the desired gene. Most of the foreign genes so far transferred into plants, which are expressed and have been inherited in a Mendelian fashion, have been genes from other plant species, genes from plant viruses, and bacterial genes encoding enzymes and toxins. However, genes from insects and from higher animals have so far been used only to give transient expression (Weising et al. 1988). There are numerous well-established techniques for plant transformation, including direct gene transfer, particle or ballistic gun, microinjection, protoplast fusion, liposome encapsidation, and *Argobacterium*-mediated transfer; these techniques have been reviewed by Potrykus et al. (1985), Weising et al. (1988) and Ream (1989). This paper focuses on the potential of plant viruses to mediate the transfer of DNA to plants and achieve foreign gene expression.

Useful Characteristics of Plant Viruses

About 700 plant viruses have been identified. Approximately 75% of these viruses have genomes of single-stranded (ss) messenger sense (+)RNA (mRNA) as their genetic material and lack any DNA intermediate in their natural life cycle (Zaitlin and Hull 1987). The largest and economically most important of the 34 plant virus groups and families currently recognized are the RNA potyviruses, with 189 members (Ward and Shukla 1991). In contrast, plant viruses with genomes of DNA are few, the two main groups being the caulimoviruses, with 12 members, and the geminiviruses, with 50 members. Historically, it is the DNA viruses that have received most attention as potential gene vectors (reviewed by Hull and Davies 1983; Davies and Stanley 1989) as DNA is more stable and far less prone to error during replication by DNA-dependent DNA polymerases with proof reading activity. Furthermore, DNA viruses usually have a nuclear phase, whereas RNA viruses are normally cytoplasmic. RNA-RNA replication strategies are highly prone to mutation, so one might expect that a mutant would be quickly lost from a biologically active system. However, since the recent advent of efficient in vitro transcription systems for RNA (Melton et al. 1984), there have been significant successes using RNA viruses as gene vectors (French et al. 1986; Gallie et al. 1987a, 1987b; Takamatsu et al. 1987).

Capsid size and morphology of plant viruses are useful characteristics for dividing viruses into taxonomic groups. However, when the genome is considered as a means of transferring foreign DNA and the intended strategy involves encapsidating the viral DNA, this capsid could prove a limiting factor. Viral genome sizes span a range of about 5×10^3 to 1×10^6 nucleotides, with the plant viral genomes falling mostly in the lower half of the range. These small viral genomes are usually encased in coats of protein or lipoprotein for protection and therefore have a limited coding capacity but must organize their own replication in a living plant cell. Packaging constraints are poorly understood for most plant viruses but isometric particles would have less scope for incorporating a large extra insertion of nucleic acid. The geminiviruses, which are quasi-isometric, consisting of two near-isometric particles associated in a twinned capsid, have a limited flexibility for encapsidating an extended genome, and this is probably restricted to about a 5% increase in the number of nucleotides before unstable multimers (chains of quasi-isometric particles) appear. Such constraints are less likely to occur in the packaging of RNA in rod-shaped particles. Despite the constraints on packaging imposed by capsid size and the limited coding capacity, all plant viruses have one or more viral coded enzymes which initiate and control autonomous replication, independent of plant chromosomes. They usually replicate to high copy number in plant cells, often in excess of 10^7 particles per cell, and will in theory amplify and express an inserted gene.

Advantages and Disadvantages of Plant Viruses as Gene Vectors

The hope of finding plant viruses that could be used as a gene vector, albeit following minor modification, may have been optimistic and was based largely on research into the molecular biology of cauliflower mosaic virus (CaMV), which goes back 15 years or more. CaMV is an isometric virus with a genome of double-stranded DNA, on which at least seven genes are recognized (Covey, 1985). However, only one of these, the gene coding for an 18K protein involved in aphid transmission, is non-essential to the survival of the virus. This gene was first replaced by the bacterial gene dihydrofolate reductase (*dhfr*) in 1984, which conferred resistance to methotrexate in cells infected by the transformed virus (Brisson et al. 1984). A major disadvantage with CaMV is the limited host range of the virus, and this applies to other members of the same group. Nonetheless the 35S CaMV promoter has been used extensively in many experimental situations to drive the expression of foreign genes. Plant viruses potentially provide essential components for specifically designing gene vectors.

One of the most widely used methods of integrating DNA into plants is *Agrobacterium*-mediated transformation, using the plasmids associated with *Agrobacterium* to transfer the DNA and stably incorporate this into plant chromosomes. Viral gene vectors would give cell infections with a high copy number compared to the low number achieved by *Agrobacterium*-mediated transfer.

Agroinoculation is also widely used to transfer viral DNA into a plant to achieve infectivity (Grimsley et al. 1986; Boulton et al. 1989). Viral gene vectors offer a quick method of infecting cells and can be used to target individual plants. Direct infection with viral vectors results in an episomal replication and phenotypic expression. However, it is also possible to construct vectors from *Agrobacterium* plasmids, viral and foreign genes which integrate but then release the replicating viral component carrying the foreign gene (Hayes et al. 1988). A major disadvantage might be the symptoms induced by a viral infection. In the same way that the plasmids of *Agrobacterium* species can be disarmed, it may be possible to ameliorate the effect of the viral genes controlling symptoms, or to select mild or non-symptomatic viruses. In practice, this may not be necessary, as plants transformed with only DNA A of tomato golden mosaic geminivirus (Rogers et al. 1986) or with gene vectors cloned from DNA A (Hayes et al. 1988) are asymptomatic. It has been reported recently that defective viral DNA ameliorates symptoms of geminivirus infection in transgenic plants (Stanley et al. 1990).

Attributes of a Putative Viral Vector

Most of our understanding of the molecular biology of plant DNA viruses comes from studying the caulimoviruses and geminiviruses and their attributes as potential viral vectors depend on an origin of replication which permits autonomous replication in cells (independent of plant chromosomes) to a high copy number. The expression of gene products depends on a strong promoter, and the presence of gene which is either additional to the normal complement or is subsituted for one not essential for the survival of the virus, such as the gene coding for the aphid transmission factor of CaMV or the coat protein gene of cassava mosaic geminivirus (Ward et al.1988). Since the sequencing and manipulation of RNA viruses are more easily accomplished using cDNA, it is also possible to define a possible strategy for a successful gene vector. A theoretical approach might be a transgene between the T-DNA borders of a transfer vector (such as pBin19), which would be consecutively a strong promoter, a foreign gene (cDNA), and a stop codon bounded by specific 5' and 3' genomic sequences. Such a cassette in a plant would probably result in a low copy number, and a low level of expression of the foreign gene in transgenic plants because (+)stranded RNA plant viruses use the (-)strand genome to synthesise subgenomic RNAs from subgenomic promoters (Miller et al. 1985). Therefore, using 5' end of the genomic RNA would presumably result in low levels of expression of the foreign gene in the transgenic plants. A more successful strategy might be to express the transgene from a (-)strand copy of the foreign gene with the viral subgenomic promoter sequence at the 5' end, which should allow tight coupling of the gene expression with provision of the viral polymerase in *trans* following a subsequent infection with the virus.

Conclusion

Cloning DNA and RNA viral fragments is now routine and with the smaller DNA viruses the entire genome may be cloned. The "foreign gene" most commonly used has often been selected from one of the many "reporter" genes (such as neomycin phosphotransferase, chloramphenicol acetyltransferase, β-glucuronidase, and *dhfr*), but the expression of another viral gene, such as the coat protein gene, has been used to show the potential of the viral vector system for studying gene function (Briddon et al. 1990). However, the true potential of the system will be realized only when desirable genes have been isolated and transferred to the crop of choice. Until that time, the use of viruses or components thereof offer the plant molecular biologist an invaluable toolbox for the development of gene vectors and for studying gene expression in plants or cells, as well as for elucidating molecular aspects of virus diseases.

Discussion

OLEMBO: What potential is there for plant transformation methods to be used as a means of production of animal products such as insulins, interferons, etc?

MARKHAM: Theoretically, there seems to be no reason why this should not be possible. Size of the gene may be a problem with current viral vectors, but with the "new" viral vectors based on TMV mentioned by Dr Fauquet, size should not be a limitation. However, I do not hav.e an example where a mammalian gene has been tried yet with a viral gene vector. The attempt to express eukaryotic genes in higher plants has met with mixed success; from more than a dozen examples, transient expression has been achieved in some — for example, firefly luciferase and mouse (*dhfr*) — but in many of the attempts to transfer mammalian genes, there has been either no expression or errors in transcription have occurred.

References

Baulcombe, D. 1989. Strategies for virus resistance. *Trends in Genetics* 5: 56-60.

Boulton, M.I., W.G. Buckholz, M.S. Marks, P. G. Markham, and J.W.Davies. 1989. Specificity of *Argobacterium*-mediated delivery of maize streak virus DNA to members of the Gramineae. *Plant Molecular Biology* 12: 31-40.

Briddon R.W., M.S. Pinner, J. Stanley, and P.G. Markham. 1990. Geminivirus coat protein gene replacement alters insect specificity. *Virology* 177: 85-94.

Brisson, J., J.R. Paszkowski, J.R. Penswick, I. Gronenborn, and T. Holn. 1984. Expression of a bacterial gene in plants by using a viral vector. *Nature (London)* 310: 511-514.

Covey, S.N. 1985.Organisation and expression of the cauliflower mosaic virus genome. Pages 121-159 in *Molecular Plant Virology* (Vol. II) edited by J.W. Davies. CRC Press, Boca Raton, Florida, USA.

Davies, J.W., and J. Stanley. 1989. Geminivirus genes and vectors. *Trends in Genetics* 5: 77-81.

French, R., M. Janda, and P. Ahlquist. 1986. Bacterial gene inserted in an engineered RNA virus: Efficient expression in monocotyledonous plant cells. *Science* 231: 1294-1297.

Gallie, D.R., D.E. Sleat, J.W. Watts., P.C. Turner, and T.M.A. Wilson. 1987a. The 5'leader sequence of tobacco mosaic virus RNA enhances the expression of foreign gene transcripts in vivo and in vitro. *Nucleic Acids Research* 15: 3257-3273.

Gallie, D.R., D.E. Sleat, J.W. Watts., P.C. Turner, and T.M.A. Wilson. 1987b. In vivo uncoating and efficient expression of foreign mRNAs packaged in TMV-like particles. *Science* 236: 1122-1124.

Grimsley, N., B. Holn, T. Holn, and R.M. Walden. 1986. Agroinfection, an alternative route for plant virus infection by using Ti plasmid. *Proceedings of the National Academy of Science, USA* 83: 3282-3286.

Hayes, R.J., I.T.D. Petty, R.H.A. Coutts, and K.W. Buck. 1988. Gene amplification and expression in plants by a replicating geminivirus vector. *Nature (London)* 334: 179-182.

Hull, R., and J.W. Davies. 1983. Genetic engineering with plant viruses, and their potential as vectors. *Advances in Virus Research* 28: 1-33.

Melton, D.A., P.A. Krieg, M.R. Rebagliati, T. Maniatis, K. Zinn, and M.R. Green.1984. Efficient in vitro synthesis of biologically active RNA hybridisation probes from plasmids containing a bacteriophage SP6 promoter. *Nucleic Acids Research* 12: 7035-7056.

Miller, W.A., T.W. Dreher, and T.C. Hall. 1985. Synthesis of brome mosaic virus subgenomic RNA in vitro by internal initiation on (-)sense genomic RNA. *Nature (London)* 313: 68-70.

Potrykus, I., R.G. Shillito, M.W. Saul, and J. Paszkowski. 1985. Direct gene transfer: State of the art and future potential. *Plant Molecular Biology Reporter* 3: 117-128.

Ream, W. 1989. *Agrobacteriun tumefaciens* and interkingdom genetic exchange. *Annual Review of Phytopathology* 27: 583-618.

Rogers, S.G., D.M. Bissaro, R.B. Horsch, R.T. Fraley, N.L. Hoffmann, L. Brand, J.S. Elmer, and A.M. Lloyd. 1986. Tomato golden mosaic virus A component DNA replicates autonomoulsy in transgenic plants. *Cell* 45: 593-600.

Stanley, J., T. Frischmuth, and S. Ellwood. 1990. Defective viral DNA ameliorates symptoms of geminivirus infection in transgenic plants. *Proceedings of the National Academy of Science, USA* 87: 6291-6295.

Takamatsu, N., M. Ishikawa, T. Meshi, and Y. Okada. 1987. Expression of bacterial chloramphenicol acetyltransferase gene in tobacco plants mediated by TMV-RNA. *EMBO Journal* 6: 307-311.

Ward, C.W., and D.D. Shukla. 1991. Taxonomy of potyviruses: Current problems and some solutions. *Intervirology* 32: 269-297.

Ward, A., P. Etessami, and J. Stanley. 1988. Expression of bacterial gene in plants mediated by infectious geminivirus DNA. *EMBO Journal* 7: 1583-1587.

Weising, K., J. Schell, and G. Kahl. 1988. Foreign genes in plants: Transfer, structure, and applications. *Annual Review of Genetics* 22: 421-477.

Wilson, T.M.A. 1989. Plant viruses: A tool-box for genetic engineering and crop protection. *BioEssays* 10: 179-186.

Zaitlin, M., and R. Hull. 1987. Plant virus-host interactions. *Annual Review of Plant Physiology* 38: 291-315.

Acknowledgements

I would like to thank all my colleagues who so generously gave their time to discuss the subject of this paper, and those who loaned the material to make its presentation possible at short notice. I would particularly like to thank Professor J.W. Davies and Drs R.W. Briddon, A.J. Maule, and J. Stanley for their suggestions and help. I would also like to thank the British Council, whose financial support made it possible for me to attend this workshop.

4.5

Gene technology and yam improvement

G. Kahl, J. Ramser, D. Kaemmer, S. Kost, I. Knobloch, R. Rompf, B. HütteL, J. Geistlinger[1], F. Weigand[2] and K. Weising[1]

Department of Biology, University of Frankfurt am Main, Siesmayerstrasse 70, D-W-6000 Frankfurt am Main, Germany[1]; Legume Improvement Program, International Center for Agricultural Research in the Dry Areas, Aleppo, Syria[2]

Abstract

The genetic improvement of yam by conventional breeding methods will increasingly be supported by the application of new techniques such as DNA fingerprinting, defence gene isolation, construction of chimeric genes, and their reintroduction into selected cultivars. The preliminary results of this approach, summarized in this paper, are encouraging.

The quality and yield of yam (*Dioscorea* spp.) in West and Central Africa are affected by viral, bacterial, and fungal diseases. Although a few disease-tolerant selections from local germplasm have been made, the effective breeding of yam using conventional crossing is largely impossible, mainly because yam itself exhibits counterproductive characters. Some 50% of its cultivars do not flower; the ovules and pollen of flowering, normally dioecious, cultivars are only weakly fertile; the small flowers make hand pollination difficult; and synchronization of flowering of both sexes is difficult. Thus, other approaches for improving yam, such as the application of gene technology, have to be adopted. In this paper we suggest a series of experiments, based on successful applications of biotechnology to other crops, which could be applied to yam cultivars.

DNA Fingerprinting

DNA fingerprinting detects a particular class of repetitive DNA in eukaryotic genomes — short sequence motifs that are tandemly arranged to form long arrays. Two characteristics qualify these sequences for DNA fingerprinting: similar motifs are dispersed throughout a particular genome (multilocus appearance); and tandemly arranged repetitive sequences normally possess a high degree of polymorphism (multiallelic appearance). Since the variability is mainly a result of different copy numbers of the basic motifs, such a sequence is called a "variable number of tandem repeats" (VNTRs). This term overlaps somewhat with "minisatellites" (for GC-rich VNTRs of 15-30 base pairs).

DNA fingerprinting starts with the isolation of genomic DNA from target plants, its restriction (fragmentation) using specific restriction endonucleases, the separation of the myriads of fragments by agarose gel electrophoresis, and the hybridization of the resulting gels to a conserved minisatellite core sequence. This procedure detects several variable loci at once and creates a "DNA fingerprint". Such fingerprints usually detect individual-specific patterns in human beings and animal species. In addition, variety- or strain-specific fingerprints are possible with plants and fungi, respectively (Weising and Kahl 1990). Since its introduction into plant biology, some 200 animal and plant species have been successfully fingerprinted with oligonucleotide probes (Epplen et al. 1991; Weising et al. 1991).

This technique has not yet been applied to yam but it has worked with other important crops, including tomato, rapeseed, tobacco, and chickpea (Weising et al. 1989; Weising et al. 1990) and with major fungal pathogens. The couple chickpea and its pathogenic fungus, *Ascochyta rabiei*, illustrates the technique's usefulness. A series of chickpea accessions and single-spored fungal isolates were fingerprinted, with the following results:

- Multiples of the simple sequence motifs CA, CT, GATA, GACA, GTG, GGAT, and TCC are present and repetitive to varying degrees in both organisms.
- The complexity, as well as the informativeness, of the fingerprint patterns depends largely on the particular repeat.
- The optimal combination of probe and species must be determined empirically for each plant or fungus; for example, $(GATA)_4$ and $(GACA)_4$ are promising, informative probes for chickpea and $(GATA)_4$ for the fungus.
- The fingerprints are somatically stable (that is, they are identical in DNA from different organs of one plant).

A major breakthrough has been the development of non-radioactive labelling techniques that allow fingerprinting even in laboratories with no radioisotope facilities, as is the case in many developing countries. One important aspect that should attract considerable interest in the new technology is that the fingerprint patterns are inherited by a co-dominant Mendelian mechanism and segregate in the F_2 generation.

In summary, DNA fingerprinting with simple repetitive sequences offers solutions to many plant breeding problems, such as determining genetic variability in, for example, yam and its major fungal pathogens, selecting doublets in germplasm collections, and cataloguing the aggressiveness of fungal pathogens and crop tolerance or resistance to these fungi. In chickpea, the use of the technique can be expanded to the development of pure chickpea lines, the introgression of resistance traits into high-yielding varieties by backcrossing, and the characterization, and perhaps isolation, of genes coding for fungal resistance.

Characterization and Improvement of Yam Defence Mechanisms

When a plant comes into contact with a fungal pathogen, it develops a series of defence reactions. One of these — the release of preformed chitinases and/or ß-1,3-glucanases from vacuoles or the induced synthesis of both enzymes (gene activation) — allows a barrier to be built against invading fungal hyphae. Since the cell walls of these hyphae in many pathogenic fungi have an inner layer of chitin polymers, surrounded by relatively massive layers of ß-1,3-glucans, the concerted action of both enzymes is thought to destroy fungal hyphae in vivo, as demonstrated in vitro. Yam bulbils and tubers both possess chitinase and ß-1,3-glucanase activity which increases after wounding (a process similar to fungal infection). Why, then, is this type of defence not working in yams? It may be that the induction of both enzymes is too slow, as is the general wound reaction of, for example, yam bulbils (Knobloch et al. 1989). This means that fungi probably overcome yam's defence system simply by growing faster. One approach towards improving the capacity of yam to cope with the fungal attack, using gene technology, would be to:

1. Isolate the chitinase and ß-1,3-glucanase genes from *Dioscorea* cultivars of interest; this is currently done in our laboratory, using genomic libraries of *D. rotundata* and *D. bulbifera*.
2. Isolate regulatory sequences (that is, promoters) of genes from yam or other plants that react rapidly (within minutes) upon wounding or fungal infection. One such genomic clone has been isolated and partly characterized; it reacts upon wounding of small or large white potato (*Solanum tuberosum* L.) tubers with a lag-phase of under 30 minutes.
3. Construct chimeric genes consisting of yam chitinase and ß-1,3-glucanase coding regions and fast wound- or fungus-induced promoters in vitro.
4. Transfer these constructs into yam protoplasts by *Agrobacterium* co-cultivation or direct gene transfer (PEG-mediated DNA uptake, electroporation, or particle-gun techniques). The latter would overcome the serious bottleneck evident in any engineering of yam — its apparent inability to regenerate whole plants from isolated protoplasts.

Agrobacterium-mediated gene transfer into yam bulbil tissue has been demonstrated in our laboratory (Schäfer et al. 1987). It is also possible to transform *D. bulbifera* protoplasts with chimeric plasmids carrying the ß-glucuronidase (*Gus*) reporter gene, which can be easily detected and quantified. A series of plant promoter-*Gus* constructs are being tested in our laboratory for their efficiency in yam protoplasts (transient expression). Moreover, a technique ("protectifer") has recently been developed to allow the protective packaging of genes in vitro before their direct transfer into target cells (Hofmann et al. 1989, in press). This novel method protects the DNA from undesirable breakdown during its passage through the target cell's membranes and cytoplasm, so that at least one intact copy of the gene of interest is integrated into the recipient's genome.

These laborious and time-consuming strategies should not only add to our fragmentary knowledge about yam's defence mechanisms against pathogenic fungi, but should also open up the way for a more selective, more effective, and more specific breeding strategy.

Conclusion

No technology developed in the past 40 years has shown as much promise as biotechnology. Workers in many fields, especially in gene technology and gene transfer into plants, have reported engineered resistance to herbicides, viruses, bacteria, and insects (Weising et al. 1988). Transgenic plants carrying new genes and expressing new traits are now being tested under field conditions. Thus, the "new technology" has already been shown to provide practical tools for plant breeding. The introduction of some of the more relevant techniques into plant breeding strategies is also requested and welcomed by important plant breeding schools (see, for example, Austin et al. 1986). We fully understand, however, that plant breeders have objections and reservations. The mass of new data, the rapid development of new techniques, the acronymic language of molecular biologists, and the massive biotechnological experimentation in an increasing number of industrial countries disturb breeders. In our view, the only way out of this dilemma is cooperation, in which breeders and molecular biologists invest fully their knowledge and capacities for mutual benefit.

Discussion

HAMILTON: What is known about the defence response of yam to fungi other than *Gromerella*?

KAHL: Some 27 fungal species attack yam, but it is not known whether these fungi, except *Glomerella*, contain β-1,3 glucans and chitin as components of their hyphael cell walls.

FAUQUET: What are the promoters "pwc" and "pwp", and are they expressed in yam?

KAHL: "Pwp" stands for promoter wound potato and "pwc" for promoter wound carrot (that is, both promoters are wound-induced).

References

Austin, R.B., R.B. Flavell, J.E. Henson, and H.J.B. Lowe. 1986. *Molecular Biology and Crop Improvement*. Cambridge University Press, Cambridge, UK.

Epplen, J.T., H. Ammer, C. Epplen, C. Kammerbauer, R. Mitreiter, L. Roewer, W. Schwaiger, V. Steimle, H. Zischler, E. Albert, A. Andreas, B. Beyermann, W. Meyer, J. Buitkamp, I. Nanda, M. Schmid, P. Nürnberg, H. Pöche, W. Sprecher, M. Schartl, K. Weising, and A. Yassouridis. 1991. Oligonucleotide fingerprinting using simple repeat motifs: Convenient, ubiquitously applicable method to detect hypervariability for multiple purposes. Pages 50-65 in *DNA Fingerprinting: Approaches and Applications* edited by T. Burke, G. Dolf, A.J. Jeffreys, and R. Wolff. Birkhäuser Verlag, Basel, Germany.

Hofmann, D., H. Zentgraf, and G. Kahl. 1989. In vitro nucleosome assembly with plant histones. *FEBS Letters* 256: 123-127.

Hofmann, D., K. Weising, H. Oelck, G. Donn, and G. Kahl. (in press). Protectifer: A novel technique to package genes for the effective transformation of plants. *Plant Molecular Biology*.

Knobloch, I., G. Kahl, P. Landre, and A. Nougarède. 1989. Cellular events during wound periderm formation in *Dioscorea bulbifera* bulbils. *Canadian Journal of Botany* 67: 3090-3102.

Schäfer, W., A. Görz, and G. Kahl. 1987. T-DNA integration and expression in a monocot crop plant after induction of *Agrobacterium*. *Nature (London)* 327: 529-532.

Weising, K., and G. Kahl. 1990. DNA fingerprinting in plants: The potential of a new method. *Biotech-Forum Europe* 7: 230-235.

Weising, K., J. Schell, and G. Kahl. 1988. Foreign genes in plants: Transfer, structure, expression, and applications. *Annual Review of Genetics* 22: 421-477.

Weising, K., F. Weigand, A. Driesel, G. Kahl, H. Zischler, and J.T. Epplen. 1989. Polymorphic GATA/GACA repeats in plant genomes. *Nucleic Acids Research* 17: 10128.

Weising, K., B. Fiala, K. Ramloch, G. Kahl, and J.T. Epplen. 1990. Oligonucleotide fingerprinting in angiosperms. *Fingerprint News* 2: 5-8.

Weising, K., J. Ramser, D. Kaemmer, G. Kahl, and J.T. Epplen. 1991. Plant DNA fingerprinting with radioactive and digoxigenated oligonucleotide probes complementary to single repetitive DNA sequences. *Electrophoresis* 12: 159-169.

5.1

RFLP technology, crop improvement, and international agriculture[1]

N.D. Young[2], D. Menancio-Hautea[3], C.A. Fatokun[4] and D. Danesh[2]

Department of Plant Pathology, University of Minnesota, St Paul, Minnesota 55108, USA[2]; Institute of Plant Breeding, University of the Philippines, Los Baños, Philippines[3]; Department of Agronomy, University of Ibadan, Ibadan, Nigeria[4]

Abstract

Genetic markers, known as restriction fragment length polymorphisms (RFLPs), are likely to have a major impact on crop improvement. Major genes for disease and pest resistance, as well as ensembles of genes controlling complex traits such as yield and quality, can be tagged with tightly linked RFLPs. Once economically important genes are tagged, individuals that carry these genes can be selected for based on the RFLP genotype. This may be essential when the desired plant phenotype is difficult or impossible to score. Undesirable chromosomal segments from one of the parents can also be selected against. To date, most RFLP research has been carried out in developed countries, but RFLPs are likely to have a major impact in the developing nations as well. Most developing countries already have a scientific foundation on which to build RFLP technology, but the use of RFLPs will require greater efforts to train scientists in RFLP techniques, improved strategies for transferring RFLP technology and equipment, adoption of techniques that do not require the use of radioactive isotopes, and an increased effort to integrate RFLPs into ongoing breeding programmes.

Restriction fragment length polymorphism (RFLP) mapping is a powerful, new tool of biotechnology that can potentially increase the effectiveness and efficiency of plant breeding. Although RFLP maps are being developed for many crops important to industrialized nations, such as maize, wheat, tomato, and soybean, there has been little effort to create RFLP maps for the "orphan" crops (Persley 1990) that are grown primarily in

1 Contribution No. 18 603 from the Minnesota Agricultural Experiment Station, USA (research conducted under Project 015).

developing countries. Moreover, there has been little work on adapting RFLP technology to suit research environments lacking a foundation in molecular genetics, which is the case throughout most of the developing world. This is unfortunate because these crops, and the people they feed, are likely to benefit the most from the application of RFLP technology.

In this paper, we outline potential applications of RFLPs to crop improvement and the conceptual basis of RFLP mapping. Tanksley et al. (1989) give a more thorough review of theory of RFLP mapping and crop improvement. We then focus on economically important genes that have been tagged with RFLPs and how gene-tagging can be used as a basis for improved selection strategies. Lastly, we examine the opportunities and challenges in applying RFLP technology to crop improvement in the international context.

Applications of RFLPs

RFLPs, like other types of genetic markers, can be used to characterize the genotype of an organism. However, RFLPs are superior to previous types of genetic markers for several reasons. Unlike morphological characteristics and isozymes, RFLPs exist in potentially unlimited numbers. They tend to be variable in most crosses, and scoring one RFLP does not interfere with scoring others. Therefore, large numbers of RFLPs can be scored in a single cross. Moreover, RFLP analysis is based on the use of purified DNA, which can be extracted from almost any type of tissue at any stage of growth. Thus, plants can be grown in any environment to provide starting material for RFLP analysis.

The most important application of RFLPs to crop improvement is in selecting individuals that carry genes of economic importance. When a gene of interest can be tagged with a tightly linked RFLP, selection can be based on scoring for the RFLP marker, rather than for the gene itself. If the phenotype is difficult to score, recessive, or impossible to monitor in the presence of other genes, the ability to use RFLPs can provide a powerful selection tool for plant breeding. Moreover, it is possible to use RFLPs to select simultaneously for desired chromosomal regions and against unwanted chromosomal segments, such as those that might be introduced while backcrossing from exotic germplasm (Young and Tanksley 1989c). This may be very important in accessing genes from unadapted varieties or wild relatives, sources considered too difficult to adapt for cultivation in the past.

Another exciting application of RFLP technology is in the analysis of complex polygenic characters controlled by several unlinked genes, which are also known as "quantitative trait loci" (QTLs). Breeding crops for polygenic traits, such as yield, has generally been more difficult than for simple, monogenic characteristics. Now it is possible, using a high density RFLP map, to identify regions of the genome that contain QTLs controlling complex characteristics by simple statistical analysis (Paterson et al. 1988). RFLPs can then simplify the analysis of such traits and provide a direct method for selecting individuals carrying desirable combinations of underlying genes.

In addition to these valuable functions in breeding and mapping, RFLPs are likely to be important to plant genetics in many other ways, such as genotype identification (DNA fingerprinting) and variety protection (Dallas 1988). RFLPs can also be used to characterize natural and controlled mating systems, levels of genotypic diversity, and phylogenetic relationships (Miller and Tanksley 1990). Moreover, they may eventually form the basis for cloning genes that are known only by phenotype (Young 1990). This is important because most current gene cloning strategies depend upon knowledge of the biochemistry of a gene's protein or mRNA product. Using a gene's RFLP map location as a basis for cloning would bypass the need for such detailed (or impossible to obtain) biochemical information.

Conceptual Basis of RFLP Mapping

To generate a map using RFLPs, cloned single copy DNA sequences that are randomly distributed throughout the entire genome are used (Botstein et al. 1980; Tanksley et al. 1989). It is not necessary that the clones come from a library of known genes or even sequences that code for proteins. Any library of single copy sequences will suffice and, indeed, non-coding sequences are sometimes best for RFLP mapping.

To carry out linkage mapping with RFLPs, one begins with an F_2 or backcross population derived from parents with differing genomes. For any region of the genome, alternate DNA sequences (alleles) can often be detected as changes in the lengths of DNA fragments generated by restriction endonuclease digestion (Figure 1 *overleaf*). This is the basis of the term "restriction fragment length polymorphisms" (RFLPs). If a sequence is then examined in the progeny using restriction enzyme analysis, that sequence will appear to segregate for the differing allelic forms of the parents. Thus, it is possible to determine the inheritance pattern for each cloned DNA sequence in the segregating population. Linkage relationships and genetic map locations can then be inferred by comparing segregation patterns for all RFLP markers. If the progeny also segregate for a gene of interest, then the location of each RFLP marker can also be determined relative to that gene (Figure 2 *overleaf*).

Mapping Plant Genes with RFLPs

The starting point for tagging genes of interest with RFLPs is a high density genetic linkage map. Such maps have been generated for many important crops, including maize (Helentjaris 1987), tomato (Tanksley et al. 1988), rice (McCouch et al. 1988), lettuce (Landry et al. 1987), potato (Bonierbale et al. 1988), wheat (Sharp et al. 1989), sorghum (Hulbert et al. 1990), brassica (Slocum et al. 1990), soybean (R.C. Shoemaker, Iowa State University, pers. comm.), and the model plant system, *Arabidopsis* (Chang et al. 1988). For most of these species, over 200 RFLP markers have been mapped, so RFLPs can be found quickly near almost any gene of interest.

Mapping disease resistance genes

Disease resistance genes were the first category of genes to be mapped extensively using RFLPs. This was partly because most characterized disease resistance genes are controlled by alleles at a single locus and most are dominant genes that are easy to assay. In lettuce, for example, RFLPs linked to four genes for resistance to lettuce downy mildew have been identified (Landry et al. 1987). In maize, RFLPs linked to resistance genes to maize dwarf mosaic virus have been mapped (McMullen and Louie 1989; Romero-Severson et al. 1989) and in soybean, RFLPs linked to several different *Phytophthora* resistance genes have been reported (R.C. Shoemaker, pers. comm). The most extensive mapping of disease resistance genes using RFLPs has been carried out in tomato. The genes that have been tagged include: *Tm1* and *Tm2* (resistance to tobacco mosaic virus), *Mi* (resistance to root-knot nematode), I, I2, I3 (resistance to *Fusarium*), Ve (resistance to *Verticillium*) and Pto (resistance to

Figure 1 Conceptual basis of RFLP analysis

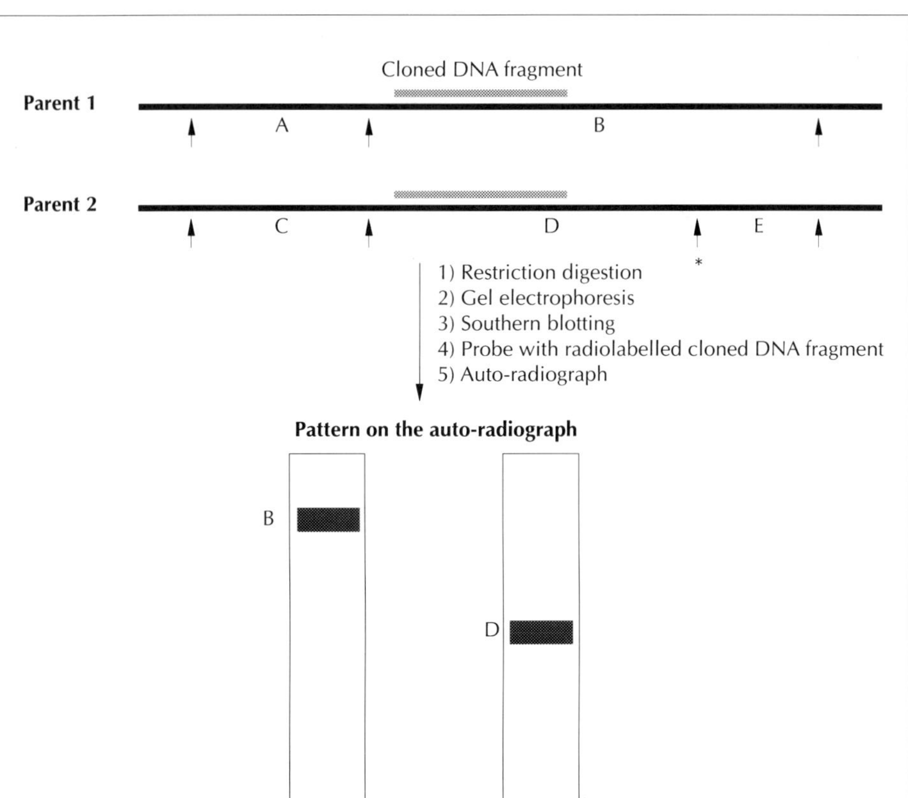

Individuals 1 and 2 are inter-fertile members of the same species or genus. The solid, horizontal lines at the top represent short segments of genomic DNA from these individuals and the arrows beneath the lines indicate restriction endonuclease sites. Since the genomic sequences of both individuals are almost identical, they have most restriction endonuclease sites in common. Occasionally, however, differences in DNA sequence lead to differences in restriction sites, denoted by an asterisk (*). Restriction enzyme digestion produces DNA fragments A to E. If these fragments are fractionated by gel electrophoresis, Southern blotted, probed with the cloned DNA sequence (grey lines), and auto-radiographed, only fragments B and D will be visible. These DNA fragments differ in size because of the new restriction site * in parent 2, and therefore migrate to different positions during electrophoresis, indicating a restriction fragment length polymorphism (RFLP).

Pseudomonas) (Young et al. 1988; Tanksley, unpubl.). Efforts are now under way to use tightly linked RFLPs to pyramid all these tomato resistance genes into a single line as quickly as possible (S.D. Tanksley, pers. comm.).

Mapping QTLs with RFLPs

While single locus, major disease resistance genes are important in crop improvement, many traits of economic importance are under the control of multiple genes that work in

Figure 2 Linkage mapping with RFLPs

	Parents		Backcross progeny							
	A	B	1	2	3	4	5	6	7	8
RFLP 1										
RFLP 2										
Disease R x n	-	+	-	+	+	-	+	+	+	-

The illustration shows the results for eight hypothetical backcross progeny from a population scored for disease resistance and two RFLP markers (1 and 2). The bands beside RFLP 1 and RFLP 2 are patterns which might be observed after probing Southern blots of the parents and backcross progeny with these RFLP markers. The results suggest that the gene controlling disease resistance is closely linked to RFLP 2 but is not linked to RFLP 1. This is indicated by the fact that there are no recombinants between the resistance phenotype and the lower band of RFLP 2. In a typical RFLP mapping project, a hundred or more progeny individuals might be scored in terms of RFLPs.

concert to produce a phenotype. Examples of such traits are overall yield, stress tolerance, and nutritional quality. One of the major initiatives in RFLP mapping is mapping the QTLs underlying such quantitatively controlled traits.

To map QTLs, a population derived from parents that differ significantly for the trait of interest is analysed with RFLPs, as described earlier. Each marker can then be tested for the likelihood that it is linked to a QTL with a significant effect on the trait of interest. In the simplest case (such as a backcross), the population can be divided into individuals that are homozygous for the RFLP allele of the recurrent parent and those that are heterozygous (RFLP alleles from both parents). The phenotypic values of these subpopulations can then be compared by analysis of variance (or regression). If there is a significant difference between the subpopulations, the RFLP may be linked to a QTL. This analysis is repeated for RFLPs distributed over the entire genome. A more rigorous statistical approach to linking QTLs to RFLPs (LOD analysis) has also been developed (Lander and Botstein 1989).

Mapping QTLs with RFLPs has now been carried out in several different systems. One example is soluble solids in tomato fruit (Paterson et al. 1988) where RFLPs linked to four QTLs controlling 44% of the phenotype were identified. In a similar study, RFLPs were used to map QTLs controlling water-use efficiency in tomato (Martin et al. 1989). In this study, RFLPs were found to be linked to three major QTLs controlling this trait. RFLPs have also been used to tag QTLs controlling the production of 2-tridecanone and density of glandular trichomes in tomato, traits that appear to be important in insect resistance (Nienhuis et al. 1987). RFLPs linked to three QTLs with major effects were uncovered and, together, they could account for 38% of 2-tridecanone production. Five independent RFLPs were found to be associated with QTLs controlling hard seededness in soybean. Together, the RFLPs could account for 71% of the phenotype of this trait (Keim et al. 1990).

Marker-Assisted Selection

In plant improvement, one of themain reasons for tagging economically important genes with RFLPs is marker-assisted selection. When a gene is mapped relative to RFLPs, selection for that gene can be based on tightly linked RFLPs, rather than phenotypic scoring. This can be extremely valuable if a trait is recessive, difficult to assay, or obscured by the expression of other traits. However, RFLPs can also be used to select against unwanted DNA from one of the parents. In a recurrent backcross breeding programme, for example, the goal might be the development of a line whose genome is predominantly that of the recurrent parent, with a small chromosomal region carrying a gene of interest retained from the donor parent. Backcross breeding programmes typically continue for at least six or more generations. Figure 3 illustrates how RFLP selection for gene(s) of interest, coupled with simultaneous selection against the genome of the donor parent, can accelerate this process significantly.

RFLPs accelerate the removal of DNA from a donor parent by selecting for individuals with rare recombination events around genes of interest, as well as those that have the smallest amount of donor DNA on unlinked chromosomes (Young and Tanksley 1989c). Computer programmes have been written to optimize the use of RFLPs in this process (Young and Tanksley 1989b). Simulations using this software indicate that a backcross breeding programme for a trait controlled by two or three genes proceeds four times faster using RFLPs than in traditional backcross breeding (Young and Tanksley 1989a).

Figure 3 RFLP-assisted backcross breeding

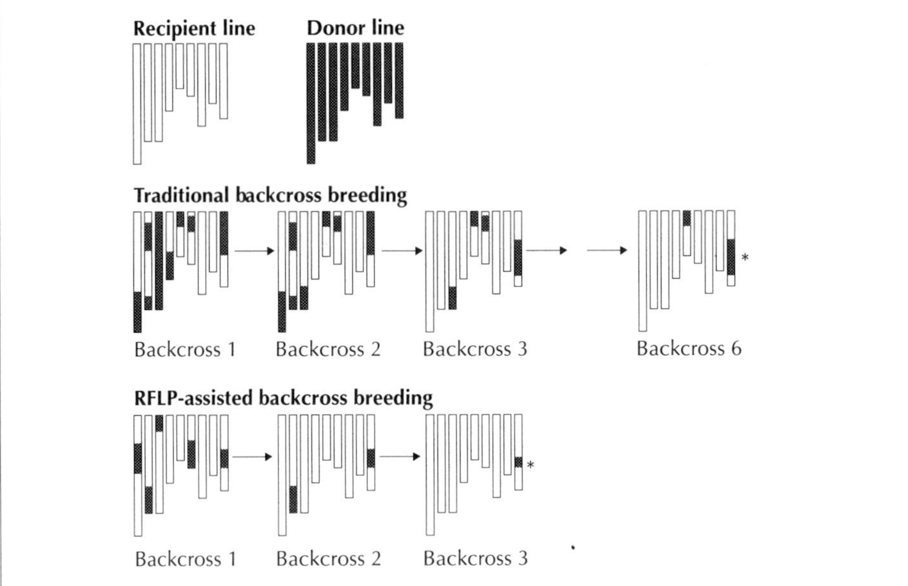

The illustration shows the increased efficiency of backcross breeding when RFLPs are used as the basis of selection. Given a medium density map of RFLPs located throughout the genome, RFLPs can be used to select for desirable crossover events near a target gene (labelled *), as well as for recovery of recurrent parent genotype on chromosomes that are not linked. Thus, an individual with a minimum amount of donor DNA can be generated in a shorter time using RFLPs than in traditional backcross breeding.

RFLP Technology in Developing Countries

Clearly, RFLP technology is beginning to have a major impact on plant improvement. However, the bulk of RFLP research has been, and continues to be, carried out in developed countries. In less developed countries there have been far fewer research or breeding projects utilizing RFLPs, although crop improvement in developing countries is likely to benefit immensely from this technology.

Opportunities provided by RFLPs

For developing countries, RFLP-assisted breeding offers many advantages over classical crop breeding strategies and even over other tools of biotechnology. RFLP technology does not require actual genetic engineering of plants, which means that RFLPs represent an "off-the-shelf" technology that does not depend upon introducing foreign genetic material into plants. Instead, RFLPs can be used to improve crops by assembling combinations of desirable genes in a shorter time compared to classical breeding, so that new varieties can be developed much faster. This will be particularly important in low-input agricultural systems where insect, microbial, and weed pests are major problems. Given the obvious need, in terms of cost and environmental impact, to reduce chemical use, RFLPs offer important opportunities to ameliorate plant disease problems. Moreover, as noted above, it is possible to use RFLPs to discern the genetic basis of complex traits. These include some of the characteristics so critical to developing countries, such as yield and stress tolerance. Using RFLPs, QTLs underlying these complex traits can be treated as simple Mendelian genes that can be manipulated far more easily in ongoing breeding programmes.

It is also important to note that developing countries grow many crop species and, often, many varieties of each crop. These countries also span nearly all possible environments, many of them far from ideal for farming. Lastly, many crop species grown in developing countries have long generation times, such as banana, cassava, yam, oil palm, and fruit trees. For all these reasons, RFLPs, with their ability to speed selection in plant breeding and shorten the breeding process, will have a major impact on variety development.

Developing countries tend to be the centres of diversity of many of the world's most important plant species. As such, they have a special role in managing world germplasm collections, and RFLPs can be a valuable component of germplasm management. Using RFLPs, accessions that are duplicates can be detected and eliminated, freeing up valuable resources. Genetic characteristics of accessions can also be catalogued and unwanted outcrossing or narrowing due to repeated seed increases monitored.

Impediments to RFLP technology

Despite the enormous promise of RFLP technology, there are many reasons why the developing world has been slow to adopt it. These problems must be addressed before RFLP technology has a significant impact outside developed countries. First of all, breeding programmes, which form the basis for implementing RFLP technology, are not well established in some developing countries and may be totally lacking for some crop species. Research projects also frequently lack efficient procurement systems and thus perishable materials, which are common in RFLP research, may often be useless before they are delivered. Moreover, RFLP mapping is currently based on the use of radioactive compounds and many countries around the world lack suitable storage and waste management systems for these potentially dangerous materials.

Just as important as technical difficulties in developing countries is limited access to information. In the early 1980s, soon after RFLPs were described, there was little awareness of the power of RFLP selection in plant breeding. This has changed, and RFLPs are now recognized worldwide as a major tool for crop improvement. Nevertheless, developing countries often have little or no access to information on new research developments and innovations. Research laboratories in the developing world frequently lack current research journals (which are often prohibitively expensive) and have poor communication facilities. However, even if this situation could be remedied, it should be noted that much of the research on RFLPs, even in industrialized countries, is carried out by private companies and therefore unavailable to research scientists at universities and national programmes.

Setting priorities

Despite these problems, many developing countries have a research base upon which to build RFLP technology. Facilities for plant tissue culture, for example, have been in place throughout the world for many years. Similarly, laboratories in many countries routinely carry out isozyme analysis and antibody-based diagnosis for plant diseases. In many ways, RFLP technology can expand from these already established resources. Moreover, the most important ingredient for implementing RFLP technology — ongoing plant breeding programmes — is in place for many important crop species.

Yet it is worthwhile to ask whether RFLP maps should be constructed for every crop species cultivated in the world. Many species are not well characterized in terms of classical genetics and in some cases breeding programmes are extremely limited and poorly funded. Given the relatively high expense of RFLP technology (Beckmann and Soller 1983), is it practical or even desirable to develop RFLP maps for these crops? Scientists, national governments, international and regional research centres, and funding agencies must set clear priorities on which systems merit significant efforts.

Implementing RFLP technology throughout the world will also require structural changes. There must be free and open exchange of plant materials, germplasm, and cloned RFLP markers, and more open and rapid exchange of information. The cost of RFLPs must be significantly reduced, perhaps with the use of the DNA amplification polymerase chain reaction (Saiki et al. 1988). At the very least, systems for networking among programmes at international and regional research centres, national institutes, and universities should be strengthened. International and regional centres, being generally more advanced in terms of research expertise, can also play an important role in training scientists from national programmes and universities. Scientists from advanced countries must also be recruited and encouraged to collaborate with those in developing nations, as a means of implementing RFLP technology in the short term, as well as transferring RFLP technology to scientists in developing countries for long-term research. Lastly, private companies must be given incentives for investing in RFLPs and other types of biotechnology research throughout the world. Private companies should be made aware of the benefits of getting a "foot in the door" by investing in developing countries. To encourage such investment, developing countries must also set clear laws and regulations on the use of biotechnology.

Conclusion

Recent advances in biotechnology offer many powerful new tools for crop improvement. Scientists are using cloned viral coat proteins to protect plants from viral pathogens (Abel et al. 1986) and cloned copies of the *Bacillus thuringiensis* toxin protein to protect them from insect pests (Vaeck et al. 1987). These, and many other applications of biotechnology, will have a major impact on crop improvement in the future. Nevertheless, they still await effective DNA transformation systems for most important crops, not to mention thorough studies of the environmental impact of releasing transgenic plants (and microorganisms). By contrast, RFLP technology brings together molecular genetics and classical plant breeding without the need, or risk, of transgenic organisms in the environment. It is in place now and the only things needed to make it generally applicable are non-radioactive probing techniques and lower costs. All countries, even those with a limited biotechnological base, should take advantage of RFLPs and marker-based breeding programmes. The benefits, in terms of faster and better crop improvement, are likely to be enormous.

References

Abel, P., R.S. Nelson, B. De, N. Hoffman, S.G. Rogers, R.T. Fraley, and R.N. Beachy. 1986. Delay of disease development in transgenic plants that express the tobacco mosaic virus coat protein gene. *Science* 232: 738-743.

Beckmann, J.S., and J. Soller. 1983. Restriction fragment length polymorphisms in genetic improvement: Methodologies, mapping and costs. *Theoretical and Applied Genetics* 67: 35-43.

Bonierbale, M.W., R.L. Plaisted, and S.D. Tanksley. 1988. RFLP maps based on common set of clones reveal modes of chromosomal evolution in potato and tomato. *Genetics* 120: 1095-1103.

Botstein, D., R.L. White, M. Skolnick, and R.W. Davis. 1980. Construction of a genetic linkage map in man using restriction fragment length polymorphisms. *American Journal of Human Genetics* 32: 314-331.

Chang, C., J.C. Bowman, A.W. DeJohn, E.S. Lander, and E.S. Meyerowitz. 1988. Restriction fragment length polymorphism linkage map for *Arabidopsis thaliana*. *Proceedings of the National Academy of Science, USA* 85: 6856-6860.

Dallas, J.F. 1988. Detection of DNA fingerprints of cultivated rice by hybridization with a human minisatellite DNA probe. *Proceedings of the National Academy of Science, USA* 85: 6831-6835.

Helentjaris, T. 1987. A genetic linkage map for maize based on RFLPs. *Trends in Genetics* 3: 217-221.

Hulbert, S.H., T.E. Richter, J.D. Axtell, and J.L. Bennetzen. 1990. Genetic mapping and characterization of sorghum and related crops by means of maize DNA probes. *Proceedings of the National Academy of Science, USA* 87: 4251-4255.

Keim, P., B.W. Diers, and R.C. Shoemaker. 1990. Genetic analysis of soybean hard seededness with molecular markers. *Theoretical and Applied Genetics* 79: 465-469.

Lander, E.S., and D. Botstein. 1989. Mapping Mendelian factors underlying quantitative traits using RFLPs linkage maps. *Genetics* 121: 185-199.

Landry, B.S., R.V. Kesseli, B. Farrara, and R.W. Michelmore. 1987. A genetic map of lettuce (*Lactuca sativa* L.) with restriction fragment length polymorphism, isozyme, disease resistance and morphological markers. *Genetics* 116: 331-337.

Martin, B., J. Nienhuis, G. King, and A. Schaefer. 1989. Restriction fragment length polymorphisms associated with water use efficiency in tomato. *Science* 243: 1725-1728.

McCouch, S.R., G. Kochert, Z. H. Yu, Z.Y. Wang, G.S. Khush, W.R. Coffman, and S.D. Tanksley. 1988. Molecular mapping of rice chromosomes. *Theoretical and Applied Genetics* 76: 815-829.

McMullen, M., and R. Louie. 1989. The linkage of molecular markers to a gene controlling the symptom response in maize to maize dwarf mosaic virus. *Molecular Plant-Microbe Interactions* 2: 309-314.

Miller, J.C., and S.D. Tanksley. 1990. RFLP analysis of phylogenetic relationships and genetic variation in the genus *Lycopersicon. Theoretical and Applied Genetics* 80: 437-448.

Nienhuis, J. T., T. Helentjaris, M. Slocum, B. Ruggero, and A. Schaefer. 1987. RFLP analysis of loci associated with insect resistance in tomato. *Crop Science* 27: 297.

Paterson, A.H., E.S. Lander, J.D. Hewitt, S. Peterson, S.E. Lincoln, and S.D. Tanksley. 1988. Resolution of quantitative traits into Mendelian factors by using a complete linkage map of restriction fragment length polymorphisms. *Nature (London)* 335: 721-726.

Persely, G.J. 1990. *Beyond Mendel's Garden: Biotechnology in the Service of World Agriculture.* CAB International, Wallingford, UK.

Romero-Severson, J., J. Lotzer, C. Brown, and M. Murray. 1989. Use of RFLPs in analysis of quantitative trait loci in maize. Pages 97-102 in *Current Communications in Molecular Biology: Development and Application of Molecular Markers to Problems in Plant Genetics* edited by T. Helentjaris, and B. Burr. Cold Spring Harbor Press, Cold Spring Harbor, New York, USA.

Saiki, R.K., D.H. Gelfand, S. Stoffel, S. Scharf, R.H. Higuchi, G.T. Horn, K.B. Mullis, and H.A. Erlich. 1988. Primer-directed enzymatic amplification of DNA with a thermostable DNA polymerase. *Science* 239: 487-491.

Sharp, P.J., S. Chao, S. Desai, and M. Gale. 1989. The isolation, characterization and application in the Triticeae of a set of wheat RFLP probes identifying each homeologous chromosome arm. *Theoretical and Applied Genetics* 78: 342-348.

Slocum, M.K., S.S. Figodre, W.C. Kennard, J.Y. Suzuki, and T.C. Osborn. 1990. Linkage arrangement of restriction fragment length polymorphic loci in *Brassica oleracea. Theoretical and Applied Genetics* 80: 57-64.

Tanksley, S.D., J.C. Miller, A.H. Paterson, and R. Bernatzky. 1988. Molecular mapping of plant chromosomes. Pages 157-173 in *Chromosome Structure and Function* edited by J.F. Gustafson, and R. Appels. Plenum Press, New York, USA.

Tanksley, S.D., N.D. Young, A. Paterson, and M. Bonierbale. 1989. RFLP mapping in plant breeding: New tools for an old science. *Bio/technology* 7: 257-264.

Vaeck, M., A. Reynaerts, H. Höfte, S. Jansens, M. DeBeuckeleer, C. Dean, M. Zabeau, M. Van Montagu, and J. Leemans. 1987. Transgenic plants protected from insect attack. *Nature (London)* 328: 33-37.

Young, N.D. 1990. Potential applications of map-based cloning to plant pathology. *Physiological and Molecular Plant Pathology* 37: 81-94.

Young, N.D., and S.D. Tanksley. 1989a. Graphics-based whole genome selection using RFLPs. Pages 123-129 in *Current Communications in Molecular Biology: Development and Application of Molecular Markers to Problems in Plant Genetics* edited by T. Helentjaris, and B. Burr. Cold Spring Harbor Press, Cold Spring Harbor, New York, USA.

Young, N.D., and S.D. Tanksley. 1989b. Restriction fragment length polymorphism maps and the concept of graphical genotypes. *Theoretical and Applied Genetics* 77: 95-101.

Young, N.D., and S.D. Tanksley. 1989c. RFLP analysis of the size of chromosomal segments retained around the *Tm-2* locus of tomato during backcross breeding. *Theoretical and Applied Genetics* 77: 353-359.

Young, N.D., D. Zamir, M.Ganal, and S.D. Tanksley. 1988. Use of isogenic lines and simultaneous probing to identify DNA markers tightly linked to the *Tm-2a* gene in tomato. *Genetics* 120: 579-585.

Acknowledgements

The authors wish to thank Dr B. Lockhart for his valuable comments on the draft of this paper. The research was supported, in part, by a grant from the Rockefeller Foundation and by GAR funds.

5.2

Assessing relationships between *Musa* species using chloroplast DNA RFLPs

N.J. Gawel[1] and R.L. Jarret[2]

Department of Horticulture, University of Georgia, 1109 Experiment Street, Griffin, Georgia 30223, USA[1]; Southern Regional Plant Introduction Station, USDA/ARS, 1109 Experiment Street, Griffin, Georgia 30223, USA[2]

Abstract

Taxonomic and phylogenetic determinations within the genus *Musa* have been established using a numerical, morphology-based scoring system. Within this system, however, the classification and relationships of some types are disputed. The application of chloroplast DNA (cpDNA) restriction fragment length polymorphism (RFLP) analysis to *Musa* taxonomy could provide valuable supplementary information about the classification of, and relationships between, *Musa* species and subspecies. Whole-cell DNA was extracted from lyophilized leaf-blade tissue and digested with various restriction enzymes, Southern blotted onto nylon membranes, and probed using radioactively labelled heterologous orchid cpDNA fragments. Phylogenies were inferred from cpDNA RFLP patterns, using Phylogenetic Analysis Using Parsimony (PAUP) software. The relationships between most species examined were as expected; however, some species, particularly *M. beccarii* and *M. basjoo*, did not conform to conventional morphology-based phylogeny.

The genus *Musa* is a vital source of food and fibre in many parts of the world; over 100 million people depend upon bananas and plantains as their principal source of carbohydrates (Rowe 1981). Yields of bananas and plantains have declined because of the spread of black sigatoka, a leaf spot disease caused by *Mycosphaerella fijiensis*. As a result, attention has become focused on the collection and conservation of *Musa* germplasm. In order to ensure that available funding is used effectively for germplasm conservation, it is essential that genetic diversity in this genus be accurately identified.

The efficient conservation of *Musa* germplasm depends upon an accurate knowledge of the degree of genetic relatedness between clones and the range of diversity present. In lieu of analysis of morphological characteristics, laboratory techniques are becoming increasingly popular. We describe here the use of heterologous chloroplast DNA (cpDNA) probes to detect cytoplasmic diversity and infer relationships between selected *Musa* species. We suggest that these, and related techniques, are likely to yield important data relative to the evolution of bananas and plantains.

Materials and Methods

DNA extraction

DNA extraction, digestion, electrophoresis, blotting, and hybridization were performed as previously described (Gawel and Jarret 1991). DNA extracts (8 µg) were digested with 20-30U of restriction enzyme (*Eco*RI, *Hind*III, *Pst*I or *Bam*HI) for 12 hours at 37°C.

Chloroplast DNA probes

Probes to determine cytoplasmic diversity were isolated from the lettuce (*Lactuca sativa*) chloroplast genome: J2, a 9.9 kb sequence from the 16S rRNA gene; J4, a 1.8 kb fragment of the 23S rRNA gene; and J12, a 10.6 kb fragment containing the large subunit of the RUBP carboxylase gene. Probes 6b, 10a, and 16 are 6.7 kb, 5.7 kb, and 4.0 kb fragments, respectively, and were isolated from an orchid (*Oncidium excavatum*) cpDNA library. Lettuce probes were obtained from Dr M. Chase (University of North Carolina, USA), and orchid probes from Dr R. Jansen (University of Connecticut, USA).

Approximately 0.5 µg of entire plasmid (containing probe insert) was labelled and used as a probe in each hybridization reaction. Each probe was random primer labelled with 70uCi 32P; intensity of incorporation was 4-8 x 10^7cpm.

Fragment positions on autoradiographs were represented as 1-0 matrices, and analysed with Phylogenetic Analysis Using Parsimony (PAUP) software (Swofford 1985). Strategy options used were SWAP = ALT, HOLD = 25, and MULPARS (max. = 100). The CONTREE programme supplied with the PAUP software was used to construct a consensus tree from all the equally parsimonious trees generated by PAUP.

Plant materials

Plant materials were obtained from Dr P. Rowe, Fundacion Hondureana de Investigacion Agricola (FHIA), La Lima, Honduras; Dr R. Ploetz, University of Florida, Homestead; and through the International Network for the Improvement of Bananas and Plantains (INIBAP) germplasm transit centre at Katholieke Universiteit Leuven, Belgium.

Results and Discussion

Species, subspecies, section, and number of chromosomes of each clone examined are presented in Table 1. Not every probe/restriction enzyme combination produced discernable polymorphisms between the clones examined.

Table 1 ***Musa* species examined using chloroplast DNA RFLPs**

Species	Section	*n*
M. balbisiana	Eumusa	11
M. basjoo	Eumusa	11
M. beccarii	Callimusa	9
M. coccinea	Callimusa	10
M. ornata	Rhodochlamys	11
M. textilis	Australimusa	10
M. velutina	Rhodochlamys	11
M. acuminata ssp. *banksii*	Eumusa	11
M. acuminata ssp. *burmannica*	Eumusa	11
M. acuminata ssp. *malaccensis*	Eumusa	11
M. acuminata ssp. *microcarpa*	Eumusa	11
M. acuminata ssp. *siamea*	Eumusa	11
M. acuminata ssp. *truncata*	Eumusa	11

Although a number of equally parsimonious trees were generated by PAUP, most of the differences between trees were due to rearrangements between and within the *M. acuminata* subspecies complex. The consensus dendrogram (Figure 1 *overleaf*) produced from the cpDNA RFLP data illustrates the differences in cpDNA between and within *Musa* species. In general, the relationships depicted in Figure 1 agree with those based on morphological data (Simmonds and Weatherup 1990). However, some do not. One explanation for this discrepancy may be that the dendrogram is based exclusively on cpDNA-derived data, while morphological characteristics are based on the expression of nuclear genes and nuclear x cytoplasmic gene interactions.

The branching pattern in Figure 1 illustrates that the *M. acuminata* subspecies clustered as expected. Within this cluster, however, cpDNA variability is evident not only between subspecies but also within subspecies; two different clones each of *M. acuminata* ssp. *truncata* and *M. acuminata* ssp. *siamea* were examined and in neither case did the clones have identical cpDNA. Intrasubspecific variability has also been documented using morphological characteristics (Simmonds 1966). As shown in Figure 1, *M. acuminata* ssp. *banksii* is placed within the *M. acuminata* subspecies cluster, supporting the assertion that it should be classified as a subspecies of *M. acuminata* and not as separate species as suggested by Argent (1976). Perhaps the use of additional probe/enzyme combinations will clarify this situation.

Although classified as a "relic species" (Simmonds 1962), the cytoplasm of *M. beccarii* was determined as being more like that of the *M. acuminata* subspecies complex than the other species examined. Morphologically, *M. beccarii* and *M. acuminata* are very distinct; they have different basic chromosome numbers ($n = 9$ and $n = 11$, respectively), they are believed to have evolved along different paths, and they are not known to hybridize (Simmonds 1962). The similarity in cpDNAmay be due to a common ancestor, or it may suggest that *M. beccarii* is aneuploid or the product of interspecific hybridization. Shepherd (1959) documents multivalent formation in meiosis of *M. beccarii*.

Figure 1 **Consensus dendrogram of *Musa* species and *M. acuminata* subspecies, based upon chloroplast DNA (cpDNA) RFLP data**

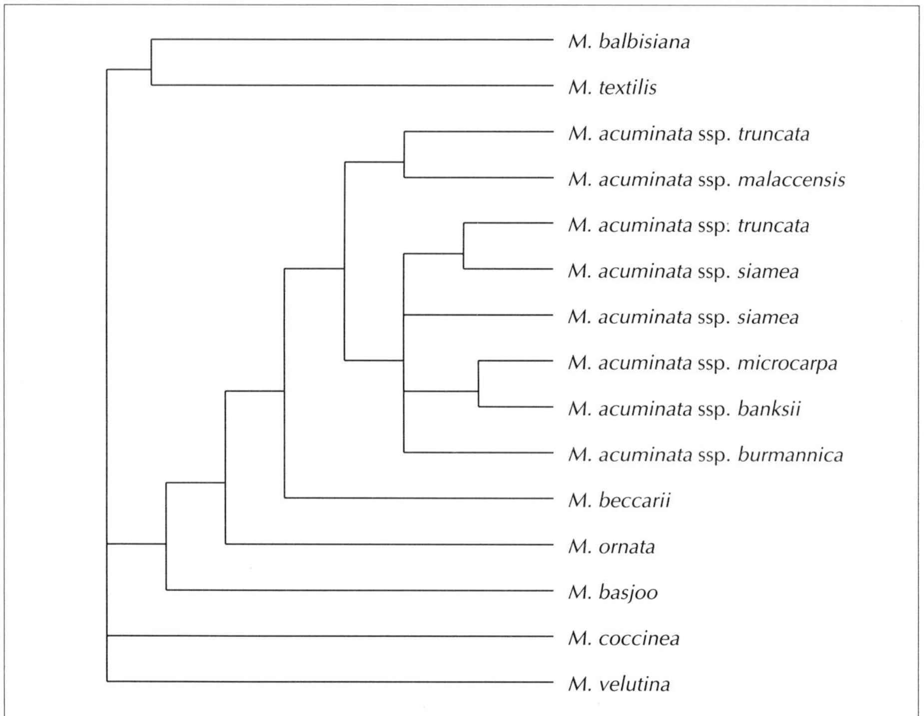

M. balbisiana and *M. textilis* also belong to different sections and have different chromosome numbers ($n = 11$ and $n = 10$, respectively). Despite these differences, these species have enough genetic similarity to readily hybridize and produce the fibre-yielding clone "Canton" (Simmonds 1966), and enough morphological similarity to be classified as closely related (Simmonds and Weatherup 1990). These similarities are reflected in the cytoplasmic relationship depicted above in Figure 1; our analysis shows the cytoplasm of *M. textilis* to be fairly close to that of *M. balbisiana*.

Numerical taxonomic analysis, based on morphological attributes (Simmonds and Weatherup 1990), places *M. basjoo* much closer to the *M. acuminata* subspecies complex than does the cpDNA-based analysis. Again, this discrepancy may be due to the basis of classification (morphological vs cytoplasmic characteristics). The relative placements of the remaining species, *M. ornata*, *M. coccinea*, and *M. velutina*, are in general agreement with those based on morphological characteristics.

The data presented here demonstrate the use of cpDNA RFLPs for phylogenetic analysis of *Musa* species and subspecies. With a few exceptions, the results are in general agreement with previously published data. Some species (*M. beccarii* and *M. basjoo*) were grouped quite differently than expected, suggesting the need for further investigation. Perhaps the use of a greater number of probe-enzyme combinations, or the analysis of genomic or mitochondrial DNA diversity, would help clarify these relationships.

References

Argent, C.G.C. 1976. The wild bananas of Papua New Guinea. *Notes of the Royal Botanic Gardens (Edinburgh)* 35: 77-114.

Gawel, N., and R.L. Jarret. 1991. Cytoplasmic genetic diversity in bananas and plantains. *Euphytica* 52: 19-23.

Rowe, P.R. 1981. Breeding an intractable crop: Bananas. Pages 66-84 in *Genetic Engineering for Crop Improvement* edited by K.O. Rachie and J.M. Lyman. Working Papers, Rockefeller Foundation, New York, USA.

Saghai-Maroof, M.A., K.M. Soliman, R.A. Jorgensen, and R.W. Allard. 1984. Ribosomal DNA spacer-length polymorphisms in barley: Mendelian inheritance, chromosomal location, and population dynamics. *Proceedings of the National Academy of Science, USA* 81: 8014-8018.

Shepherd, K. 1959. Two new basic chromosome numbers in Musaceae. *Nature (London)* 183: 1539.

Simmonds, N.W. 1962. *The Evolution of the Banana*. Longman, London, UK.

Simmonds, N.W. 1966. *Bananas* (2nd edn). Longman, London, UK.

Simmonds, N.W., and S.T.C. Weatherup. 1990. Numerical taxonomy of the cultivated bananas. *New Phytology* 11: 567-571.

Swofford, D.L. 1985. *PAUP-Phylogenetic Analysis using Parsimony, Version 2.4.0*. Illinois Natural History Survey, Champaign, IL, USA.

Acknowledgements

This research was funded by the International Board for Plant Genetic Resources (IBPGR), whose support is gratefully acknowledged.

5.3

RFLP mapping in cowpea[1]

N.D. Young[2], C.A. Fatokun[3], D. Danesh[2] and D. Menancio-Hautea[4]

Department of Plant Pathology, University of Minnesota, St Paul, Minnesota 55108, USA[2]; Department of Agronomy, University of Ibadan, Ibadan, Nigeria[3]; Institute of Plant Breeding, University of the Philippines, Los Baños, the Philippines[4]

Abstract

Linkage maps based on restriction fragment length polymorphisms (RFLPs) are beginning to have a major impact on crop improvement. Recognizing the many important applications of RFLPs in plant breeding, the authors of this paper are constructing an RFLP linkage map for cowpea, a major legume crop in Africa and throughout the developing world. The results indicate that DNA clones from several legume genera can be used to detect RFLPs in cowpea. While the level of DNA sequence polymorphism in this crop is relatively low, it is adequate for RFLP mapping. In a preliminary analysis of eight RFLP markers in an F_2 population of 58 individuals, three markers showed skewed segregation ratios, two markers were found to be linked to one another, and another appeared to be tightly linked to a major gene controlling pod length.

Restriction fragment length polymorphisms (RFLPs) are DNA-based genetic markers. They can be used to construct high density linkage maps in plants because they can be detected in potentially unlimited numbers, tend to be moderately or highly variable, and are free of pleiotropic effects. RFLPs make it possible to map the locations of genes of economic importance precisely, enabling breeders to select desirable individuals in plant breeding programmes. They can also provide detailed genotypic information on individual plants or groups of plants for use in germplasm management and patenting improved crop varieties.

1 Contribution No. 18 943 from the Minnesota Agricultural Experiment Station, USA (research conducted under Project 015).

Linkage maps based on RFLPs have been constructed for several important crops (see Young et al., Paper 5.1, pp 221-230 this volume). Less attention has been paid to crops grown mainly in developing countries, such as cowpea (*Vigna unguiculata* (L.) Walp.) for which no RFLP map is yet available. Nonetheless, cowpeas are a major source of vegetable protein for people and livestock in developing countries, particularly in Africa where the crop suffers severe losses because of insect and microbial pests (Singh and Allen 1979). As RFLPs provide a method for faster and more effective selection during plant breeding, an RFLP map for cowpea would have a significant effect on its genetic improvement.

In this paper, we report on preliminary work to construct an RFLP map for cowpea that will eventually consist of over 100 DNA markers. Our work shows that DNA clones from a wide variety of related leguminous species can be used as sources of RFLPs for cowpea. Our results also indicate that the level of DNA sequence variation in cowpea is relatively low, but still adequate for RFLP mapping. We also present segregation and linkage data for a few of the cowpea RFLP markers that we have characterized to date, including evidence that one of the RFLPs is closely linked to a quantitative trait locus (QTL) controlling pod length.

Methods

Experimentally, RFLP mapping is based on hybridizing cloned DNA sequences with "blots" that contain restriction enzyme-digested DNA from a population of related individuals (such as an F_2 or backcross population). For each cloned DNA sequence, the pattern of DNA bands for each progeny individual can be used in segregation analysis to construct a genetic linkage map. Thus, two major components are required for RFLP mapping: a set of putative RFLP markers, usually a library of cloned DNA sequences derived from the organism of interest; and purified DNA from a related group of individuals. More detailed information on the conceptual basis of RFLP mapping can be found in Tanksley et al. (1989), Young (1990) and Young et al. (Paper 5.1, pp 221-230 this volume).

DNA isolation from cowpea

Because large amounts of DNA are required for both parents and progeny individuals, we adopted a fast and simple technique for DNA isolation, based on the method described by Dellaporte (1983). Briefly, this technique begins with homogenization of frozen leaves, followed by high-salt precipitation of proteins and polysaccharides. The supernatant is then extracted with an organic solvent, treated with ribonuclease, and the DNA precipitated from the aqueous phase with isopropanol. The DNA, although not pure at this stage, is still suitable for restriction enzyme digestion and RFLP mapping.

Purified DNA is then digested with one of several restriction endonucleases, which recognize specific four- or six-base pair nucleotide sequences and cleave the DNA at or near that site. Because recognition is highly specific, genomic DNA from any individual is cut into a unique and reproducible set of DNA molecules of varying sizes. The digested DNA can then be separated by size, using agarose gel electrophoresis, and it is transferred to a permanent filter (for subsequent DNA hybridization) by Southern blotting (Southern 1975).

Construction of a cowpea genomic DNA library

As a source of putative RFLP clones, we constructed a library of single- and low-copy DNA sequences from cowpea. Highly repetitive sequences are unsuitable for RFLP mapping because they yield banding patterns that are too complex to interpret. Single- and low-copy libraries are usually constructed using one of two methods, either cloning cDNA sequences derived from purified messenger RNAs or cloning random genomic sequences enriched for single and low copy DNA sequences.

Our library is a random genomic library of cowpea DNA. To construct this library, we isolated large amounts of high molecular weight cowpea genomic DNA from dark-grown sprouts, digested the DNA with the methylation-sensitive restriction enzyme *Pst*I, and then fractionated the digested DNA by sucrose-gradient centrifugation. Because single- and low-copy sequences in plants tend to be under-methylated (Miller and Tanksley 1990), this process yielded a DNA fraction highly enriched in single- and low-copy DNA sequences. The size-selected DNA was then ligated into the plasmid vector pUC-19, transformed into the *Escherichia coli* strain DH5a, and a total of 200 clones grown for further study.

In addition to the cowpea genomic library constructed in our laboratory, we have also analysed DNA clones from several other leguminous species to see if they would be suitable for RFLP analysis in cowpea. We constructed a genomic library from the related *V. radiata* (mung bean), using a strategy identical to that described above. In addition, Dr R. Shoemaker (Iowa State University, USA) provided genomic clones from *Glycine max* (soybean), and Dr E. Valleijos (University of Florida, USA) from *Phaseolus vulgaris* (common bean).

Segregation analysis and linkage mapping

When parents of an F_2 or backcross population contain different alleles of a cloned DNA sequence, progeny segregate for the differing allelic forms, as indicated by different-sized DNA bands derived from the parents. We have been using an F_2 population derived from a cross between the cultivated cowpea line IT 84S-2246-4 and the non-cultivated cowpea NI 963, provided by Dr G. Myers (International Institute of Tropical Agriculture).

Before segregation analysis could begin, putative RFLP markers were surveyed against DNA from the parents. This indicated the level of DNA sequence variation between the parents and showed which restriction enzyme would be best for mapping a given clone in the progeny. We surveyed individual DNA clones by labelling each clone with the radio-isotope, 32P-dCTP, using the random hexamer method (Feinberg and Vogelstein 1983). The labelled DNA clone was then hybridized with a Southern blot containing restriction enzyme-digested DNA from both parents (up to seven different restriction enzymes). For each clone, the restriction enzyme (if any) that yielded bands with the clearest difference between the parents was chosen and used to digest the F_2 progeny. Once we identified the best restriction enzyme for mapping a given DNA clone, we hybridized the clone to a blot containing DNA from 58 F_2 plants that had been digested with the appropriate enzyme.

For each F_2 individual, RFLP results were used to infer whether the individual was homozygous for IT 84S-2246-4, homozygous for NI 963, or a heterozygote of the two. We coded this data for each RFLP into a numeric form and entered results into a computer for linkage analysis. This process was repeated for each DNA clone. To prepare a linkage map, we used the computer programme MAPMAKER (Lander et al. 1987), which calculates the log of the odds-ratio (LOD score) for multipoint linkage among RFLP markers.

Results

Testing different sources of putative RFLPs

DNA clones from four genomic libraries have been tested for their suitability for RFLP mapping in cowpea. In addition to clones from the cowpea *Pst*I genomic library that we constructed, clones from similar genomic libraries of soybean, common bean, and mung bean have also been tested. These experiments consisted of radiolabelling several clones from each library and hybridizing the clones to blots containing restriction-enzyme digested DNA from the two parents of our F_2 population. Hybridization signals were evaluated to determine whether the clone hybridized to a single (or low) copy sequence, or to a sequence which is highly repeated in the cowpea genome. As noted earlier, only clones that gave a single- or low-copy signal would be useful for RFLP mapping. Clones were also evaluated for DNA polymorphisms (differences in banding patterns) between the parents (Table 1).

Table 1 Summary of clones examined for RFLP mapping

Source of clones	Number tested	Single- or low-copy (%)	Polymorphic (%)
Soybean	19	79	40
Cowpea	11	71	43
Mung bean	5	100	0
Common bean	4	75	0

To date, 39 DNA clones have been tested by hybridization with blots of cowpea DNA. The results for one of the hybridization experiments (using a soybean DNA clone as the radiolabelled probe) are shown in Figure 1. Almost every DNA clone, irrespective of its source, exhibited a clear, prominent signal with cowpea DNA. This indicates that there is significant sequence homology between the legume taxa we analysed and cowpea. Because the number of clones that have been tested is still very low, it is not possible to conclude whether there are significant differences in the frequency at which high copy clones occur within different libraries. It is also not possible to infer whether any library is more likely to show polymorphisms between the parents of our F_2 population.

Segregation analysis of RFLP markers

Of the 39 DNA markers analyzed, eight have been hybridized with all 58 F_2 individuals in the mapping population. The results of one of the segregation experiments, in which the RFLP shown in Figure 1 was hybridized to a DNA blot of all F_2 individuals, are shown in Figure 2 (*overleaf*). Similar hybridization signals (although different banding patterns)

Figure 1 **Restriction enzyme survey of soybean clone, pA64, against cowpea DNA**

Genomic DNA was prepared from the cowpea lines IT 84S-2246-4 and NI 963, digested with five different restriction enzymes, fractionated by agarose gel electrophoresis, and Southern blotted onto nylon membrane. The DNA blot was then probed with a radiolabelled copy of clone pA64. Two lanes are shown for each enzyme (ERI and *Eco*RI; ERV and *Eco*RV; Hae and *Hae*III; Hin and *Hind*III; Bam and *Bam*HI). The left lane in each pair is IT 84S-2246-4 and the right lane is NI 963. Polymorphisms can be observed for all enzymes (as indicated by differences in banding patterns between the cowpea lines) except *Bam*HI.

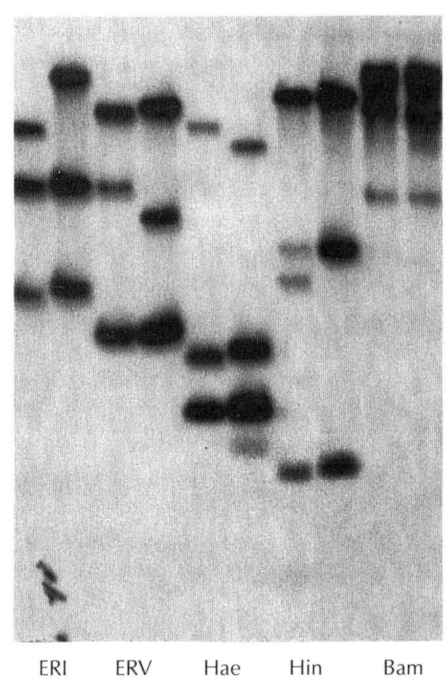

were obtained for each of the eight RFLPs analysed. Among the eight RFLPs, two (p77 and p106) were found to be linked and are presumably located on the same chromosome. The crossover frequency between these RFLPs was 23%.

We also determined whether the RFLPs segregated according to the typical 1:2:1 ratio expected for co-dominant markers in F_2 populations. Segregation results for all eight RFLPs were evaluated using the chi-square test. Of the eight RFLPs characterized, five showed no significant deviation from the expected 1:2:1 ratio. The other three RFLPs, which differed significantly from expectation ($P < 0.01$), consisted of two markers that were skewed in the direction of the cultivated cowpea parent and one skewed in the direction of the wild parent. Since the three RFLPs that show significant skewing are unlinked, there must be at least three regions of the genome which exhibit this phenomenon. Such skewing in the direction of one parent during RFLP mapping has been observed before (Young et al. 1988). In cowpea, we do not yet know the cause of the segregation distortions that were observed.

Quantitative trait loci analysis with RFLPs

One of the most powerful applications of RFLP mapping in crop improvement is the dissection of complex genetic traits into simple Mendelian genes. The parents of our F_2 population differ in many morphological characteristics (as well as in disease and pest resistance). Although a complete analysis of these complex traits must await the completion

Figure 2 Restriction enzyme analysis of DNA from a cowpea F_2 population

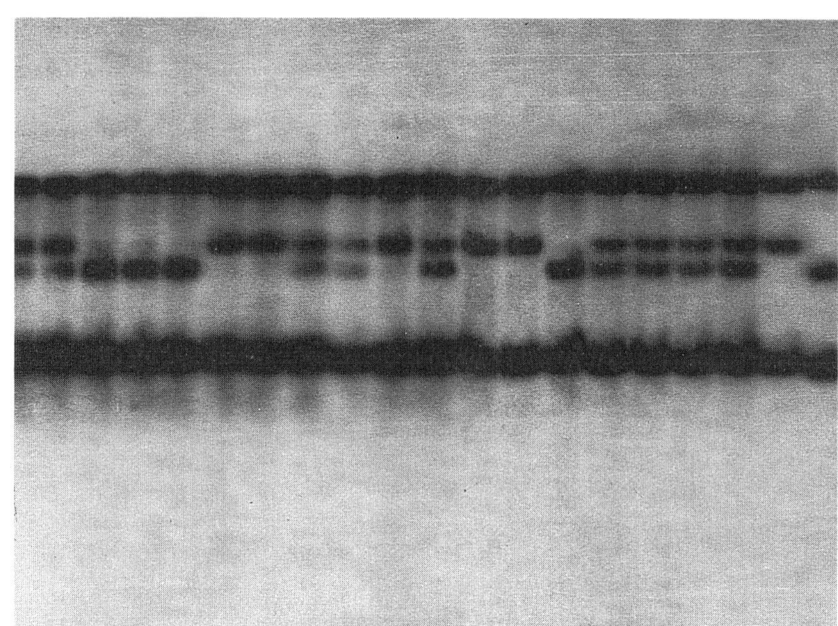

DNA was extracted from 58 F_2 progeny derived from a cross between IT 84S-2246-4 and NI 963. DNA was then digested with *Eco*RV, fractionated by gel electrophoresis, and blotted onto nylon membrane. The DNA blot was then probed with a radiolabelled copy of clone pA64. The results for 20 F_2 individuals are shown. For each individual, the presence of only the upper band (of the two middle bands) indicates that that individual is homozygous for the pA64 allele from IT 84S-2246-4, while the presence of only the lower band indicates that the individual is homozygous for the allele from NI 963. The presence of both bands indicates that the individual is a heterozygote.

of the RFLP map, even with just eight RFLP markers it has been possible to detect linkage between one of the RFLPs and pod length. In simple regression analysis, RFLP marker pM185b was found to be statistically significantly associated with this characteristic (Figure 3). Since the r^2 for the regression line was estimated to be 0.259, it can be inferred that a significant proportion of the variation for this characteristic can be accounted for by pM185b and that a gene with a major effect on pod length may be located nearby.

Future Research

A "framework" RFLP map for cowpea

The long-range goal of our research is development of a medium-density RFLP map for cowpea, of between 100 and 200 markers. Our results indicate that DNA clones from a variety of legumes can be used as RFLPs in cowpea and that the level of DNA sequence in

Figure 3 Regression analysis of pM185b genotype versus pod length phenotype

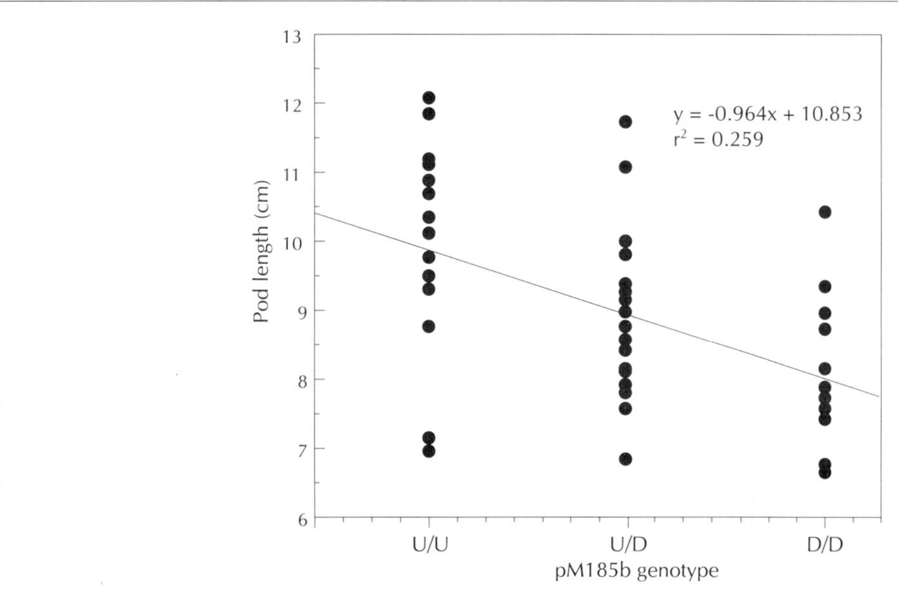

A total of 58 F_2 progeny from a cross between IT 84S-2246-4 and NI 963 were analysed in terms of pod length and RFLP genotype (for eight mapped RFLP markers). In single factor analysis of variance (ANOVA), the inherited genotype for marker pM185b was found to be significantly related to observed pod length ($P = 0.0003$). The results indicate that pod length is reduced by 0.964 cm for each additional dose of the NI 963 allele of pM185b (indicated by "D" for *dekindtiana*) and increased by the same amount for each IT 84S-2246-4 allele (indicated by "U" for *unguiculata*) that an F_2 individual inherits.

cowpea, although relatively low, will be adequate for eventually constructing a linkage map. Because so many DNA clones will have to be tested in order to construct the map, it is important that our method for DNA isolation is simple and rapid. It is also vital that our DNA blots can be re-probed five or more times and that hybridization signals we obtain can easily be analyzed after only one or a few days exposure.

Estimating the number of clones required for RFLP mapping

Although these results are preliminary, they do indicate the total number of clones needed to construct a medium density RFLP map for cowpea. As the final goal is a linkage map with RFLP markers spaced about 10 centimorgans on average, over 100 total RFLPs will need to be mapped. Of the 39 DNA clones analysed (from all four sources), only 73% were single- or low-copy sequences; of these, only 42% exhibited DNA polymorphisms that would be useful for segregation analysis. This indicates that only 30% of DNA clones that are tested will be suitable for RFLP mapping, a value similar to other inbred leguminous genera (R.C. Shoemaker, pers. comm.; E. Valleijos, pers. comm.). This analysis also indicates that, to complete the cowpea RFLP map, at least 300 total DNA clones must eventually be analysed.

Tagging genes of economic importance with RFLPs

Once constructed, the "framework" cowpea RFLP map should accelerate attempts to tag genes of interest with tightly linked markers. Rather than randomly seeking markers that lie near a target gene, scientists will be able to step along each chromosome efficiently, testing a minimum number of markers for linkage. This will be very important in tagging genes of interest because statistical theory indicates that attempts to find linked markers using a random strategy require more than three times as many markers to be tested. Also, in previous work on tomato it was possible to group several RFLPs for hybridization as a single pooled set (Young et al. 1988). If this strategy can be applied to cowpea in conjunction with a complete linkage map, it may be possible to test markers spanning whole arms of chromosomes for linkage all at once. Thus, RFLPs tightly linked to a gene of interest could be identified with as few as 20-30 hybridization experiments.

While we build a framework linkage map for cowpea, we will be identifying RFLPs near genes that segregate in the current F_2 population. As noted earlier, we have already identified an RFLP that appears to be near a major gene controlling pod length. The parents of this population also differ in some other morphological traits, so it is likely that RFLPs near genes controlling other important traits will eventually be mapped. Moreover, this F_2 population is known to segregate for resistance to maruca (pod borer), and is likely to segregate for resistance to several viruses, pod scab, anthracnose, brown blotch, bruchids, aphids, leafhoppers, and root knot nematode (G.O. Myers, pers. comm.). Thus, markers near these important pest and disease resistance genes will be pinpointed as the RFLP map is constructed, and these will be immediately available for use in cowpea improvement.

Discussion

NG: How accurate is it for elimination of duplicate germplasm accessions and selection of a core collection that can represent the diversity of the whole collection?

YOUNG: RFLPs, particularly "hypervariable" DNA probes, have the potential to identify putative duplicates in a germplasm collection. However, RFLPs give only an indication of the probability that two accessions are identical; they cannot give an unequivocal answer. Accuracy can be increased by testing with more markers, but this costs more. In developing a core collection which represents the diversity of the whole collection, RFLPs can indeed optimize selection of subgroups that span variation in terms of DNA sequence, but it is still unknown whether choosing genotypes on the basis of DNA sequence will adequately represent phenotypic variation from the collection.

RAO: Do you think that during the scoring of QTLs traits it is practically possible to identify the "quota" of participation of different alleles to the expression of the character?

YOUNG: When using RFLPs to analyse QTLs, it is possible to estimate the relative contribution of each QTL to the total phenotype, based on linked RFLPs. However, there are two different variables that affect the accuracy of the RFLP-based estimates. On the one hand, RFLPs can miss the presence of a QTL if there has been too much recombination. On the other hand, an RFLP can be very tightly linked to a QTL that contributes only a tiny fraction of the total phenotype.

KIM: In a polygene system using RFLP, can you detect types of gene action — additive, dominance, and epistasis — and can you detect stability and productivity of genes?

YOUNG: It is entirely feasible to determine the nature of the effect of a gene (on a polygenic trait) using RFLPs. For each RFLP marker, statistical tests indicate whether the gene acts in a dominant, recessive, additive, or even epistatic manner by comparing means and variance of the different RFLP genotypic classes. Regarding the second question, RFLPs can be used to test gene stability and productivity. If stability is defined as similar effect in different environments and or different genetic backgrounds, RFLP analysis of appropriate populations (in appropriate backgrounds) would indicate stability. This requires significant effort and some experiments are under way now. In terms of productivity, the relative contribution to a polygenic phenotype that can be explained with a specific RFLP is indicated by statistical analysis.

HASEGAWA: Please comment, in terms of time and effort, on the relationship between the theory of utility of RFLP markers pertaining to the linkage of the trait and the physical marker of the degree of polymorphism in the cultivated species, density of the map, etc.

YOUNG: For RFLP markers to be useful in breeding and selection, putative RFLPs must show clear differences (in banding pattern) between parents; they must also be relatively tightly linked to the gene of interest. Useful polymorphisms between parents will be most common in relatively wide crosses, at least with current technology. Thus, RFLPs will be most applicable to wide crosses for now. In terms of linkage between RFLPs and target genes, the closer the better, but it is generally agreed that a map of about one RFLP/10-20 centimorgans will be effective. Even more important is to have two markers that flank a gene and to carry out selection on the basis of both flanking markers.

PERRINO: It has been suggested that RFLPs be used to identify duplicates in germplasm collections in order to reduce the number of accessions to be preserved in a genebank. I would like to know your opinion if the size of the collection is, say, 10 000 accessions.

YOUNG: Hypervariable RFLP probes would definitely assist in identifying duplicates in a germplasm collection. As a foundation, one would need to identify and characterize one or more hypervariable RFLPs. Once appropriate RFLPs are developed, analysis of a germplasm collection of 10 000 accessions would be entirely feasible. Assuming a simple DNA extraction procedure is effective for the organism of interest, RFLP analysis would require approximately 1 year. A germplasm manager should remember, however, that RFLPs indicate only the likelihood that two accessions are identical; they cannot give unequivocal and final answers.

SINGH: Your proposal to use RFLPs for backcross breeding is very attractive. It should not only cut down time but also reduce the number of steps in backcrossing. How soon will this technique be available for routine use by breeders? Also, what are the needs in terms of manpower development and laboratory facilities?

YOUNG: For some crops, RFLP maps are already available; these include maize, tomato, potato, wheat, lettuce, brassica, and sorghum. In these crops, RFLPs could be used to assist ongoing breeding projects immediately. However, except for maize, narrow crosses would not be suitable for RFLP analysis. In terms of new crops for which a linkage map is yet to be developed, it takes approximately 1-2 years to construct a map, assuming a suitable F_2 or backcross population is available (including cowpea, cassava, and plantain). The amount of manpower/laboratory effort required for marker-assisted breeding, complete "RFLP-typing" 100 individuals requires DNA extraction, electrophoresis, blotting, and nucleic acid hybridization; probably 6 months and US $10,000 or more in expendible materials. Remember, however, that more effective selection, and fewer generations, are eventually required in breeding using RFLPs. Thus, overall costs of traditional versus RFLP-assisted breeding should be considered.

OKORO: What are the possibilities of transferring insecticidal proteins from resistant rice plants to gall-midge susceptible rice plants released in Nigeria, using the RFLP technique being developed by scientists like you?

YOUNG: As a high density RFLP map already exists for rice, it is likely that this technology can assist in breeding for gall midge. However, RFLPs are applicable to naturally occurring genetic resistance, which must be identified in crossable genotypes. Many resistance genes may act through mechanisms other than the production of insecticides. To identify scientists from developed countries interested in this project, the key would be to identify a funding source, considering the high cost of RFLP research.

FILIPPONE: I would like to address a more general question: What about the use of non-radioactive probes to be applied to RFLP?

YOUNG: Technology for non-radioactive nucleic acid probes has improved significantly in the past 2-3 years. One important improvement has been the development of enhanced chemiluminescent techniques. However, non-radioactive techniques are still more expensive (two- or threefold) than radioactive probes. The technique also requires precise experimentation — it is far from " low tech."

References

Dellaporte. 1983. A plant DNA minipreparation: Version II. *Plant Molecular Biology Reporter* 1: 19-21.

Feinberg, A.P., and B. Vogelstein. 1983. A technique for radio-labeling DNA restriction fragments to high specific activity. *Analytical Biochemistry* 132: 6-13.

Lander, E.S., P. Green, J. Abrahamson, A. Barlow, M.J. Daly, S.E. Lincoln, and L. Newburg. 1987. MAPMAKER, an interactive computer program for constructing genetic linkage maps of experimental and natural populations. *Genomics* 1: 174.

Miller, J.C., and S.D. Tanksley. 1990. Effect of different restriction enzymes, probe source, and probe length on detecting restriction fragment length polymorphism in tomato. *Theoretical and Applied Genetics* 80: 385-389.

Singh, S.R. and D.J. Allen. 1979. *Cowpea Pests and Diseases*. IITA Manual Series No. 2. IITA, Ibadan, Nigeria.

Southern, E. 1975. Detection of specific sequences among DNA fragments separated by gel electrophoresis. *Journal of Molecular Biology* 98: 503-517.

Tanksley, S.D., N.D. Young, A.H. Paterson, M.W. Bonierbale. 1989. RFLP mapping in plant breeding: New tools for an old science. *Bio/technology* 7: 257-264.

Young, N.D. 1990. Potential applications of map-based cloning to plant pathology. *Physiological and Molecular Plant Pathology* 37: 81-94.

Young, N.D., D. Zamir, M.W. Ganal, and S.D. Tanksley. 1988. Use of isogenic lines and simultaneous probing to identify DNA markers tightly linked to the *Tm-2a* gene in tomato. *Genetics* 120: 579-585.

Acknowledgements

The authors wish to thank Dr B. Lockhart for his valuable comments on the draft of this paper. The research was supported, in part, by a grant from the Rockefeller Foundation and by GAR funds.

5.4

RFLP marker-assisted breeding for maize virus resistance

R.C. Pratt[1], M.D. McMullen[2] and R. Louie[3]

Department of Agronomy, Ohio State University, Ohio Agricultural Research and Development Center, 1680 Madison Avenue, Wooster, Ohio 44691-4096, USA[1]; Corn and Soybean Research Unit (CSRU), Agricultural Research Service (ARS), United States Department of Agriculture (USDA)[2]; Department of Plant Pathology, Ohio State University, Ohio Agricultural Research and Development Center, 1680 Madison Avenue, Wooster, Ohio 44691-4096, USA[3]

Abstract

The use of restriction fragment length polymorphisms (RFLPs) has been proposed as a means of enhancing the success of plant breeding. This paper discusses the rationale involved in deciding whether or not to use RFLPs to assist the improvement of yield and of host plant resistance to two maize viral pathogens. At present, a cost-effective argument probably cannot be made for the use of RFLPs to assist selection for yield. For maize dwarf mosaic virus (MDMV), a tightly bracketed major gene for resistance has been mapped to the short arm of chromosome 6. However, there is not a compelling argument to use RFLPs to assist in the selection for this trait because artificial inoculation is effective, resistance is largely dominant, and modifiers apparently control the expression of resistance in different susceptible genetic backgrounds. No resistance factors have been described for resistance to maize chlorotic dwarf virus (MCDV). Because of the effort and expense involved in screening for MCDV resistance, it may be more cost effective to establish and use RFLP markers tightly linked to loci conferring MCDV resistance in breeding MCDV-resistant lines.

Indirect selection entails selecting markers (secondary traits) in lieu of genetically correlated economic traits (primary traits). Prospective markers may be correlated with phenotypic traits (such as a height rather than total weight), or biochemical (for example, isoenzyme) or molecular (for example, RFLP) markers. In molecular marker-based indirect selection, we use markers that are physically linked to loci governing the expression of primary traits rather than secondary traits which are genetically correlated with primary traits. Limiting our consideration of genetic correlation to linkage between the secondary and primary traits allows us to directly examine the factors that determine the efficiency of indirect selection.

Indirect selection will be superior to direct selection when a substantially higher selection intensity can be practised on the secondary versus the primary trait, or if the heritability of the secondary trait exceeds that of the primary trait (Falconer 1981). For the purposes of this discussion, we will define heritability as the proportion of total phenotypic variance that is attributable to the average effects of genes and not to genotype x environment interaction or random environmental influences.

With regard to the first criterion, indirect selection will be favoured chiefly when technical difficulties in selecting the primary trait are encountered. Difficulties might arise, for example, when the primary character is difficult to measure with precision (which results in low heritability), when it is expressed only in later stages of the life cycle, or when it is more costly to measure than the secondary trait (Falconer 1981). Tight linkage between an RFLP marker (heritability approximately 1.0) and a primary trait will result in a combined heritability approaching 1 (0.9-1.0) for the marker-trait region. This will fulfill the second criterion because primary traits (which are determined by their phenotype) will usually have a heritability value considerably less than 1.

RFLP-trait associations are usually established by crossing parents which differ markedly for the trait and subsequently scoring segregating populations of progeny to ascertain correlations between the expression of the trait and the parental markers in the progeny. The statistical interpretation of the results can be accomplished in several ways, which may be more or less prone to under- or over-estimating true associations (Knapp et al. 1990). The inherent assumption in establishing these associations has been that the contributions of loci governing a trait will be of sufficient magnitude that they can be detected, and that the relative contribution of a given locus is essentially constant across environments. Variability in the expression of a quantitative trait locus (QTL) in different environments would reduce the average effect and heritability of that locus. Not all RFLP trait-associations are completely free of environmental interaction and their establishment may be better performed in multiple environments (Beckmann and Soller 1986; Bubeck et al. 1990). Thus, as the relative contribution of a particular locus to the expression of the trait decreases, and its dependency on specific environmental conditions for expression increases, the probability that usable marker-trait associations can be established will decline.

RFLP-Assisted Breeding of Maize

RFLP-assisted breeding offers unique advantages because the maize genetic map is well established and the species displays considerable genomic variability. These factors allow for expedient establishment of polymorphisms between even closely related parents. RFLPs are co-dominant and they can be scored in the individual progeny of crosses between contrasting parental lines without affecting one's ability to measure agronomic traits (Helentjaris et al. 1985). Also, many molecular markers for maize are already available for mapping studies and an increasingly more extensive RFLP map is being developed.

Should we try to use RFLPs for the enhancement of traits governed by few, or many, loci? Should we use them to solve seemingly intractable problems or should we simply strive to improve the selection efficiency of traits we can already improve through conventional means? These are important questions because of the facilities and associated costs of RFLP techniques. One should also consider the present need for the use of radioisotopes. Deciding whether or not to use RFLPs will be strongly influenced by a breeding team's experience with the particular economic traits.

Yield is generally the first priority in maize breeding programmes. It is controlled by many genes. Perhaps molecular markers could contribute to enhanced selection efficiency of maize yield (Tanksley et al. 1989). So far, it has been difficult to identify loci that contribute more than 5% of total yield using isozyme markers (Stuber et al. 1987). This result may be due partly to the incomplete coverage of the genome offered by isozyme markers. Selecting for yield in an open-pollinated population is unwieldy, and selection for yield in hybrids infers the efficient selection of not only yield potential, but also of combining ability. Recent results indicate that RFLP-based genetic distance measures are of limited use in predicting heterotic performance of single crosses between unrelated lines (Melchinger et al. 1990). RFLPs will play a much more important role in facilitating the selection of "yield" loci when the cost of RFLP analysis substantially decreases. For the present, the use of RFLPs should be most productive when directed toward traits controlled predominantly by one or a few genes.

Maize Viruses

Maize viruses cause economically important diseases in maize worldwide (Gordon et al. 1982). In the majority of cases, the most economical, least disruptive, and environmentally sound approach to controlling viral diseases is the planting of resistant hybrids or varieties. The development of virus-resistant maize lines is an ongoing goal of public and private USA and international corn breeding programmes (Scott 1982; Eberhart 1983; Findley et al. 1984; CIMMYT 1988; Efron et al. 1989). Despite this considerable effort in breeding for host plant resistance, the genetic bases of resistance to most maize virus diseases are incompletely understood. Detailed knowledge about the number and chromosomal positions of genes controlling virus resistance would allow the development of RFLP-based breeding strategies and would also enhance classical breeding strategies for the improvement of host plant resistance.

Host plant resistance to maize dwarf mosaic virus

Maize dwarf mosaic virus (MDMV) is the most widespread and prevalent viral pathogen of maize in the USA (Gordon et al. 1981). Breeding for MDMV resistance has been undertaken since the occurrence of widespread epiphytotics in the early 1960s. Field screening in areas of natural disease occurrence has suffered from variable time of infection, mixed infections, and variable incidence of the disease. This can be overcome by using large-scale artificial inoculation procedures (Louie 1983, 1986). Sustained breeding efforts using artificial inoculation have improved the MDMV tolerance of commercial USA maize hybrids (Eberhart 1982).

Early genetic studies of MDMV resistance attempted to fit the progeny of resistant x susceptible crosses into symptom response classes predicted by various models of gene number and dominance effects (Findley et al. 1977; Scott and Rosenkranz 1982; Roane et al. 1983; Mikel et al. 1984). Even when the inbred Pa405 (perhaps the best known source of resistance to MDMV) was used as the resistant parent, models involving one or two (Findley et al. 1977), three (Mikel et al. 1984), and five (Rosenkranz and Scott 1984) resistance genes were reported.

Recently, RFLP analysis has resulted in the identification and mapping of a major gene in Pa405 that confers resistance to MDMV strain A (McMullen and Louie 1989). This gene, designated *Mdm*1 (maize dwarf mosaic virus resistance, gene 1), is located near the centromere of chromosome 6 and is tightly linked to, and located between, the RFLP marker loci UMC-85 and BNL6.29. Thus, the gene is bracketed in a distance far less than the 20cM marker bracket recommended for gene introgression by Soller and Plotkin-Hazan (1977).

*Mdm*1, or another gene closely linked to it, is also involved with resistance to four additional strains of MDMV (Louie et al. 1991). Recent publications by Roane et al. (1989a, 1989b) and Scott (1989) have reported the presence of a major gene for MDMV resistance on chromosome 6 from a number of different resistant inbreds.

Influence of genotypic background on expression of *Mdm*1

In backcrosses to yM14 (a white endosperm M14 inbred line), *Mdm*1 delayed symptom appearance and reduced symptom severity. Most plants with *Mdm*1 gave a symptomless response in both field and greenhouse trials. Although *Mdm*1 delayed symptom appearance in backcrosses to the inbred K55, almost all plants exhibited virus-induced symptoms, ranging from limited chlorotic streaks to a delayed generalized mosaic. There was also a difference in the expression of the resistance from Pa405 in the F1s; Pa405 x yM14 plants were symptomless, but most Pa405 x K55 plants exhibited delayed symptoms. Thus, the genotype of the susceptible parent affects the expression of resistance.

Similar to the influence of K55 in the hybrid (Pa405 x K55), many agronomically elite inbreds (W.R. Findley, pers. comm.) do not enable the expression of a high degree of dominance of MDMV resistance when crossed with Pa405. Breeding to confer complete MDMV resistance on many inbreds is difficult. It appears that genotype-specific modifiers play an important role in the expression of resistance, and thus unless a few "universal" modifiers can be identified, it may become necessary to identify and "tag" specific modifiers for every genetic background one wants to convert to full MDM resistance. This would greatly reduce the utility of indirect selection for improved MDM resistance.

Host plant resistance to maize chlorotic dwarf virus

Maize chlorotic dwarf (MCD) is an important disease in the southern, south-eastern, and mid-Atlantic regions of the USA. Attempts to improve the resistance of maize to maize chlorotic dwarf virus (MCDV) have been hampered by the lack of immune or highly tolerant germplasm, the inability to mechanically transmit MCDV, and the paucity of knowledge regarding the fundamental genetic basis of tolerance to the pathogen. Unlike MDMV, MCDV is obligately transmitted by leafhoppers and this increases the difficulty in screening for resistance. The result of these difficulties has been slow improvement of MCDV tolerance in USA maize hybrids; today, there exists only a handful of hybrids with tolerance to MCDV. Were it possible to establish RFLP-trait associations for MCDV tolerance, the need for insect-mediated screening of each breeding cycle might be reduced. This could result in a more expedient and economical improvement of MCDV resistance.

Early studies on the genetic basis of maize virus resistance suffered from confusion in data due to the frequent presence of both MDMV and MCDV in the field (Findley et al. 1984). Additionally, maize genotypes display a wide range of symptoms even when only MCDV is present (Louie et al. 1974). Because of this, in most early studies plants were scored as either positive or negative for virus disease symptoms. This approach obviously did not foster differentiation between symptom types, symptom severity, or the association of specific symptom responses to a specific virus infection or to the degree of tolerance.

There are conflicting interpretations of the genetic basis of resistance to MCDV. For example, Rosenkranz and Scott (1987) ascribed tolerance to additive gene action, whereas Naidu and Josephson (1976) suggested that a degree of dominance is associated with tolerance. A lack of genotypes with complete resistance to MCDV has hampered attempts to interpret data from the progeny of resistant x susceptible crosses.

There is a clear need to identify sources of resistance and to define the genetic control of the differential inbred responses to MCDV. Determination of the chromosomal regions controlling specific symptomatic responses would also allow the use of RFLP-assisted selection for resistance to MCDV. If marker-assisted selection proves to be effective, the reduced need for insect-vectored screening would probably be a favourable cost trade-off.

Conclusion

One of the major questions about the use of RFLP analysis in plant breeding is whether the characterization of loci identified as controlling a trait in one cross will also be a sufficient description of the genetic basis of that trait in other crosses. Could the same trait associations used to select for virus resistance in one instance be used in other crosses? While this appears true for major gene resistance to MDMV conferred by *Mdm*1, genotypic modification of the level of resistance may complicate selection for complete MDMV resistance in varied backgrounds.

The choice of the trait to pursue in using marker-assisted strategies is critical. Economic traits, such as yield and MDMVresistance, are relatively inexpensive and straightforward to select for, so a cost/benefit analysis is unlikely to warrant the application of RFLP technology as it exists today. In these cases, a clearer understanding of the genetic basis of the trait may affect the selection strategy used. Some traits that are costly or technically difficult to select for (such as MCDV resistance) may justify the application of RFLP technology. Effective and expedient improvement of these traits may be realized at a reasonable cost. Additionally, when high priority is placed on timeliness (for example, the "crash conversion" of an inbred line by backcrossing in one to several genes), marker-assisted breeding may be worthwhile because one can simultaneously select for the trait of interest and against the undesirable donor background.

To conclude, we would state that the more difficult the traits are to select by conventional methods, the stronger will be the argument to entertain the use of RFLP marker-assisted breeding. As one goes from traits with high heritability, and economical, straightforward selection to traits with low heritability, and difficulty or expense in selection, the potential benefit of RFLPs becomes considerably greater. The nature of the traits, and whether or not one can reveal those loci which are primary contributors to the phenotypic expression of the desirable trait, will determine the degree to which RFLPs will be successful markers. If evaluation of the trait is complicated and inconsistent because of the complexity in achieving, for example, uniform disease infection or in creating the proper environment for disease expression, then RFLPs will be very useful. If the economic trait is difficult to select for because it is governed by many genes displaying minor individual contributions, and it is much influenced by genotype interactions with, and effects of, the environment, the multiple linkages with RFLP markers will be hard to detect and establish. If, after considerable effort and expense, a moderate number of the most influential regions can be identified and bracketed with markers, then selection efficiency will be enhanced even under poor environmental conditions for trait expression. It should be clear that for traits that can be easily evaluated, conventional approaches are probably more cost effective.

Discussion

ROSA-RAO: Would you like to comment or report on the use of non-radioactive probes in RFLP work?

PRATT: There are certainly many applications for the use of non-radioactive probes, and I would expect their use to increase in the future. Non-radioactive labelling is very effective in detecting RFLPs in the organelle genomes, and it is widely used by plant pathologists for detection of specific pathogens.

PERSLEY: Is there only one maize RFLP map, or are there several maps? Are these all available to maize breeding programmes in Africa?

PRATT: Mapping of publically available probes is being undertaken primarily by the United States Department of Agriculture, Agricultural Research Service at the University of Missouri, Columbia, Missouri, USA. These probes, and their map positions, are updated and published annually in the newsletter of the maize genetics cooperative. The probes are available to all researchers (to the best of my knowledge). RFLP maps generated by private concerns are under proprietary restrictions, and their availability will probably be limited through contractual arrangements, or not at all.

AKORODA: How does one handle screening given variable levels of disease, since the disease varies in mixes and intensities?

PRATT: Marker-assisted breeding relies on establishing associations between, for example, RFLPs and symptoms expression. The initial associations should be established under the most controlled conditions possible. Firm establishment of these associations across environments will allow one to have greater confidence if one wishes to rely on marker-based selection only. The stability of marker-trait associations may vary, depending on the intensity of disease pressure and the particular genetic background being examined.

OLEMBO: Have there been any field traits for maize streak virus, since this is an important disease in certain African countries, such as Kenya?

PRATT: I know that IITA has a very active programme in this area. The maize virus disease research group at Ohio State University has conducted laboratory studies and student training in MSV-host plant relationships with IITA. We are precluded from conducting greenhouse and field studies in the USA by quarantine regulations. We are interested in applying RFLP technology to further elaborate the genetic basis of resistance to MSV in cooperation with IITA. Field studies will certainly be an important part of that work.

OGBE: Techniques for gene transfer have been well developed, but there are no genes to work with. Will RFLP help in locating these genes?

PRATT: RFLP technology can be a very important tool in locating the position of genes of interest. Once located, techniques such as "chromosome walking", which have been performed successfully in mammalian systems, may be feasible in certain plant systems in the near future.

KAHL: Of the non-radioactive techniques, our laboratory prefers digoxygenic-labelled probes to biotinylated probes. Enhanced chemiluminescence does not reproducibly give results. However, due to its higher resolving power, radioactive probing is certainly the best up to now.

References

Beckman, J.S., and M. Soller. 1986. Restriction fragment length polymorphisms and genetic improvement of agricultural species. *Euphytica* 35: 111-124.

Bubeck. D.M., M.M. Goodman, W.D. Beavis, and D. Grant. 1990. Quantitative trait loci controlling resistance to gray leaf spot in maize. In *Agronomy Abstracts, American Society of Agronomy* 194.

CIMMYT. 1988. *The CIMMYT 1988 Annual Report*. CIMMYT, Mexico DF, Mexico.

Eberhart, S.A. 1982. Developing virus resistant commercial maize hybrids. Pages 258-261 in *Proceedings, 2nd International Maize Virus Disease Colloquium and Workshop* edited by D.T. Gordon, J.K. Knoke, L.R. Nault, and R.M. Ritter. Ohio Agricultural Research and Development Center, Wooster, USA.

Efron, Y., S.K. Kim, J.M. Fajemisin, J.H. Mareck, C.Y. Tang, Z.T. Dabrowski, H.W. Rossel, and G. Thottappilly. 1989. Breeding for resistance to maize streak virus: A multidisciplinary approach. *Plant Breeding* 103: 1-36.

Falconer, D.S. 1981. *Introduction to Quantitative Genetics*. Longman, London, UK.

Findley, W.R., R. Louie, J.K. Knoke, and E.J. Dollinger. 1977. Breeding corn for resistance to virus in Ohio. Pages 123-128 in *Proceedings, First International Maize Virus Disease Colloquium* edited by L.E. Williams, D.T. Gordon, and L.R. Nault. Ohio Agricultural Research and Development Center, Wooster, USA.

Findley, W.R., R. Louie, and J.K. Knoke. 1984. Breeding corn for resistance to corn viruses in Ohio. Pages 56-67 in *Proceedings, Annual Corn Sorghum Research Conference*. Publication 399. American Seed Trade Association, Washington DC, USA.

Gordon, D.T., O.E. Bradfute, R.E. Gingery, J.K. Knoke, R. Louie, L.R. Nault, and G.E. Scott. 1981. Introduction: History, geographical distribution, pathogen characteristics, and economic importance. Pages 1-12 in *Virus and Virus-Like Diseases of Maize in the United States* edited by D.T. Gordon, J.K. Knoke, and G.E. Scott. Southern Cooperative Series Bulletin 247. Ohio Agricultural Research and Development Center, Wooster, USA.

Gordon, D.T., J.K. Knoke, L.R. Nault, and R.M. Ritter (eds). 1982. *Proceedings, 2nd International Maize Virus Disease Colloquium and Workshop* edited by D.T. Gordon, J.K. Knoke, L.R. Nault, and R.M. Ritter. Ohio Agricultural Research and Development Center, Wooster, USA.

Helentjaris, T., G. King, M. Slocum, C. Siedenstrang, and S. Wegman. 1985. Restriction fragment polymorphisms as probes for plant diversity and their development as tools for applied plant breeding. *Plant Molecular Biology* 5: 109-118.

Knapp, S.J., W.C. Bridges Jr., and D. Birkes. 1990. Mapping quantitative trait loci using molecular marker linkage maps. *Theoretical and Applied Genetics* 79: 583-592.

Louie, R. 1983. Transmission of maize dwarf mosaic virus with solid-stream inoculum. *Plant Disease* 67(12): 1328-1331.

Louie, R. 1986. Effects of genotype and inoculation protocols on resistance evaluation of maize to maize dwarf mosaic virus strains. *Phytopathology* 76: 769-773.

Louie, R.A. J.K. Knoke, and D.T. Gordon. 1974. Epiphytotics of maize dwarf mosaic and maize chlorotic dwarf diseases in Ohio. *Phytopathology* 64: 11.

Louie, R., W.R. Findley, J.K. Knoke, and M.D. McMullen. 1991. Genetic basis of resistance in maize to five maize dwarf mosaic virus strains. *Crop Science* 31: 14-18.

McMullen, M.D., and R. Louie. 1989. The linkage of molecular markers to a gene controlling the symptom response in maize to maize dwarf mosaic virus. *Molecular Plant Microbe Interactions* 2: 309-314.

Melchinger, A.E., M. Lee, K.R. Lampkey, and W.L. Woodman. 1990. Genetic diversity for restriction fragment length polymorphisms: Relation to estimated genetic effects in maize inbreds. *Crop Science* 30: 1033-1040.

Mikel, M.A., C.J. D'Arcy, A.M. Rhodes, and R.E. Ford. 1984. Genetics of resistance of two dent corn inbreds to maize dwarf mosaic virus and transfer of resistance into sweet corn. *Phytopathology* 74: 467-473.

Naidu, B., and L.M. Josephson. 1976. Genetic analysis of resistance to the corn virus disease complex. *Crop Science* 16: 167-172.

Roane, C.W., S.A. Tolin, and C.F. Genter. 1983. Inheritance of resistance to maize dwarf mosaic virus in the maize inbred line Oh7B. *Phytopathology* 73: 845-850.

Roane, C.W., S.A. Tolin, and H.S. Aycock. 1989a. Genetics of reaction to maize dwarf mosaic virus strain A in several maize inbred lines. *Phytopathology* 79: 1364-1368.

Roane, C.W., S.A. Tolin, H.S. Aycock, and P.J. Donahue. 1989b. Association of *Mmd*1, a gene conditioning reaction to maize dwarf mosaic virus, with genes conditioning endosperm color (y1) and type (su2) in maize. *Phytopathology* 79: 1368-1372.

Rosenkranz, E., and G.E. Scott. 1984. Determination of the number of genes for resistance to maize dwarf mosaic virus strain A in five corn inbred lines. *Phytopathology* 74: 71-76.

Rosenkranz, E., and G.E. Scott. 1987. Type of gene action in the resistance to maize chlorotic dwarf virus in corn. *Phytopathology* 77: 1293-1296.

Scott, G.E. 1982. Breeding for and genetics of virus resistance in field corn. Pages 248-254 in *Proceedings, 2nd International Maize Virus Disease Colloquium and Workshop* edited by D.T. Gordon, J.K. Knoke, L.R. Nault, and R.M. Ritter. Ohio Agricultural Research and Development Center, Wooster, Ohio, USA.

Scott, G.E., and E. Rosenkranz. 1982. A new method to determine the number of genes for resistance to maize dwarf mosaic in maize. *Crop Science* 22: 756-761.

Scott, G.E. 1989. Linkage between maize dwarf mosaic virus resistance and endosperm color in maize. *Crop Science* 29: 1478-1480.

Soller, M., and J. Plotkin-Hazan. 1977. The use of marker alleles for the introgression of linked quantitative alleles. *Theoretical and Applied Genetics 51: 133-137.*

Stuber, C.W., M.D. Edwards, and J.F. Wendel. 1987. Molecular marker-facilitated investigations of quantitative trait loci in maize. II. Factors influencing yield and its component traits. *Crop Science* 27: 639-648.

Tanksley, S.D., N.D. Young, A.H. Paterson, and M.W. Bonierbale. 1989. RFLP mapping in plant breeding: new tools for an old science. *Bio/technology* 7: 257-264.

Acknowledgements

We wish to thank B. McBlain and R. Gingery for their helpful comments on the manuscript. This paper is a cooperative contribution of the Ohio State University, Ohio Agricultural Research and Development Center (OSU-OARDC) and the USDA Agricultural Research Service (ARS); it was submitted as Journal Article 35-91 by the OSU-OARDC. State and federal funds for the work described in this paper were provided to the OSU-OARDC and the USDA-ARS.

5.5

The use of chloroplast DNA analysis in yam phylogeny reconstruction[1]

R. Terauchi

International Institute of Tropical Agriculture, PMB 5320, Ibadan, Nigeria

Abstract

The enormous polymorphism and plasticity of yam has made it a difficult plant to classify using conventional taxonomic methods. Recently developed techniques of molecular genetics (DNA analysis) offer opportunities for elucidating the phylogeny and origins of yam species. This knowledge is important for the effective breeding and conservation of yams.

The protein and mineral contents of yam are high (Ayensu and Coursey 1972), and thus more emphasis should be placed on this crop as a means of addressing the food crisis and malnutrition problems in tropical regions. Among the many cultivated species, economically important ones are *Dioscorea alata* L. and *D. esculenta* (Lour.) Burk. from South-East Asia, *D. rotundata* Poir. and *D. cayenensis* Lamk. from West Africa, and *D. trifida* L. from the Caribbean. Four other economically important species are *D. bulbifera* L., *D. opposita* Thunb., *D. hispida* Dennst., and *D. dumetorum* (Knuth) Pax.

Despite their economic importance, very little is known about the origin and phylogeny of most yam species. This is because the yam group presents difficult material to work with using conventional taxonomic methods. Phenotypic characteristics are highly plastic (Chikwendu and Okezie 1989). Chromosomes are minute, and the chromosome number, which is usually critically examined for taxonomic purposes, is very high and unstable, even within a single plant (Terauchi 1990). Recently developed techniques in molecular genetics (DNA analysis) offer considerable help in elucidating the phylogeny and origins of plant species (Avise et al. 1987; Palmer 1987; Palmer et al. 1988). The power of DNA analysis

1 Contribution No. IITA/91/CP/27 from the International Institute of Tropical Agriculture, PMB 5320, Ibadan, Nigeria.

lies in its direct evaluation of genotypes rather than of expressed phenotypes. DNA analysis of yam should help us to resolve the origins and phylogeny of yam. This knowledge is important for contemporary breeding and conservation, as well as from the botanical and historical perspectives. When identified, the close wild relatives of yams are likely to be useful for introducing desirable genes into cultivars using various breeding methods — in direct crosses, as bridging species with more distant or incompatible relatives, or as resources for genetic engineering. This paper presents some methods and results of yam phylogenetic research, based on DNA (especially chloroplast DNA) analysis.

Chloroplast DNA

Plant species have three discrete genomes: the chloroplast genome (cpDNA), the mitochondrial genome (mtDNA), and the nuclear genome. Among them, cpDNA has been most extensively used for plant phylogenetic studies, for the following reasons (Palmer 1987):

- Chloroplast DNA is present in higher copy number than other kinds of genomes, and thus is easily detected.
- Genome size is relatively small (120-217 kb) in land plants, making it convenient to handle.
- The structure of cpDNA is simple and usually not heterogenous within an individual, so that secure homologous comparisons can be made among a wide variety of species.
- The cpDNA genome is inherited from only one parent and thus there is no recombination between different cpDNAs.

This last point allows us to consider cpDNAs as attributes for the operational taxonomic unit (OTU) in reconstructing maternal phylogeny, as is the case for most angiosperms.

Methods of Chloroplast DNA Analysis in Yams

Restriction endonuclease analysis is usually employed to compare DNA structure. DNA is extracted, digested with restriction endonucleases, subjected to electrophoresis in an agarose gel, transferred to a nylon membrane (Southern transfer), and detected by hybridization with well-defined cloned DNA fragments obtained from one of the taxa under study. A difference in restriction profile thus found between two individuals is called a restriction fragment length polymorphism (RFLP).

DNA extraction

DNA extraction is the most difficult and critical step in the experiment. Yam leaves usually contain large amounts of polysaccharides, which interfere with restriction enzyme digestion and electrophoresis. For most yam species, the minipreparation method described by Dellaporta et al. (1983), followed by DNA purification with CTAB (Murray and Thompson 1980), gives good results.

DNA probes

A cpDNA clone bank of *D. bulbifera* was constructed with its *Bam*HI fragments, using the cpDNA of one accession from Madagascar (Terauchi et al. 1989). From this bank, four clones, B5 (9.4 kb), B6 (9.0 kb), B8 (6.55 kb), and B16 (2.3 kb), were selected as the probes. In addition, three clones of the *Sal*I fragments of *D. opposita* cpDNA, S2 (26.0 kb, DO#2), S5-6 (26.4 kb, DO#1), and S4-7 (30.0 kb, DO#3) (Terauchi et al. 1991), were used as the heterologous probes to detect cpDNA variation in the species to be analysed. Figure 1 gives their location in the physical maps of the respective cpDNAs. The probe DO#1 covers the small single copy region entirely, DO#2 corresponds to most parts of the inverted repeat, and DO#3, with the four *Bam*HI probes (B5, B6, B8, and B16), covers about 70% of the large single copy region. Altogether, these probes cover 80% of the total chloroplast genome. They were sequentially hybridized with the DNAs from the yam material to be analysed.

Figure 1 The *Bam*HI restriction map of *Dioscorea bulbifera* cpDNA (accession DB1) and *Sal*I map of *D. opposita* cpDNA

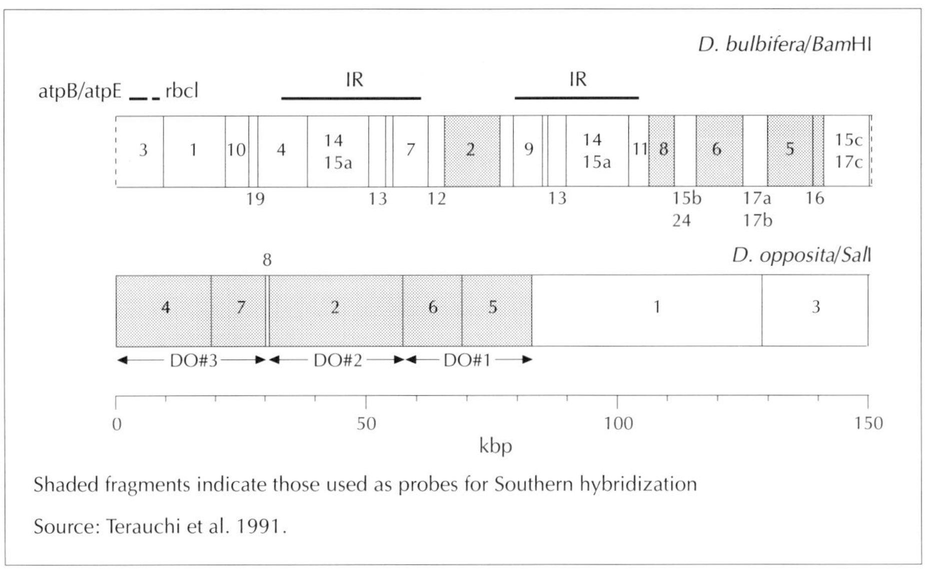

Shaded fragments indicate those used as probes for Southern hybridization

Source: Terauchi et al. 1991.

Non-radioactive detection method of Southern hybridization

In most cases, a radioisotope, ^{32}P, has been used to label DNA probes for Southern hybridization. The half-life of ^{32}P is 14 days, and it is not easy to obtain ^{32}P with high activity in Africa because the transport system is not very efficient. Even if active isotope is available, much caution and some skill are needed to handle it without being exposed to harmful radioactivity. Recently, non-radioactive methods for hybridization detection have been developed, such as the Dig-ELISA (Boehringer) and the enhanced chemiluminescence (ECL) (Amersham) methods. For detection of DNAs present in many copies, the Dig-ELISA method gives sufficient signals. DNA probes, once prepared, can be used several times for different hybridization experiments.

Phylogeny of Aerial Yam

Aerial yam, *D. bulbifera*, is a highly polymorphic species distributed widely in Africa and Asia. RFLP analysis of cpDNAs from 15 accessions collected from Africa and Asia was carried out as described above (Terauchi et al. 1991), using nine restriction enzymes. Ten variable sites, located in the large and small single-copy regions, were uncovered among the 15 accessions, of which six showed base substitutions and four carried length mutations. Based on these results, chloroplast genomes of the 15 accessions could be classified into nine types. The results indicate the following taxonomic relationships: African and Asian chloroplast genomes diverged from each other earliest in time; the E-type chloroplast genome, occurring in the south-eastern edge of the Asian continent, appears to be the most ancient among all the Asian chloroplast genomes; and four chloroplast genomes, found in Asian insular regions, are probably derived independently from the E-type genome (Figure 2).

Figure 2 Geographical distribution and evolutionary relationship of chloroplast genomes (nine types, A-I) found in *Dioscorea bulbifera*

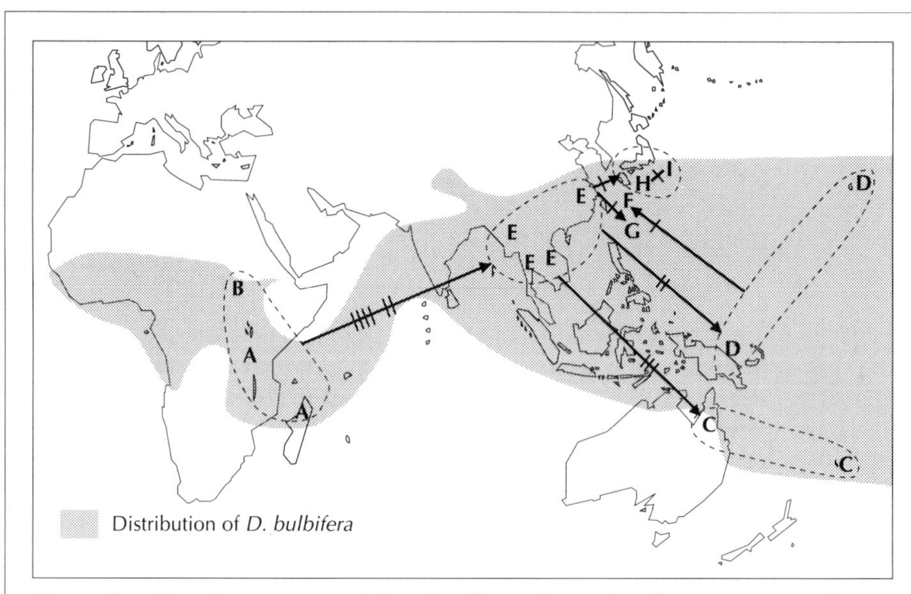

The number of bars drawn across each branch indicates the number of mutations assumed to have occurred in it.

Source: Terauchi et al. 1991.

Phylogeny of Guinea Yam

Guinea yam (*D. rotundata-D. cayenensis* complex) consists of two phenotypically distinguishable plant types: white yam (white tuber flesh, short growth period), and yellow yam (yellow tuber flesh, long growth period). Although the typical plants of white and yellow yams can be easily recognized, there is a wide spectrum of gradation between them, and thus their taxonomy has been a subject of much controversy. Some taxonomists treated them as

Figure 3 Summary of the phylogenetic relationships between the five chloroplast genomes of Guinea yam (*Dioscorea cayenensis-rotundata*) and its wild relatives

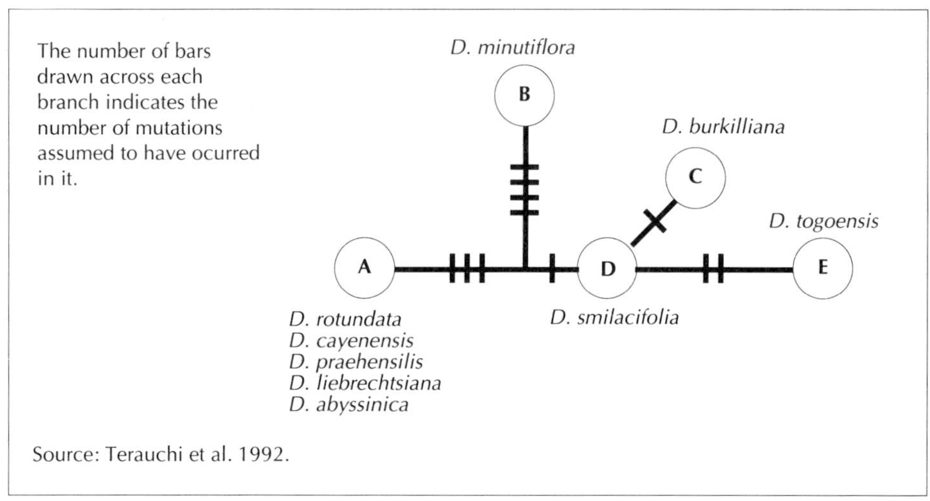

Source: Terauchi et al. 1992.

two discrete species, *D. rotundata* Poir. and *D. cayenensis* Lamk. (Hutchinson and Dalziel 1931; Burkill 1960; Ayensu 1971; Akoroda and Chheda 1983; Onyilagha and Lowe 1985); others consider them as two subspecies or varieties of one species, *D. cayenensis* (Chevalier 1936; Miège 1968, 1979; Martin and Rhodes 1978). Hamon and Toure (1990) proposed to treat this species complex as the *Dioscotea cayenensis-rotundata* complex. There is no consensus as to the origin and phylogenetic relationships between these two types. On morphological grounds, Miège (1982) proposed *D. praehensilis* Benth., *D. sagittifolia* Pax., *D. lecardii* De Wild., *D. liebrechtsiana* De Wild., and *D. abyssinica* Hochest ex. Knuth as the possible wild progenitors of Guinea yam. Other wild species closely related to Guinea yam include *D. minutiflora* Engl., *D. smilacifolia* De Wild., *D. burkilliana* J. Miège, *D. mangenotiana* J. Miège, and *D. togoensis* Knuth (Chevalier 1936; Miege 1982).

RFLP analysis was conducted using seven accessions of typical *D. rotundata*, five accessions of typical *D. cayenensis*, two types between *D. rotundata* and *D. cayenensis*, two accessions each from *D. praehensilis*, *D. minutiflora*, *D. smilacifolia*, *D. abyssinica*, and *D. togoensis*, and one accession each from *D. liebrechtsiana* and *D. burkilliana* (Terauchi et al., 1992). By applying the DNA probes noted above to DNAs that were digested with eight restriction enzymes, five cpDNA types (types A-E) were disclosed. CpDNA type A was carried by five species, *D. rotundata*, *D. cayenensis*, *D. praehensilis*, *D. liebrechtsiana*, and *D. abyssinica*; cpDNA type B, C, D, and E were carried by *D. minutiflora*; *D. burkilliana*, *D. smilacifolia*; and *D. togoensis*, respectively. An outline of the relationships between these five cpDNA types is depicted in Figure 3.

Additional RFLP analysis using nuclear ribosomal DNA, suggested that white yam, *D. rotundata*, had originated directly from one of three wild species, *D. praehensilis*, *D. liebrechtsiana*, and *D. abyssinica*. Yellow yam, *D. cayenensis*, had probably originated from a hybrid between one of the species of cpDNA type A and one of the four species *D. minutiflora*, *D. burkilliana*, *D. smilacifolia*, and *D. togoensis*. The former species must have contributed as the female parent and the latter as the male parent.

References

Akoroda, M.O., and H.R. Chheda. 1983. Agro-botanical and species relationships of Guinea yams. *Tropical Agriculture* 60: 242-247.

Ayensu, E.S. 1971. Dioscoreales. Pages 1-182 in *Anatomy of the Monocotyledons* edited by C.R. Metcalfe. Oxford University Press, Oxford, UK.

Ayensu, E.S., and D.G. Coursey. 1972. Guinea yams — The botany, ethnobotany, use and possible future of yams in West Africa. *Economic Botany* 26: 301-318.

Avise, J.C., J. Arnold, R.M. Ball, E. Bermingham, T.Lamb, J.E. Neigel, C.A. Reeb, and N.D. Saunders. 1987. Intraspecific phylogeography: The mitochondrial DNA bridge between population genetics and systematics. *Annual Review of Ecology and Systematics* 18: 489-522.

Burkill, I.H. 1960. The organography and the evolution of the Dioscoreaceae, the family of the yams. *Journal of the Linnaean Society (Botany)* 56: 319-412.

Chevalier, A. 1936. Contribution à l'étude de quelques espèces africaines du genre *Dioscorea. Bulletin du Musée National de l'Histoire Naturelle, Paris* 8: 520-551.

Chikwendu, V.E., and C.E.A. Okezie. 1989. Factors responsible for the ennoblement of African yams: Inferences from experiments in yam domestication. Pages 344-357 in *Foraging and Farming* edited by D.R. Harris and G.C. Hillman. Unwin Hyman, London, UK.

Dellaporta, S.L., J. Wood, and J.B Hicks. 1983. A plant DNA minipreparation: Version II. *Plant Molecular Biology Reporter* 1: 19-21.

Hamon, P., and B. Toure. 1990. The classification of the cultivated yam (*Dioscorea cayenensis-rotundata* complex) of West Africa. *Euphytica* 47: 179-187.

Hutchinson, J., and J.M. Dalziel. 1931. *Flora of West Tropical Africa* (Vol. 2) (1st edn). Crown Agents, London, UK.

Martin, F.W., and A.M. Rhodes. 1978. The relationship of *Dioscorea cayenensis* and *D. rotundata. Tropical Agriculture* 55: 193-206.

Miège, J. 1968. Dioscoreaceae. Pages 144-154 in *Flora of West Tropical Africa* (Vol. 3, Part 1) (2nd edn) edited by F.W. Hepper. Crown Agents, London, UK.

Miège, J. 1979. Sur quelques problems taxonomiques posès par *Dioscorea cayenensis* et *D. rotundata.* Pages 109-113 in *Taxonomic Aspects of African Economic Botany.* Proceedings, Plenary Meeting of AETFAT, Las Palmas, Gran Canaria.

Miège, J. 1982. Appendice: Notes sur les espèces *Dioscorea cayenensis* Lamk. et *D. rotundata* Poir. Pages 367-375 in *Yams/Ignames* edited by J. Miège and S.N. Lyonga. Oxford University Press, Oxford, UK.

Murray, M.G., and W.F. Thompson. 1980. Rapid isolation of high-molecular-weight plant DNA. *Nucleic Acids Research* 8: 4321-4325.

Onyilagha, J.D., and J. Lowe. 1985. Studies on the relationships of *Dioscorea cayenensis* and *Dioscorea rotundat*a cultivars. *Euphytica* 35: 733-739.

Palmer, J.D. 1987. Chloroplast DNA evolution and biosystematic uses of chloroplast DNA variation. *American Naturalist* 130: S6-S29.

Palmer, J.D., R.K. Jansen, H.J. Michaels, M.W Chase, and J.R Manhart. 1988. Chloroplast DNA variation and plant phylogeny. *Annals of the Missouri Botanic Gardens* 75: 1180-1206.

Terauchi, R. 1990. Genetic diversity and phylogeny of two species of the genus *Dioscorea.* PhD thesis, Kyoto University, Kyoto, Japan.

Terauchi, R., T. Terachi, and K.Tsunewaki. 1989. Physical map of chloroplast DNA of aerial yam, *Dioscorea bulbifera* L. *Theoretical and Applied Genetics* 78: 1-10.

Terauchi, R., T. Terachi, and K. Tsunewaki. 1991. Intraspecific variation of chloroplast DNA in *Dioscorea bulbifera* L. *Theoretical and Applied Genetics* 81: 461-470.

Terauchi, R., V.A. Chikaleke, G. Thottappilly, and S.K. Hahn. 1992. Origin and phylogeny of Guinea yams as revealed by RFLP analysis of chloroplast DNA and nuclear ribosomal DNA. *Theoretical and Applied Genetics* 83: 743-751.

5.6

Isozyme analysis and its application in plant breeding[1]

R. Asiedu

International Institute of Tropical Agriculture, PMB 5320, Ibadan, Nigeria

Abstract

Plant breeders face the difficult task of having to select for traits that are often under complex genetic control and subject to edapho-climatic influences. Electrophoresis of protein extracts from plant tissue, using different kinds of support media and buffer systems, allows separation of the multiple forms of enzymes (isozymes) on the basis of charge and/or molecular size. The resulting polymorphisms are useful as genetic markers. Isozyme analysis is a powerful technique for estimating genetic variability, identifying cultivars and germplasm accessions, confirming hybrids, tracking alien chromosome segments, tagging genes, and screening backcross populations more efficiently.

Many of the characteristics of interest to plant breeders cannot be scored on an individual plant basis, and thus examination of populations is necessary. Quite often, the plants need to be grown to maturity before the trait can be scored with certainty. Breeders occasionally rely on morphological characteristics (markers) known to be associated with some of the more subtle traits, or they take advantage of markers established through cytological techniques, such as differential chromosome staining. There is another class of markers — biochemical/molecular markers — which have many advantages over other marker types. These are based on DNA polymorphisms or polymorphisms among structural genes coding for proteins. At the DNA level, fingerprinting and/or analysis of restriction fragment length polymorphisms (RFLPs) may be conducted. Protein markers can further be categorized as storage protein or isozyme markers.

1 Contribution No. IITA/91/CP/28 from the International Institute of Tropical Agriculture, PMB 5320, Ibadan, Nigeria.

The major advantages of biochemical/molecular markers are that they are numerous, widely applicable to various tissues of the plant, rapid to assay for, and co-dominant in inheritance, making it easy to distinguish homozygotes from heterozygotes. Moreover, such markers are free from the deleterious associations or epistatic/pleiotropic effects which characterize morphological markers.

Methodology

The traditional definition of an isozyme or isoenzyme as the multiple molecular form of an enzyme having the same catalytic specificity but probably differing slightly in kinetic properties (Markert and Moller 1959) has undergone some revision, as discussed by Kephart (1990). Theoretically, isozyme markers cannot be as numerous as DNA markers because not all modifications at the DNA level are expressed in the gene products. However, analysis of the former will usually have an advantage in terms of cost.

The basic steps in isozyme analysis involve the extraction of enzymes from plant tissue, the application of the extract to a solid support medium, electrophoretic separation, enzyme activity staining, and the interpretation of the observed patterns.

The efficiency of extraction of enzymes and maintenance of their integrity before and during electrophoresis are important to isozyme analysis. Buffers of various complexities have been used with or without protectants such as 2-mercaptoethanol, dithiothreitol, polyvinylpyrrolidone, and EDTA, or detergents such as Triton X-100 (Kephart 1990). The plant species and the particular tissue to be used influence the choice of buffer. It is usually preferable to keep samples cool during the grinding and centrifugation phases. It is possible to use distilled water for extraction under some circumstances or to avoid prior extraction altogether, as demonstrated in isoelectric focusing for analysis of glucose phosphate isomerase in wheat where half-grains were applied directly to the gel surface (Chojecki et al. 1983). Appropriate methods for grinding the samples vary widely, depending on available resources, number of samples, crop species, and the innovativeness of the researcher. These include squeezing of extracts from leaves in an eppendorf tube, grinding in a mortar, and various mechanical devices.

Support media that have been used for electrophoresis of isozymes include filter paper, agar, cellulose acetate, starch, agarose, and polyacrylamide. The last four have been the most popular in recent times, each with advantages and some limitations (Houwing and van Dreven 1986). The pore size of the matrix determines the extent of molecular sieving during electrophoresis. This size is controlled by the concentration of starch, agarose, or acrylamide, which usually ranges between 6% and about 15%. In the case of acrylamide, the level of cross-linkage is also important and this depends on the ratio of acrylamide to bis-acrylamide (Maurer 1971).

The common methods of isozyme separation have been isoelectric focusing and conventional electrophoresis in a native or non-dissociating medium. In the former, molecular sieving is minimal and molecules move in the electric field accross a pH gradient until they reach their isoelectric points where they stop (Righetti 1983). In the latter, the protein molecules adopt charges due to the pH of the medium and move in the electric field at a mobility determined by the size and sign of their charge as well as their molecular size (Maurer 1971). The electrical power supply to the run is usually a compromise between the desire for a quick separation that would give high resolution of bands and the need to avoid overheating and distortions due to excessive voltage. Usually the best conditions for separation are determined empirically, depending on factors such as gel concentration and dimensions, ionic strength and pH of electrode and gel buffers, temperature, heat sensitivity of the enzyme, and (in the case of electrofocusing) the pH range of the gel. Broad guidelines are available, which are useful for many crops (Shields et al. 1983; Hussain et al. 1988).

Gels may be set up as rods in glass tubes or as slabs. The former are usually run vertically whereas the latter may be run horizontally as well, depending on what electrophoresis cell is available. Slab gels facilitate comparison of several samples simultaneously. Continuous or discontinuous buffer and pH systems may be used. Continuous systems are simpler but discontinuous systems are often used for higher resolution (Maurer 1971).

Detection of isozymes in the support medium after separation follows action by the enzyme on an appropriate substrate(s) provided in solution and the interaction of the products with diazonium salts (such as fast blue BB) and other compounds that precipitate as insoluble dyes. The reaction may also be coupled to a reduction of tetrazolium salts into insoluble formazan dyes through mediation by nicotinamide adenine dinucleotide (or its phosphate) and phenazine methosulphate (Vallejos 1983). Many recipes have been published for the different enzymes (Shaw and Prasad 1970; Vallejos 1983). Interpretation of the pattern obtained can be simple where a few well-separated bands are involved or it may be complicated enough to make investment in a densitometer worthwhile, especially if many runs are involved. Depending on the objectives of the researcher, it is important to verify the genetic basis of observed patterns through isozyme analysis of progenies. Shields et al. (1983) provided examples of the genetic interpretations of electrophoretic patterns.

Applications

Isozyme analysis has had many useful applications in plant genetics and breeding (Tanksley and Orton 1983; Asiedu et al. 1990; Kephart 1990). Within the limits of the correlation for variability at isozyme loci and variability at loci for agronomic traits, isozyme polymorphisms have been useful indicators of the diversity of genotypes (Brown et al. 1978; Asiedu et al. 1988).

The identification of cultivars and germplasm accessions through isozyme analysis offers far more precision than using morphological descriptors alone. It facilitates establishment of genetic purity, location of duplicate accessions, and identification of somaclonal variants (Hussain et al. 1987; Ramirez et al. 1987; Ryan and Scowcroft 1987).

Confirmation of the authenticity of hybrids can be difficult in the absence of reliable genetic markers. Isozyme loci have been used effectively for this purpose (Asiedu and Mujeeb-Kazi 1987).

Alien chromosome segments carrying loci of important traits have been traced through large populations of breeding material using isozyme markers, at a pace that would be difficult to match using other techniques. The most celebrated example is the *1BL/1RS* translocation in some wheat lines, which carries genes for rust resistance as well as a locus for the isozyme glucose phosphate isomerase (GPI). Stickiness of dough in bread making has been associated with the presence of the translocation, making it important for wheat breeders in Europe, Australia, and the Americas to detect the translocation in their materials. The method of choice, for most programmes, has been to assay for GPI using isoelectric focusing (Chojecki et al 1983).

It has been demonstrated that the use of isozymes markers can reduce the number of backcrosses required in the transfer of traits into a cultivar through repeated selection of progenies with a higher proportion of recurrent parent alleles (Tanksley et al. 1981). Isozyme analyses could be used to test the normality of gene flow in segregating generations of interspecific crosses (Rick 1969), since alleles do not often segregate according to Mendelian ratios in such crosses.

Arguably the most exciting use of an isozyme marker is for the tagging of a plant gene of economic interest, following establishment of a close linkage between the locus of the isozyme structural gene and that of the economic/agronomic trait. An example is the use of the endopeptidase locus in wheat to screen for resistance to eyespot disease (Doussinault et al. 1983; Riley 1989).

Kephart (1990) lists many other instances in which breeders and geneticists have used isozyme analysis for various purposes, including the elucidation of clonal diversity in apomicts, breeding and mating systems, insect x plant interactions, and phenotype x environment interactions.

It is important to note, however, that isozyme analysis will operate most effectively in a breeding programme when used in a complementary fashion with other markers in an effort to gain maximum control and understanding of the normally complex genetic traits breeders have to work with.

References

Asiedu, R., and A. Mujeeb-Kazi. 1987. Diagnostic applications of isozyme markers in wheat *(Triticum* spp.) wide crosses. *Agronomy Abstracts* (1987): 54.

Asiedu, R., A. Mujeeb-Kazi, C. Ayala, and V. Rosas. 1988. Genetic variability among some accessions of intermediate wheat-grass (*Thinopyrum intermedium*): Implications for wheat wide-cross breeding. *Agronomy Abstracts* (1988): 72.

Asiedu, R., J.M. Fisher, and C.J. Driscoll. 1990. Resistance to *Heterodera avenae* in the rye genome of triticale. *Theoretical and Applied Genetics* 79: 331-336.

Brown, A.H.D., E. Nevo, D. Zohary, and O. Dagan. 1978. Genetic variation in natural populations of wild barley (*Hordeum spontaneum*). *Genetica* 49: 97-108.

Chojecki, A.J.S., M.D. Gale, L.M. Holt, and P.I. Payne. 1983. The intrachromosomal mapping of a glucose phosphate isomerase structural gene, using allelic variation among stocks of Chinese Spring wheat. *Genetic Research* 41: 221-226.

Doussinault, G., A. Delibes, E. Sanchez-Monge, and F. Garcia-Olmedo. 1983. Transfer of a dominant gene for resistance to eyespot disease from a wild grass to hexaploid wheat. *Nature (London)* 303: 698-700.

Houwing, A., and F. van Dreven. 1986. A proposed method of polyacrylamide gel electrophoresis in acid environments applied to gliadins of wheat grains. *Euphytica* 36: 55-60.

Hussain, A., W. Bushuk, H. Ramirez, and W. Roca. 1987. Identification of cassava (*Manihot esculenta* Crantz) cultivars by electrophoretic patterns of esterase isozymes. *Seed Science and Technology* 15: 19-22.

Hussain, A., W. Bushuk, H. Ramirez, and W. Roca. 1988. *A Practical Guide for Electrophoretic Analysis of Isoenzymes and Proteins in Cassava, Field Beans and Forage Legumes.* CIAT/University of Manitoba, Canada.

Kephart, S.R. 1990. Starch gel electrophoresis of plant isozymes: A comparative analysis of techniques. *American Journal of Botany* 77: 693-712.

Markert, C., and F. Moller. 1959. Multiple forms of enzymes: Tissue, ontogenetic, and species specific patterns. *Proceedings of the National Academy of Science, USA* 45: 753-763.

Maurer, H.R. 1971. *Disc Electrophoresis and Related Techniques of Polyacrylamide Gel Electrophoresis.* Walter de Gruyter, Berlin, Germany.

Ramirez, H., A. Hussain, W. Roca, and W. Bushuk. 1987. Isozyme electrophoregrams of sixteen enzymes in five tissues of cassava (*Manihot esculenta* Crantz) varieties. *Euphytica* 36: 39-48.

Rick, C.M. 1969. Controlled introgression of chromosomes of *Solanum pennelli* into *Lycopersicon esculentum*: Segregation and recombination. *Genetics* 62: 753-768.

Righetti, P.G. 1983. *Isoelectric Focusing: Theory, Methodology and Applications.* Elsevier, Amsterdam, The Netherlands.

Riley, R. 1989. Plant biotechnologies in developing countries: The plant breeder's perspective. Pages 85-94 in *Plant Biotechnologies for Developing Countries.* Proceedings, CTA/FAO Symposium, 26-30 June 1989, Luxembourg. CTA/FAO, Wageningen, the Netherlands.

Ryan, S.A., and W.R. Scowcroft. 1987. A somaclonal variant of wheat with additional ß-amylase isozymes. *Theoretical and Applied Genetics* 73: 459-464.

Shaw, C.R., and Prasad R. 1970. Starch gel electrophoresis of enzymes — A compilation of recipes. *Biochemical Genetics* 4: 297-320.

Shields, C.R., T.J. Orton, and C.W. Stuber. 1983. An outline of general resource needs and procedures for the electrophoretic separation of active enzymes from plant tissue. Pages 443-468 in *Isozymes in Plant Genetics and Breeding* (Part A) edited by S.D. Tanksley and T.J. Orton. Elsevier, Amsterdam, The Netherlands.

Tanksley, S.D., H. Medina-Filho, and C.M. Rick. 1981. The effect of isozyme selection on metric characters in an interspecific backcross of tomato — Basis of an early screening procedure. *Theoretical and Applied Genetics* 60: 291-296.

Tanksley, S.D., and T.J. Orton (eds). 1983. *Isozymes in Plant Genetics and Breeding.* Elsevier , Amsterdam, The Netherlands.

Vallejos, E. 1983. Enzyme activity staining. Pages 469-515 in *Isozymes in Plant Genetics and Breeding* (Part A) edited by S.D. Tanksley and T.J. Orton. Elsevier, Amsterdam, The Netherlands.

5.7

Equipping a laboratory for RFLP analysis

R.L. Jarret[1] and N. Gawel[2]

Southern Regional Plant Introduction Station, United States Department of Agriculture (USDA)/Agricultural Research Service (ARS),1109 Experiment Street, Griffin, Georgia 30223, USA[1]; Department of Horticulture, University of Georgia, 1109 Experiment Street, Griffin, Georgia 30223 USA[2]

Abstract

In most instances, laboratory facilities for RFLP analysis can be created by modifying existing facilities. Space requirements include sufficient laboratory bench space, an isolated area for handling radionucleotides, and a photographic darkroom. The equipment needs listed in this paper will vary with the specific research objectives.

Few, if any, of the pieces of laboratory equipment listed in Tables 1, 2 and 3 could be considered specific for the detection of restriction fragment length polymorphisms (RFLPs). Most items on these lists can be found in laboratories conducting molecular-level research in numerous disciplines, including plant pathology, genetics, and biology. However, for the purpose of preparing a compete list of the equipment needed for RFLP analysis, items have been grouped according to their general availability. Equipment and facilities that tend to be available in laboratories and institutes performing both basic and applied research have been listed separately (Table 1) from items more likely to be found in association with basic research programmes (Tables 2 and 3); access to the latter may be more limited.

All the equipment listed in Table 1 (*overleaf*)may be considered essential for RFLP analysis (Blumberg 1987). However, certain items may be substituted. For example, a crushed ice maker is not essential. Shaved ice or ice cubes work equally well and may be obtained locally from various sources. A chemical fumehood provides not only an area for safe handling of volatile organic solvents, but also serves as a containment area for manipulations with radionucleotides. The autoclave should ideally be of sufficient size to accept large flasks or carboys for sterilization. Water of molecular biology grade may be obtained from a glass still or by ultrafiltration (reverse osmosis) and subsequent deionization.

Table 1 General facilities and equipment recommendations

Item
Autoclave and sterilization tape
Balances (analytical and top-loading)
Bunsen burner
Chemical fumehood
Dissecting tools (scalpels, scissors, forceps, etc.)
Freezer
Glassware washer or facilities for washing glass
Heating/magnetic stirrer plates
Ice maker and ice buckets
Laboratory glassware, including large graduated cylinders, erlenmeyer flasks, beakers, etc.
pH meter
Refrigerator
Timers
Water purification system or glass still

Specialized Facilities and Equipment

The equipment listed in Table 2 is more specific for molecular-level research, including RFLP analysis. While some items in this category are more readily substituted than those listed in Table 1, all are essential, or highly desirable, for RFLP analysis. For short-term projects, it may be more efficient to share or borrow items on this list, if possible. If a long-term programme involving detection of RFLPs is planned, it is probably worthwhile investing in the listed pieces of equipment or substitutes for them.

We routinely use a freeze-dryer to lyophilize leaf tissue prior to DNA extraction. This allows us to accumulate samples for analysis over a period of time. Alternatively, DNA may be extracted directly from freshly harvested tissues or from tissues stored in liquid nitrogen. During the extraction process, yields may be increased by grinding samples to a fine powder in liquid nitrogen with a mortar and pestle. A waterbath, screw-cap test tubes resistant to organic solvents, a vortex mixer, and test-tube racks may be required, depending on the extraction technique used.

Investigators using RFLPs can expect to spend a considerable amount of time selecting and maintaining bacterial cultures. A laminar-flow hood, bacterial inoculating loops, test tubes, a platform shaker, and a 37°C incubator or temperature-controlled room are desirable. Glycerol cultures of transformed bacteria can be safely stored for up to a year at -20°C or for much longer periods at -70°C. A non-self-defrosting freezer is required for storage of bacterial cultures and restriction enzymes. In addition to storage of bacterial glycerol cultures, a -70°C freezer is also useful in reducing film exposure time during autoradiography (Maniatis et al. 1982).

A high-speed, refrigerated centrifuge is indispensable for DNA extractions and recovery of cells from large-scale bacterial cultures. On the other hand, a microcentrifuge is essential for the small-scale isolation of plasmid DNA, for recovery of DNA following ethanol

Table 2 Specialized facilities and equipment needs

Item
Bacterial inoculating loops
Darkroom with supplies for X-ray film development
Electrophoresis power supplies (high and medium voltage ranges) and gel apparatus, including mini-gel and larger gel box sizes with combs of various sizes
Dewer flask(s) or other isothermic container(s)
High speed centrifuge (e.g., Beckman J2-21) with medium capacity rotor (e.g., Beckman JA-20)
Heating block (to hold microcentrifuge tubes)
Incubator(s) (37°C) or temperature controlled room
Laminar-flow hood
Microcentrifuge
Microfiltration apparatus
Micropipetters (1-20 µl, 10-100 µl, and 100-1000 µl)
Peristaltic pumps and tubing
Plastic and pyrex (or glass) baking dishes
Racks to hold test tubes and microcentrifuge tubes
Radiation safety equipment, including safety glasses, laboratory coats, waste-disposal storage containers, plexiglas shields, geiger counter, etc.
Rotary platform shaker
Scintillation counter
Spectrophotometer with quartz cuvettes
Transilluminator
Vacuum oven or UV-crosslinker
Waterbath (ambient temperature to boiling)
Vortex mixer
X-ray film cassettes, intensifying screens, and film hangers

precipitation, and for DNA purification. An ultracentrifuge (Table 3 *overleaf*) may or may not be necessary for the preparation of DNA sufficiently pure for digestion with restriction endonuclease. Nevertheless, an ultracentrifuge can be extremely useful for other purposes (Maniatis et al. 1982). We use a spectrophotometer to assay DNA preparations for quantity and quality.

Following digestion, DNA fragments are separated on horizontal agarose gels. These gels are generally run between 50 and 100V constant in Tris-EDTA-acetic acid (TEA) buffer, and appropriate power supplies are required. A range of gel box sizes and combs is recommended; they are well worth the investment. Peristaltic pumps are used for buffer recirculation. No special equipment is required for the transfer of DNA fragments to the solid support. However, prior to blotting, the investigator may wish to examine the digestion on a transilluminator after staining the gel with ethidium bromide (a carcinogen). After blotting, membranes are baked in a vacuum oven or treated by shortwave UV light.

The detection of RFLPs is accomplished by labelling a DNA sequence with ^{32}P and allowing that sequence to hybridize with homologous sequences on the membrane. The amount of ^{32}P incorporated is determined with a scintillation counter. Alternatively, a crude estimate can be obtained with a geiger counter. Labelled sequences may be separated from

Table 3 Optional facilities or equipment

Item/facility	Possible uses
Cold room	DNA extraction
Dot-blot apparatus	quantification of insert DNA
Fraction collector	sample isolation
Freezer (-7°C)	bacterial culture storage and autoradiography
Glassware oven	mRNA isolation
Gradient former	organelle isolation
Hand-held UV lamp	DNA detection
Large-capacity rotor for high-speed centrifuge	organelle isolation, maxi-preps
Lyophilizer	useful for tissue preparation
Microwave oven	gel preparation
Refractometer	DNA quantification
Scanning densitometer	quantification of signal
Shaking waterbath(s)	DNA extraction, hybridization
Sonicator	shearing DNA
Spin vac	sample concentration
Ultracentrifuge with swinging bucket and fixed-angle rotors	DNA purification

Table 4 Representative examples of principle expendable supplies and their possible uses

Item	Uses
Agarose	gel electrophoresis
Chromatography columns	isolation of labelled probe
Micropipet tips	various
Microcentrifuge tubes	various
Nylon membrane (or nitrocellulose)	Southern blotting
Paper towels	Southern blotting
Petri dishes	bacterial plating, culture
Pipets (disposable)	various
Plastic or latex gloves	various
Plastic wrap	wrapping membranes
Radionucleotide (^{32}P)	probe labelling
Restriction endonucleases	DNA digestion
Whatman 3 mm filter paper	Southern blotting
X-ray film	autoradiography

free radionucleotides on disposable chromatography columns (Sephadex G-50). In our laboratory, membranes are placed in plastic or glass baking dishes and pre-hybridizations and hybridizations are performed in a 65°C incubator; the incubator also serves as a platform shaker. A sonicator is useful for shearing of salmon-sperm DNA, which is generally included in hybridization solutions; however, the DNA may also be sheared manually.

Hybridized sequences are detected by placing membranes on X-ray film in autoradiography film cassettes, equipped with intensifying screens. Film is exposed as required at -20°C or -70°C and then developed.

Expendable disposable supplies (Table 4) can represent a large expense during RFLP analysis. Chief among these supplies are microcentrifuge tubes and plastic micropipet tips. If a large number of samples are to be analysed, restriction enzymes, nylon membrane, agarose, and radionucleotide can be quite expensive.

Handling Radionucleotides

Great care must be exercised during the receipt, handling, and disposal of radionucleotides. Each manipulation involving ^{32}P should be planned carefully in advance, to minimize exposure to the user and to reduce the potential for spillage on, or contamination of, adjacent objects. Liquid waste can be collected in 5-gallon plastic jerricans and transferred to 30 or 50 gallon steel drums with removable tops for storage during radioactive decay. Waste disposal is an institute-specific issue (Zoon 1987) and all national, state, and local guidelines should be strictly adhered to.

Discussion

JARRET: As I mentioned during my presentation, it is difficult to discuss the cost of equipping an RFLP laboratory in general terms since the cost will vary dramatically in response to the available equipment and facilities. To initiate a laboratory and equip it completely, I would estimate a cost of about US $200 000 to US $300 000 at present.

References

Blumberg, D.D. 1987. Equipping the laboratory. Pages 3-19 in *Guide to Molecular Cloning Techniques* edited by S.L. Berger and A.R. Kimmel. Academic Press, San Diego, USA.

Maniatis, T., E.F. Fritsch, and J. Sambrook. 1982. *Molecular Cloning: A Laboratory Manual.* Cold Spring Harbor Laboratory, Cold Spring Harbor, New York, USA.

Zoon, R.A. 1987. Safety with ^{32}P- and ^{35}S-labeled compounds. Pages 25-29 in *Guide to Molecular Cloning Techniques* edited by S.L. Berger and A.R. Kimmel. Academic Press, San Diego, USA.

6.1

Virus-engineered resistance: Concepts, efficacy, and stability

C. Fauquet[1] and R.N. Beachy

Department of Biology, Washington University, St Louis CB 1137, Missouri 63130, USA (affiliated to the Institut Français de Recherche Scientifique pour le Développement en Coopération, ORSTOM[1])

Abstract

This paper reviews four strategies being used in research on virus resistance through genetic engineering in terms of their concepts, efficacy, and stability. The most recent strategy is the use of ribozymes, which involves using catalytic RNAs to cause the cleavage of viral RNA molecules. Reports of in vitro experiments have documented cleavage of viral RNAs but there are no reports of in vivo experiments. The second strategy involves an attempt to block the translation of viral RNAs by expressing genes that encode sequences complementary to viral genes. Only a few cases have been reported, and little success achieved. The third strategy is to block virus replication, including the use of competitor sequences, subgenomic sequences, and the expression of satellite RNA. This strategy has been applied with some success in the field. The fourth and most promising strategy involves integrating a gene encoding viral coat protein (CP) into the plant genome. An increasing number of examples of the use of this strategy have been reported, and resistance specificity and efficacy have been evaluated for viruses belonging to 10 groups; furthermore, several successful field experiments have been conducted. The paper summarizes CP-mediated resistance and puts forward hypotheses on the possible mechanisms of resistance.

The recent development of gene transfer technologies has made it possible to transfer useful traits to a number of crop plants. Among the various possibilities, conferring of resistance to viruses has probably been the most successful application of plant genetic engineering. When used to complement current breeding programmes, these technologies can help control plant viruses and reduce their impact on crop productivity. Over the past 5 years, a variety of molecular strategies have been used in attempts to control viruses. Some of these are still under investigation in laboratories; others have reached the field testing stage.

This paper reviews these various strategies, discussing the state of development, the efficacy, and the stability of each of them. Particular emphasis is placed on coat protein (CP) mediated resistance because of the success of this strategy and the large amount of data and examples now available.

Ribozyme Strategy

A new approach to achieving virus resistance is the use of autocatalytic RNA cleaving molecules, known as 'ribozymes' (Cech 1986; Kim and Cech 1987). In viroid RNAs, such as avocado sunblotch viroid, and satellite RNAs, such as the satellite of tobacco ring spot virus (TobRSV), there is the possibility of self-cleavage during replication (Buzayan et al. 1986; Hutchins et al. 1986; Prody et al. 1986; Forster and Symons 1987). The sites of cleavage are intramolecular and presumably occur when the RNA molecule is in the correct configuration, thereby activating the cleavage reactions. Cleavage is effective on the positive and negative strand of the RNA; it is highly specific and is associated with conserved sequence domains. Several studies have been conducted to determine the optimal in vitro conditions of cleavage (Haseloff and Gerlach 1988; Gerlach 1989). Genes encoding sequences bearing specific virus cleavage sites have been integrated into transgenic plants and should generate sequence specific endonuclease activities. Constructs have been made to inactivate various viruses, including tobacco mosaic virus (TMV) and barley yellow dwarf virus (BYDV) (Gerlach 1989) but, to date, no in vivo results have been published.

Translation Strategy

Translation strategy involves the integration into the plant genome of sequences generating complementary sequences to viral RNA that interfere with the translation of viral genes by hybridization of the coding sequence. It has been reported that synthesis of complementary RNA (antisense RNA) can reduce the accumulation of gene products in both procaryotes and eukaryotes (Ecker and Davis 1986; Green et al. 1986; Rothstein et al. 1987). It is likely that antisense RNAs anneal with sense RNAs to form a double strand complex, which is rapidly degraded or which inhibits translation of the RNA. Several viral CP antisense constructs, including TMV, cucumber mosaic virus (CuMV), and potato virus X (PVX), have been integrated into plants and the plants tested for resistance to infection (Cuozzo et al. 1988; Hemenway et al. 1988; Powell et al. 1989). In all cases, resistance has been reported against infection by the homologous virus but only at low virus inoculum concentrations. In addition to CP antisense sequences, other genes have been tested with the antisense strategy, but to date little or no resistance to virus infection has been reported (Beachy et al. 1987; Rezaian et al. 1988).

Replication Strategy

One of the first steps in virus multiplication is the replication of the viral genome and it seems logical that blocking this phase should be an efficient way to protect plants. By replication strategy we mean the blockage of the replication of the virus. Two approaches to blocking a virus infection have been considered, the antisense and the sense approaches.

Antisense approach

The replication strategy using the antisense approach attempts to block the replication of a virus by hybridization of complementary sequences to the replicase viral gene or to sequences recognized by the replicase during replication. This strategy is at a very preliminary stage of investigation but is presented because of promising in vitro results and the possibility that it may be applicable when other approaches fail. In transient assays with protoplasts of wheat, the antisense sequence of the first 250 nucleotides of the replicase gene of the geminivirus wheat dwarf mosaic virus (WDMV) completely inhibited virus replication (Gronenborg 1990). A second example involved the geminivirus tomato golden mosaic virus (TGMV). The complete antisense sequence of the replicase gene of the TGMV was integrated into the tobacco genome and several lines were reported to exhibit a level of resistance when challenged with varying concentrations of TGMV (Lichtenstein and Buck 1990). This approach is an interesting alternative but must be further tested before it can be considered as a useful and practical strategy. A final example concerns turnip yellow mosaic virus (TYMV), where antisense sequences corresponding to the tRNA-like structure of the 3' extremity of the TYMV RNA have been shown to strongly inhibit replicase activity (Cellier et al. 1990). Transgenic plants that produce such sequences are under evaluation.

Sense approach

The second approach to reduce replication involves the expression of "sense" viral sequences.

Competitor RNA

Sense sequences comprising the above-mentioned tRNA-like structure of TYMV have been used to compete with similar viral sequences and thus reduce virus replication activity. In vitro studies have demonstrated such competition (Morch et al. 1987; Cellier et al. 1990;) and in vivo experiments are currently in progress to confirm these results. In contrast, a similar approach used with TMV seemed not to induce any resistance (Powell et al. 1990).

Subgenomic DNA

Some viruses produce subgenomic molecules during virus infection; for example, several geminiviruses produce subgenomic DNA molecules of the B component. Insertion of one copy of such DNA of the African cassava mosaic virus (ACMV) into the tobacco genome reduced disease symptoms when the plants were challenged with ACMV (Frischmuth et al. 1990; Stanley et al. 1990). Symptom amelioration is associated with a general reduction in the level of viral DNA, including B DNA which is responsible for symptoms, while the subgenomic DNA is specifically amplified. This type of resistance is specific to ACMV; other geminiviruses are unable to amplify the subgenomic of ACMV.

Satellite RNA

Another approach to conferring protection against viruses is to induce the expression of virus satellite (SAT) RNAs. SAT RNAs are associated with several viruses and are dependent upon a helper virus for their replication and spread in the infected plant. It has been reported that the presence of SAT RNAs in CuMV-infected tobacco reduces disease symptoms (Mossop and Francki 1979; Collmer et al. 1983). Similarly, in tobacco plants, infection with a mixture of TobRSV and SAT TobRSV led to an amelioration of symptoms (Gerlach et al. 1986). Transgenic plants that express these satellite sequences were tested for disease development and viral replication with the corresponding virus. Amelioration of symptoms and reduced virus replication was the result in both cases (Gerlach et al. 1987; Harrison et al. 1987; Jacquemond et al. 1988). However, when the CuMV/SAT RNA expressing plants are infected with a related but different cucumovirus, there is symptom amelioration but no reduction in virus replication.

For both systems, some systemically infected plants developed symptoms although they were less severe than those in the control plants. Recently, this strategy has been applied to tomato (Tien et al. 1990; Tousch et al. 1990), and has proved to be efficient for the reduction of symptoms in both greenhouse and field experiments (Tien et al. 1990). Not all satellite sequences provide symptom attenuation and sometimes they can cause necrosis; the sequences responsible for severe symptoms are reduced to a few nucleotides (Devic et al. 1990; Jaegle et al. 1990). There is a risk that amplifying a satellite in transgenic plants may result in some of the molecule reverting to a necrotic form, causing dramatic symptoms when naturally infected by the helper virus. This possibility will greatly limit the use of the SAT RNA strategy unless further studies demonstrate a high degree of stability in the system.

TMV replicase strategy

Recently, a new source of genetically engineered resistance has been reported, involving the transformation of plants with non-structural viral genes. Tobacco plants transformed with the TMV 54 kDa gene, which is derived from a portion of the replicase complex, are immune to extremely high concentrations of TMV virions or RNA (up to 500 μg/mL) of the strain U1 (Golemboski et al. 1990). This immunity is extremely specific to the strain U1, or a mutant YS1/1, of TMV and susceptible to the other strains, including the related U2 strain.

Coat Protein Strategy

Among the different strategies for controlling viruses by genetic engineering, the CP strategy currently appears to be the most promising. Many examples have been published and a number of experiments are in progress. Efficiency in terms of protection, stability, and specificity has been evaluated for several cases. The type of resistance and the mechanisms of action of the CP-mediated resistance have been investigated and there is now a large amount of information available. Laboratory experiments and field tests with various crops have been carried out and the first commercial use of this type of resistance has been forecast for 1995.

Definition, concept, and production of coat protein-mediated resistance

CP-mediated resistance refers to the resistance to virus infection caused by the expression of a CP gene in transgenic plants. The expression of a CP gene confers resistance to the virus from which the CP gene was derived and to related viruses. Resistance is not due to somaclonal variation caused by the transformation event or the tissue culture procedure, and it is stably inherited by subsequent generations.

The principle of the CP strategy is the expression by the plant of the viral CP gene integrated into the plant genome. Construction of the chimaeric gene should include the selection of an appropriate transcriptional promoter to induce the expression of the CP gene at sufficient levels and with sufficient tissue specificity to produce disease resistance. Several different transcriptional promoters have been used and the one that has proved most effective is the CaMV 35S. This promoter leads to high levels of mRNA and protein in most of the plants in which it has been tested. Furthermore, an enhanced 35S promoter, pE35S, which is produced by duplication of an upstream regulatory sequence (Kay et al. 1987) causes even higher levels of gene expression. The coding region used for the gene is obtained by deriving double-stranded DNA from the virus genome. When necessary, specific mutations of the gene may be used to increase the translation of any mRNA, following the consensus sequence rules described by others (Kozak 1988). These and other changes can increase the amount of CP produced in transgenic plants. The third part of the chimaeric genes is a sequence to confer transcript termination and polyadenylation. Little evidence has been published to date to indicate that a specific 3' end is preferable in transgenic plants. The 3' ends used for most chimaeric genes expressed in plants have been taken from the T-DNA region of the Ti plasmid (Powell et al. 1986, 1990; Nelson et al. 1988).

Assessing disease resistance involves the inoculation of plants with viruses that express the CP gene (CP+) and those that do not (CP-). A comparison of numbers of sites of infection, disease incidence, development of disease symptoms, and accumulation of virus is generally used to evaluate resistance.

In order to use populations of plants that are identical in age, growth rate, and size, R1 or successive generations of plants are used. Prior to inoculating seedlings with a virus, the segregation of the introduced gene is usually determined by an immunological reaction to detect the CP or by following the expression of a gene that is introduced with the CP gene.

Efficacy of coat protein-mediated resistance

The efficacy of CP-mediated resistance is indicated by the number of examples where resistance has been achieved using this technique, by the spectrum of specificity of protection, and by the type of resistance achieved.

Specificity of resistance

Since 1986, the date of the first publication describing the CP strategy (Powell et al. 1986), there have been reports of CP-mediated resistance involving a variety of viruses and host plants. A list of published and unpublished reports is presented in Table 1. It includes viruses belonging to 10 virus groups and hosts that include members of the Chenopodiaceae, Leguminosae, and Solanaceae. A detailed review of CP-mediated resistance has been recently published (Beachy et al. 1990).

Spectrum of resistance

A study on the spectrum of resistance of transgenic plants expressing a CP gene was carried out on tobamoviruses (Anderson et al. 1989; Nejidat and Beachy 1990). In these studies, CP(+) tobacco lines that expressed the U1 TMV CP gene were inoculated with other tobamoviruses. Based upon comparisons of amino acid sequences and/or amino acid composition of virus CPs (Gibbs 1986), tobamoviruses have degrees of relatedness to TMV ranging from 85% to 39%. Infection by TMV, ToMV, pepper mild mosaic virus (PMMV), and tomato mild green mosaic virus (TMGMV) was inhibited by 95-98%, by 80-95% for ondontoglossum ringspot virus (ORSV), and by 40-60% for RMV and sunnhemp mosaic virus (SHMV) Nejidat and Beachy 1990). On the basis of these studies, it was concluded that viruses that are related to TMV on the basis of CP by greater than 50-60% are less able to infect resistant lines than are less closely related tobamoviruses. TMV CP(+) plant lines were also inoculated with members of different virus groups including CuMV, alfalfa mosaic virus (AlMV), PVX, and potato virus Y (PVY). There was no resistance against infection by any of the viruses on inoculated leaves of CP(+) plants; however, there were somewhat reduced rates of systemic spread of CuMV, PVX, and PVY in CP(+) compared with CP(-) plants (Anderson et al. 1989).

Experiments on CP protection against potyviruses, which are particularly important because many economically important plant viruses belong to this virus group, have shown that expression of the CP gene of soybean mosaic virus protected tobacco plants from infection by tobacco etch virus (TEV) and potato virus Y (PVY) (Stark and Beachy 1989). The CP gene sequences of these potyviruses are about 55-60% homologous. In the case of tobraviruses, heterologous protection is also effective for viruses having about 60% homology (van Dun and Bol 1988). The specificity of CP-mediated resistance is restricted to members of the same virus group and within a group. Results to date suggest the need for more than 60% amino acid sequence homology of CPs to result in heterologous protection in several of the virus groups tested.

Multiple manifestations of resistance

Resistance to inoculation

In each of the examples of CP-mediated resistance described above, resistance was manifested in several features. First, there was a reduction in the numbers of sites of infection on inoculated leaves. Fewer starch lesions were produced after inoculation with

Table 1 Examples of coat protein (CP) mediated resistance in transgenic plants

Virus group	CP gene	Transgenic plant	Virus resistance	Reference
Tobamovirus	TMV	tobacco	TMV	Powell et al. 1986
	"	"	ToMV	Nelson et al. 1988
	"	"	PMMV	Nejidat and Beachy 1990
	"	"	TMGMV	"
	"	"	ORSV	"
	"	tomato	TMV	Nelson et al. 1988
	"	"	ToMV	"
	ToMV	tomato	ToMV	Sanders et al. (in press)
Tobravirus	TRV	tobacco	TRV	van Dun and Bol 1988
	"	"	PEBV	"
Carlavirus	PVM	potato	PVM	Wefels et al. 1990
	PVS	tobacco	PVS	McKenzie and Tremaine 1990
Potexvirus	PVX	tobacco	PVX	Hemenway et al. 1988
	"	potato	PVX	Lawson et al. 1990
Potyvirus	PVY	potato	PVY	"
	SMV	tobacco	PVY	Stark and Beachy 1989
	"	"	TEV	"
	ZYMV	tobacco	ZYMV	Namba et al. 1990
	WMV II	tobacco	WMV II	"
	PRSV	tobacco	PRSV	Ling et al. 1990
	"	"	TEV	"
Furovirus	BNYVV	beet (protoplast)	BNYVV	Kallerhoff et al. 1990
AlMV group	AlMV	tobacco	AlMV	Loesch-Fries et al. 1987
	"	"	"	Tumer et al. 1987
	"	"	"	van Dun et al. 1987
	"	tomato	"	Tumer et al. 1987
	"	alfalfa	"	Hill et al. 1991
Cucumovirus	CuMV	tobacco	CuMV	Cuozzo et al. 1988
	"	"	"	Nakayama et al. 1990
	"	tomato	CuMV	Cuozzo et al. 1988
Ilarvirus	TSV	tobacco	TSV	van Dun et al. 1988
Luteovirus	PLRV	potato	PLRV	Tumer et al. 1990
	"	"	"	Kawchuk et al. 1990

Note: a AlMV = alfalfa mosaic virus; BNYVV = beet necrotic yellow vein virus; CuMV = cucumber mosaic virus; ORSV = ondontoglossum ringspot virus; PEBV = pea early browning virus; PLRV = potato leafroll virus; PMMV = pepper mild mosaic virus; PRSV = papaya ringspot virus; PVM = potato virus M; PVS = potato virus S; PVX = potato virus X; PVY = potato virus Y; SMV = soybean mosaic virus; TEV = tobacco etch virus; TMGMV = tobacco mild green mosaic virus; TMV = tobacco mosaic virus; ToMV = tomato mosaic virus; TRV = tobacco rattle virus; TSV = tobacco streak virus; WMV II = watermelon mosaic virus II; ZYMV = zucchini yellow mosaic virus.

PVX on CP(+) tobacco plants than on CP(-) plants (Hemenway et al. 1988), and there were fewer chlorotic lesions caused by TMV infection on tobacco plants that expressed the TMV CP gene than on those that did not (Powell et al. 1986). Similarly, the numbers of necrotic local lesions caused by TMV infection on CP(+) Xanthi *nc* tobacco local lesion were between 95 and 98% lower than on CP(-) plants (Nelson et al. 1987). These experiments indicate that the expression of a CP gene causes a reduction in the number of sites where infection is established upon inoculation.

Resistance to virus spread within the plant

The second manifestation of resistance in CP-engineered plants is a reduced rate of systemic disease development throughout the CP(+) plants. Thus if inoculation results in infection on the inoculated leaves, the likelihood that the infection will become systemic is considerably lower in CP(+) plants than in CP(-) plants. Grafting studies in which a stem segment of a transgenic TMV(CP+) tobacco plant was inserted between the rootstock and apex of a wild tobacco, have demonstrated that the CP(+) segment prevented the virus from moving to the upper part of the grafted plant. Thus CP may play a role in the long distance movement of a virus and consequently resistance has a component that affects systemic spread of the infection, at least in the TMV-tobacco system (Wisniewski et al. 1990).

Resistance to symptom expression

A third manifestation of resistance is a reduced rate of disease development on systemic hosts that are CP(+). In all examples of CP(-) mediated resistance, CP(+) plant lines were less likely to develop systemic disease symptoms than those that were CP(-). Several plant lines that expressed the PVX CP gene did not become systemically infected when inoculated with high levels of virus (Hemenway et al. 1988). Similar results have been reported for CP-mediated resistance against CuMV (Cuozzo et al. 1988), TMV (Powell et al. 1986), and other viruses.

Resistance to virus multiplication

A last manifestation of resistance is lower accumulation of virus in CP(+) compared with CP(-) plant lines. The enzyme-linked immunosorbent assay (ELISA) and semi-quantitative western or immuno dot-blots have been used to quantify virus accumulation in inoculated leaves and other plant parts in most examples of CP-mediated resistance (Powell et al. 1986; Nelson et al. 1987; Cuozzo et al. 1988; Hemenway et al. 1988; Lawson et al. 1990). In some cases of resistance, plants accumulated little or no virus after inoculation (Hemenway et al. 1988; Lawson et al. 1990), and could be considered to be immune to infection under the conditions of the tests.

Manifestations of CP-mediated resistance can usually, but not always, be overcome by inoculating with relatively high concentrations of virus. A virus concentration of 10 µg/mL of TMV almost breaks the CP-mediated resistance to TMV in a system where 0.01 µg/mL causes disease in CP plants (Powell et al. 1986); 50 µg/mL is needed to overcome CP resistance to AlMV, PVX, PVY, and TEV (Tumer et al. 1987; Hemenway et al. 1988; Stark and Beachy 1989; Lawson et al. 1990). In many cases, resistance is largely overcome by inoculation with RNA rather than virions, with the exception of PVX CP(+) lines of tobacco and potato (Hemenway et al. 1988; Lawson et al. 1990).

Most CP-mediated resistance has been assessed by mechanical inoculation with the virus, but the most important criterion for virus resistance is its effectiveness under natural modes of contamination (that is, in vegetative propagation) and via natural vectors. Information regarding these points is limited but significant. In the case of dually engineered resistance against PVX and PVY in potato (Lawson et al. 1990), it has not been possible to recover either of the viruses in the tubers. At least one line of potato has shown a good level of resistance against aphid inoculation of PVY. Potato leafroll virus (PLRV), a member of the luteovirus group, is non-mechanically transmissible and all CP-engineered potato plants challenged by using aphid inoculation (Kawchuk et al. 1990; Tumer et al. 1990) have demonstrated some degree of resistance.

Stability of resistance

CP-mediated resistance is, in most cases, monogenic and is inherited by the subsequent generations, as with any other genetic trait. Consequently, the genetic stability of this resistance gene is expected to be the same as any other gene. The biological stability of such resistance can be questioned, but no answer can be provided until it is used under natural conditions. One may argue that a single point mutation can change a vital amino acid in the expressed CP and consequently alter the resistance, but the probability of this occurring will not be greater than for any other monogenic resistance gene. Furthermore, we know that CP-mediated resistance is effective for viruses differing by up to 40% in their CP sequences. In order to test the stability of the system, TMV CP(+) lines of tomato have been inoculated 15 times, successively, with the same isolate of virus, with the idea of selecting TMV molecules able to overcome the protection; however, no resistance-breaking strain has yet been identified (White and Beachy, pers. comm.).

Field experiments with coat protein-mediated resistant plants

There have been several field tests of virus-engineered resistant plants. Tomato plants expressing the TMV and ToMV CP genes have been tested in the field for several years; potato plants expressing the PVX and PVY CP genes have also been tested in the field.

Tobacco plants that expressed the CP gene of AlMV were field tested in Wisconsin in 1988 (K.J. Krahn, pers. comm.). CP(+) plants developed disease more slowly or not at all compared to CP(-) plants; symptom development was correlated with virus accumulation. At 85 days after inoculation, only 9% of the CP(+) plants had developed a systemic infection, while 93% of the CP(-) plants had systemic infection.

The first field test with TMV-resistant tomato plants (cultivar VF36) was conducted in 1987 (Nelson et al. 1988). CP(+) plants, mechanically inoculated with TMV, exhibited a delay in the development of disease symptoms or did not develop symptoms. Only 5% of the CP(+) plants developed disease symptoms by fruit harvest, while 99% of the inoculated CP(-) plants developed symptoms. Lack of visual symptoms was associated with lack of virus accumulation. Fruit yields of infected plants decreased 26-35% compared to healthy plants, whereas yields from the CP(+) line equalled those of uninoculated CP(-) plants.

To determine if the TMV CP gene conferred protection against infection by field isolates of ToMV, tests were conducted in 1988 in Florida and Illinois, USA. In Florida, progeny that were homozygous for the TMV CP gene and VF36 CP(-) plants were challenged with a Florida isolate of ToMV (Naples C). The field test in Illinois was conducted to determine whether expression of the TMV CP gene in tomato would protect against a number of different strains of TMV and ToMV. The TMV CP gene conferred resistance against ToMV-Naples C infection under Florida field conditions and against two strains of TMV under Illinois field conditions. Only weak protection was conferred against infection by the ToMV strains under Illinois field conditions (Sanders et al., in press).

To enhance protection against ToMV, plants were produced that expressed a CP gene derived from ToMV-Naples C. These lines were evaluated under field conditions in Illinois along with the control tomato line UC82B, with lines expressing the TMV CP gene, and with lines expressing both TMV and ToMV CP genes. The TMV CP(+) lines were resistant to TMV infection as shown in the earlier field test; however, they were less resistant to infection by ToMV-Naples C. The tomato lines expressing ToMV CP gene were highly resistant to infection by ToMV-Naples C. Plants that expressed both TMV and ToMV CP genes were equally well protected against TMV and ToMV.

Field tests were recently conducted with Russet Burbank potato plants expressing the CP genes of PVX and PVY (Lawson et al. 1990). PVY causes significant yield depression in potato and, in combination with PVX, it produces the severe disease rugose mosaic. To determine if expression of PVX and PVY CP genes would protect potato plants from the synergistic effects of PVX and PVY infection in the field, plants propagated from Russet Burbank CP(-) and from plant lines expressing both CP genes were inoculated with both PVX and PVY and transplanted into the field (Kaniewski et al. 1990). Plants from four CP(+) lines were significantly protected from infection by PVX. However, three of the lines were not protected from infection by PVY when simultaneously inoculated by both viruses. Plants of one line, however, were highly resistant to both PVX and PVY, as predicted from growth chamber tests. Tuber yields at maturity in non-inoculated plots were the same for all lines. In contrast, tuber yields of all inoculated lines were markedly reduced, except the line resistant to both viruses, which was unaffected by virus inoculation.

Conclusion

In the past few years a profusion of techniques to control viruses by genetic engineering have emerged. Most of them have demonstrated potential for conferring resistance in plants and provide hope for future crop improvement. Several techniques, such as the satellite and CP strategies, have already been applied in field experiments, and their potential to provide a source of resistance to viruses has been confirmed.

The ribozyme strategy is interesting because of its potential applicability to all viruses and its high specificity, but it needs to be demonstrated under in vivo conditions. The satellite strategy, although very effective, is strictly limited to viruses having satellites, which are very few in the plant virus world; its usage is questionable because of the risk that would accompany widespread use. The CP-mediated resistance for controlling viruses is the most safe, efficient, and widely documented type of engineered resistance. Several field experiments have been conducted with various crops, and CP-mediated protection has been shown to be effective in all cases. The specificity of CP-mediated resistance is broader than any other type of virus resistance, as plants can be protected against viruses of the same group sharing up to 60% in their CP sequence. CP transgenic plants are resistant to high concentrations of virus inoculum, and in one case even RNA inoculum. Resistance against vector inoculation and inoculation via vegetative propagation have also been demonstrated. The resistance generated by the CP strategy is of a multiple type, reducing the number of infection sites, symptoms, virus multiplication, and long-distance spread through plants. All these manifestations could be the result of a single mechanism or multiple effects of the CP on different targets. Despite the number of studies conducted to understand the mechanisms of action of the CP strategy, the question of how the CP confers resistance to engineered plants remains unanswered.

Discussion

ENE-OBONG: The major insect pest of cassava in West Africa is probably *Zonocerus variegatus*. Most plants are known to possess certain secondary compounds that protect them from insects. Obviously, cassava is not resistant to this insect, which means that *Z. variegatus* contains enough enzymes to break down the HCN released by linamarin bioconversion. Protection of cassava against this insect should be a good topic for study.

References

Anderson, E.J., D.M. Stark, R.S. Nelson, P.A. Powell, N.E. Tumer, and R.N. Beachy. 1989. Transgenic plants that express the coat protein genes of tobacco mosaic-virus or alfalfa mosaic-virus interfere with disease development of some non-related viruses. *Phytopathology* 79: 1284-1290.

Beachy, R.N., D.M. Stark, C.M. Deom, M.J. Oliver, and R.T. Fraley. 1987. Expression of sequences of tobacco mosaic virus in transgenic plants and their role in disease resistance. Pages 169-180 in *Tailoring Genes for Crop Improvement*. Plenum Press, New York, USA.

Beachy, R.N., S. Loesch-Fries, and N.E. Tumer. 1990. Coat protein-mediated resistance against virus infection. *Annual Review of Phytopathology* 28: 451-474.

Buzayan, J.M., W.L. Gerlach, and G. Breuning. 1986. Satellite tobacco ringspot virus RNA: A subset of the RNA sequence is sufficient for autolytic processing. *Proceedings of the National Academy of Science, USA* 83: 8859-8862.

Cech, T.M. 1986. The chemistry of self-splicing RNA and RNA enzymes. *Science* 236: 1532-1539.

Cellier, F., B. Zaccomer, M.D. Morch, A.L. Haenni, and M. Tepfer. 1990. Strategies for interfering in vivo with replication of turnip yellow mosaic virus. Page 123 in *Proceedings, Eighth International Congress of Virology, 26-30 August 1990, Berlin, Germany.* International Union of Microbiological Societies, Free University of Berlin, Germany.

Collmer, C.W., M.E. Tousignant, and J.M. Kaper. 1983. Cucumber mosaic virus associated RNA 5: X. The complete nucleotide sequence of a CARNA 5 incapable of inducing tomato necrosis. *Virology* 127: 230-234.

Cuozzo, M., K.M. O'Connell, W. Kaniewski, R.X. Fang, N.H. Chua, and N.E. Tumer. 1988. Viral protection in transgenic plants expressing the cucumber mosaic virus coat protein or its antisense RNA. *Bio/technology* 6: 549-557.

Devic, M., M. Jaegle, and D. Baulcombe. 1990. Cucumber mosaic virus satellite RNA (Y strain): Analysis of sequences which affect systemic necrosis on tomato. *Journal of General Virology* 90 : 1443-1449.

Ecker, J.R., and R.W. Davis. 1986. Inhibition of gene expression in plant cells by expression of antisense RNA. *Proceedings of the National Academy of Science, USA* 83: 5372-5376.

Forster, A.C., and R.H. Symons. 1987. Self-cleavage of plus and minus RNAs of a virusoid and a structural model for the active sites. *Cell* 49: 211-220.

Frischmuth, T., S. Ellwood, and J. Stanley. 1990. Amplification of geminivirus subgenomic DNA causes symptom amelioration in transgenic *Nicotiana benthamiana*. Page 455 in *Proceedings, Eighth International Congress of Virology, 26-30 August 1990, Berlin, Germany*. International Union of Microbiological Societies, Free University of Berlin, Germany.

Gerlach, W.L., H.P. Haseloff, M.J. Young, and G. Bruening. 1989. Use of plant virus satellite RNA sequences to control gene expression. Pages 177-184 in *Viral Genes and Plant Pathogenesis* edited by T.P. Pirone and I.G. Shaw. Springer-Verlag, New York, USA.

Gerlach, W.L., J.M. Buzayan, I.R. Scheider and G. Breuning. 1986. Satellite tobacco ringspot virus RNA: Biological activity of DNA clones and their in vitro transcripts. *Virology* 151: 172-185.

Gerlach, W.L., D. Llewellyn, and J. Haseloff. 1987. Construction of a plant disease resistance gene from the satellite RNA of tobacco ringspot virus. *Nature (London)* 328: 802-805.

Gibbs, A. 1986. Tobamovirus classification. Pages 168-180 in *The Plant Viruses (Vol. 2): The Rod-Shaped Plant Viruses*. Plenum Press, New York, USA.

Golemboski, D.B., G.P. Lomonossoff, and M. Zaitlin. 1990. Plants transformed with a tobacco mosaic virus nonstructural gene sequence are resistant to the virus. *Proceedings of the National Academy of Science, USA* 87: 6311-6315.

Green, P.J., O. Pines, and M. Inouye. 1986. The role of antisense RNA in gene regulation. *Annual Review of Biochemistry* 55: 569-597.

Gronenborg, B. 1990. Wheat dwarf virus vectors, a way to deliver chimaeric transposons to cereals. *Journal of Cellular Biochemistry* (Supplement 14E 1990): 160.

Harrison, B.D., M.A. Mayo, and D.C. Baulcombe. 1987. Virus resistance in transgenic plants that express cucumber mosaic virus satellite RNA. *Nature (London)* 334: 799-802.

Haseloff, J., and W.L. Gerlach. 1988. Simple RNA enzymes with new and highly specific endoribonuclease activities. *Nature (London)* 334: 585-591.

Hemenway, C., R.X. Fang, W.K. Kaniewski, N.H. Chua, and N.E. Tumer. 1988. Analysis of the mechanism of protection in transgenic plants expressing the potato virus X coat protein or its antisense RNA. *EMBO Journal* 7: 1273-1280.

Hill, K.K., N. Jarvis-Eagan, E.L. Halk, K.J. Krahn, L.W. Liao, R.S. Mathewson, D.J. Merlo, S.E. Nelson, K.E. Rashka, and L.S. Loesch-Fries. 1991. The development of virus resistant alfalfa, *Medicago sativa* L. *Bio/technology* 9(4): 373-377.

Hutchins, C.J., P.D. Rathjen, A.C. Forster, and R.H. Symons. 1986. Self cleavage of plus and minus RNA transcripts of avocado sunblotch viroid. *Nucleic Acids Research* 14: 3627-3640.

Jacquemond, M., A. Amselem, and M. Tepfer. 1988. A gene coding for a monomeric form of cucumber mosaic virus satellite RNA confers tolerance to CuMV. *Molecular Plant Microbe Interactions* 1: 311-316.

Jaegle, M., M. Devic, M. Longstaff, and D. Baulcombe. 1990. Cucumber mosaic virus satellite RNA (Y strain): Analysis of sequences which affect mosaic symptoms on tobacco. *Journal of General Virology* 90: 1905-1912.

Kallerhoff, J., P. Perez, D. Gérentes, C. Poncetta, S. Bentahar, and J. Perret. 1990. Sugar beet transformation for resistance to rhizomania. In *Proceedings, Eighth International Congress of Virology, 26-30 August 1990, Berlin, Germany*. International Union of Microbiological Societies, Free University of Berlin, Germany.

Kaniewski, W.K., C. Lawson, B. Sammons, L. Haley, J. Hart, X. Delannay, and N.E. Tumer. 1990. Field resistance of transgenic Russet Burbank potato to effects of infection by potato virus X and potato virus Y. *Bio/technology* 8: 750-754.

Kawchuk, L.M., R.R. Martin, and J. Mcpherson. 1990. Resistance in transgenic plants expressing the potato leafroll luteovirus coat protein gene in transgenic potato. *Molecular Plant Microbe Interactions* 3: 301-307.

Kay, R., A. Chan, M. Daly, and J. McPherson. 1987. Duplication of CaMV 35S promoter sequences creates a strong enhancer for plant genes. *Science* 236: 1299-1302.

Kim, S.H., and R.T. Cech. 1987. Three dimensional model of the active site of the self-splicing rRNA precursor of tetrahymena. *Proceedings of the National Academy of Science, USA* 84: 8788-8792.

Kozak, M. 1988. Leader length and secondary structure modulate mRNA function under condition of stress. *Molecular Cell Biology* 6: 2737-2744.

Lawson, C., W.K. Kaniewski, L. Haley, R. Rozman, C. Newell, P.R. Sanders, and N.E. Tumer. 1990. Engineering resistance to mixed virus infection in commercial potato cultivar: Resistance to potato virus X and potato virus Y in transgenic Russet Burbank. *Bio/technology* 8: 127-134.

Lichtenstein, C.L., and K.W. Buck. 1990. Expression of antisense RNA in transgenic tobacco plants confers resistance to geminivirus infection. *Journal of Cellular Biochemistry* (Supplement 14E 1990): 23.

Ling, K., S. Namba, C. Gonzalves, J.L. Slighton, and D. Gonzalves. 1990. Cross-protection against tobacco etch virus with transgenic tobacco plants expressing papaya ringspot virus coat protein gene. In *Proceedings, Eighth International Congress of Virology, 26-30 August 1990, Berlin, Germany.* Internat. Union of Microbiological Societies, Free University of Berlin, Germany.

Loesch-Fries, L.S., D. Merlo, T. Zinnen, L. Burhop, K. Hill, K. Krahn, N. Jarvis, S. Nelson, and E. Halk. 1987. Expression of alfalfa mosaic virus RNA4 in transgenic plants confers virus resistance. *EMBO Journal* 6: 1845-1851.

McKenzie, D.J., and J.H. Tremaine. 1990. Transgenic *Nicotiana debneyii* expressing viral coat protein are resistant to potato virus S infection. *Journal of General Virology* 71: 2167-2170.

Morch, M.D., R.L. Joshi, T.M. Denialt, and A.L. Haenni. 1987. A new sense RNA approach to block viral RNA replication in vitro. *Nucleic Acids Research* 15: 4123-4129.

Mossop, D.W., and R.I.B. Francki. 1979. Comparative studies on two satellite RNAs of cucumber mosaic virus. *Virology* 95: 395-404.

Nakayama, M., T. Yoshida, T. Okuno, and I. Furusawa. 1990. Protection against cucumber mosaic virus and its RNA infection in transgenic tobacco plants expressing coat protein and antisense RNA of the virus. Page 457 in *Proceedings, Eighth International Congress of Virology, 26-30 August 1990, Berlin, Germany.* International Union of Microbiological Societies, Free University of Berlin, Germany.

Namba, S., K. Ling, C. Gonzalves, J.L. Slighton, and D. Gonzalves. 1990. Comparative expression of coat protein genes of PRV, ZYMV and WMWII in transgenic tobacco plants. In *Proceedings, Eighth International Congress of Virology, 26-30 August 1990, Berlin, Germany.* International Union of Microbiological Societies, Free University of Berlin, Germany.

Nejidat, A., and R.N. Beachy. 1990. Transgenic tobacco expressing a tobacco mosaic coat protein gene are resistant to some tobamoviruses. *Molecular Plant Microbe Interactions* 3: 247-251.

Nelson, R.S., A.P. Powell, and R.N. Beachy. 1987. Lesions and virus accumulation in inoculated transgenic tobacco plants expressing the coat protein gene of tobacco mosaic virus. *Virology* 158: 126-132.

Nelson, R.S., S.M. McCormick, X. Delannay, P. Dubé, J. Layton, E.J. Anderson, M. Kaniewska, R.K. Proksch, R.B. Horsch, S.G. Rogers, R.T. Fraley, and R.N. Beachy. 1988. Virus tolerance, plant growth, and field performance of transgenic tomato plants expressing coat protein from tobacco mosaic virus. *Bio/technology* 6: 403-409.

Powell, A.P., R.S. Nelson, B. De, N. Hoffmann, S.G. Rogers, R.T. Fraley, and R.N. Beachy. 1986. Delay of disease development in transgenic plants that express the tobacco mosaic virus coat protein gene. *Science* 232: 738-743.

Powell, P.A., D.M. Stark, P.R. Sanders, and R.N. Beachy. 1989. Protection against tobacco mosaic virus in transgenic plants that express TMV antisense RNA. *Proceedings of the National Academy of Science, USA* 86: 6949-6952.

Powell, P.A., P.R. Sanders, N.E. Tumer, and R.N. Beachy. 1990. Protection against tobacco mosaic virus infection in transgenic plants requires accumulation of capsid protein rather than coat protein RNA sequences. *Virology* 175: 124-130.

Prody, G.A., J.T. Bakos, J.M. Buzayan, I.R. Schneider, and G. Breuning. 1986. Autolytic processing of dimeric plant virus satellite RNA. *Science* 231: 1577-1580.

Rezaian, M.A., K.G.M. Skene, and J.G. Ellis. 1988. Antisense RNAs of cucumber mosaic virus in transgenic plants assessed for the control of the virus. *Plant Molecular Biology* 11: 463-471.

Rothstein, S.J., J. DiMaio, M. Strand, and D. Rice. 1987. Stable and heritable inhibition of the expression of nopaline synthase in tobacco expressing antisense RNA. *Proceedings of the National Academy of Science, USA* 84: 8439-8443.

Sanders, P.R., W.K. Kaniewski, L. Haley, B. LaVallée, X. Delannay, and N.E. Tumer. (in press). Field trials of transgenic tomatoes expressing the tobacco mosaic or tomato mosaic coat proteins. *Phytopathology.*

Stanley, J., T. Frischmuth, and S. Ellwood. 1990. Defective viral DNA ameliorates symptoms of geminivirus infection in transgenic plants. *Proceedings of the National Academy of Science, USA* 87: 6291-6295.

Stark, D.M., and R.N. Beachy. 1989. Protection against potyvirus infection in transgenic plants: Evidence for broad spectrum resistance. *Bio/technology* 7: 1257-1262.

Tien, P., S.Z. Zhao, X. Wang, G.J. Wang, C.X. Zhang, and S.X. Wu. 1990. Virus resistance in transgenic plants that express the monomeric gene of cucumber mosaic virus (CuMV) satellite RNA in greenhouse and field. Page 123 in *Proceedings, Eighth International Congress of Virology, 26-30 August 1990, Berlin, Germany.* International Union of Microbiological Societies, Free University of Berlin, Germany.

Tousch, D., M. Jacquemond, and M. Tepfer. 1990. Behaviour towards cucumber mosaic virus of transgenic tomato plants expressing CuMV-satellite RNA genes. Page 483 in *Proceedings, Eighth International Congress of Virology, 26-30 August 1990, Berlin, Germany.* International Union of Microbiological Societies, Free University of Berlin, Germany.

Tumer, N.E., K.M. O'Connell, R.S. Nelson, P.R. Sanders, and R.N. Beachy. 1987. Expression of alfalfa mosaic virus coat protein gene confers cross-protection in transgenic tobacco and tomato plants. *EMBO Journal* 6: 1181-1188.

Tumer, N.E., W.K. Kanieswki, C. Lawson, L. Haley, and P.E. Thomas. 1990. Engineering resistance to potato leafroll virus in transgenic Russet Burbank potato. Page 122 in *Proceedings, Eighth International Congress of Virology, 26-30 August 1990, Berlin, Germany.* International Union of Microbiological Societies, Free University of Berlin, Germany.

van Dun, C.M.P., and J.F. Bol. 1988. Transgenic tobacco plants accumulating tobacco rattle virus coat protein resist infection with tobacco rattle virus and pea early browning virus. *Virology* 167: 649-652.

van Dun, C.M.P., J.F. Bol, and L. van Vloten-Doting. 1987. Expression of alfalfa mosaic virus and tobacco rattle virus coat protein genes in transgenic tobacco plants. *Virology* 159: 299-305.

van Dun, C.M.P., B. Overduin, L. van Vloten-Doting, and J.F. Bol. 1988. Transgenic tobacco expressing tobacco streak virus or mutated alfalfa mosaic virus coat protein does not cross-protect against alfalfa mosaic virus infection. *Virology* 164: 383-389.

Wefels, E., A. Courtpozanis, F. Salamini, and W. Rohde. 1990. Analysis of transgenic potato lines expressing the coat proteins of potato virus Y (PVY) or potato virus M (PVM). Page 454 in *Proceedings, Eighth International Congress of Virology, 26-30 August 1990, Berlin, Germany.* International Union of Microbiological Societies, Free University of Berlin, Germany.

Wisniewski, L.A., P.A. Powell, R.S. Nelson, and R.N. Beachy. 1990. Local and systemic movement of tobacco mosaic virus (TMV) in tobacco plants that expresses the TMV coat protein gene. *Plant Cell* 2: 559-567.

Acknowledgements

Work on this publication has been made possible by the support of the Institut Français de Recherche Scientifique pour le Développement en Coopération (ORSTOM), Washington University, and the Rockefeller Foundation.

6.2

Cassava viruses and genetic engineering

C. Fauquet[1], D. Bogusz, P. Chavarriaga, C. Franche, C. Schopke and R.N. Beachy

Department of Biology, Washington University, St Louis CB 1137, Missouri 63130, USA (affiliated to the Institut Français de Recherche Scientifique pour le Développement en Coopération, ORSTOM[1])

Abstract

Cassava is affected by a number of viruses, of which the African cassava mosaic virus (ACMV) is the most damaging in Africa, and the cassava common mosaic virus (CCMV) in South America. In 1986, a new application of genetic engineering, coat protein (CP) mediated resistance, was demonstrated as an efficient way of controlling plant virus diseases. A joint programme entitled the International Cassava-Trans Project (ICTP) and being implemented jointly by the Institut Français de Recherche Scientifiquc pour lc Développement en Coopération (ORSTOM) and Washington University aims to apply this technique to cassava in order to reduce the impact of virus infection on cassava production. Because these viruses also infect *Nicotiana benthamiana*, this plant is used as a model for establishing the best molecular strategy for driving resistance against these two viruses. Gene constructions, including those containing ACMV and CCMV CP coding sequences, have been made and transgenic *N. benthamiana* lines obtained. These lines are under investigation for CP expression and resistance to the corresponding virus; preliminary results of these studies are presented. In order to test the gene constructs, transient assays, with a marker gene, have been established to demonstrate the ability of cassava cells to express these constructs. A procedure for the regeneration of cassava plants from somatic embryos has been optimized and the first results of transformation with the particle gun and *Agrobacterium tumefaciens* are presented.

The objective of the International Cassava-Trans Project (ICTP) is to produce by genetic engineering cassava plants resistant to two cassava viruses: the African cassava mosaic virus (ACMV) and the cassava common mosaic virus (CCMV). These viral diseases were chosen because of their economic importance in Africa and South America, respectively (Fauquet and Beachy 1989). The basis of the technique chosen is the integration of genes encoding viral proteins in the plant genome, in order to reduce viral replication and thus to limit the impact of infection on plant production. The technique used here is the coat protein (CP) strategy, which consists of integrating the viral CP gene into the plant genome. This strategy was first applied to express a gene encoding tobacco mosaic virus (TMV) CP in tobacco; the transgenic tobacco plants obtained were resistant to TMV (Powell et al. 1986). Since 1986 the CP strategy has been successfully applied to control a number of different viruses in several plants (Beachy et al. 1990).

Among the different examples of resistance produced by genetic engineering, an important criterion in the majority of cases is a stable and high level of expression of the inserted gene. The level and pattern of gene expression is greatly dependent upon the chimaeric gene construct used, on the site of insertion of the foreign gene in the plant genome, and on the number of inserted genes. The two last factors are not under the control of the investigator.

On the other hand, it is possible to study the influence of each part of the chimaeric gene on the expression of the gene: the transcriptional promoter, the leader sequence of the messenger RNA, the coding region, the untranslated region, and the termination sequence. It is, therefore, possible to evaluate the role of each part of the construct by generating constructs that differ from each other in each region of the gene. Large numbers of transgenic plants are regenerated following transformation with different constructs and the plants are checked for the presence of the foreign gene, mRNA expression, and CP accumulation prior to challenging the progeny of the transformant with the virus.

In the ICTP, the most significant difficulty is cassava regeneration and transformation. Consequently, indirect methods are necessary to optimize the gene constructs, and to study their expression. In our case, this can be done with *Nicotiana benthamiana* because it is a host for both cassava viruses and it can be readily transformed and regenerated. Furthermore, we can carry out transient assays with the particle gun. The *N. benthamiana* model can tell us if the chosen strategy is efficient for controlling these viruses, and the transient assays can tell us which gene construct is functional and which promoter has the highest level of expression in cassava cells.

Logistics of the International Cassava-Trans Project

The ICPT is a result of cooperation between the Institut Français de Recherche Scientifique pour le Développement en Coopération (ORSTOM) and Washington University, St Louis, USA. It aims to coordinate the institutes and agencies interested in the application of biotechnology to cassava, especially the development of viral resistance in cassava. ICPT researchers are interacting with European and American laboratories working on different aspects that relate to the project, including virology, molecular biology, and regeneration of cassava.

The project is closely related to projects of the international institutes of the Consultative Group on International Agricultural Research (CGIAR) which have a mandate for cassava improvement, namely Centro Internacional de Agricultura Tropical (CIAT) and International Institute of Tropical Agriculture (IITA). The project is also related to national institutes or universities of several developing countries; for example, a young Ivorian has joined the ICPT team for 2 years. Recently, the United States Agency for International Development (USAID) and the Monsanto Company have provided support for a post-doctoral African scientist, and it is anticipated that this researcher will collaborate with the ICPT team in the future. The project receives funds and technical support from a number of different agencies, including ORSTOM, the Rockefeller Foundation, Deutsche Gesselleschaft für Technische Zusammenarbeit (GTZ), USAID, Monsanto Company, and the Technical Centre for Agricultural and Rural Cooperation, the Netherlands (CTA) (Fauquet and Beachy 1989).

The Tobacco Model with ACMV and CCMV

Detection of the coat proteins of ACMV and CCMV

One critical requirement in the proposed studies is the detection of CPs in transgenic plants. The level of expression of CP can be very low (Beachy et al. 1990), and it is essential to develop techniques that enable detection of these low levels.

Two methods are currently used, the enzyme-linked immunosorbent assay (ELISA) technique and the western blot technique. For the ACMV, both have been used; the ELISA technique can detect up to 10µg of virus per well, which corresponds to a level of expression of 0.01% of total protein content in transgenic plants. The western blot technique can detect as low as 1µg of capsid protein (0.002% of total protein content) but with a relatively high background. For CCMV, the antibodies are much more efficient and the methods can readily detect levels of virus and CP of 1 and 0.1µg (0.001 and 0.0002% of total protein content for ELISA and western blot techniques, respectively), without significant background due to the plant proteins.

Constructs with ACMV and CCMV coat proteins

The gene constructs and vector should have two major characteristics: the ability to integrate into the plant genome at a high efficiency; and a constitutive expression (that is, in all cell types throughout the life of the plant) in the transgenic plants, which can interfere with the viral infection. The binary vector, which is derived from the Ti plasmid of *Agrobacterium tumefaciens*, has been chosen because of its high efficiency of gene transfer into the plant genome. Most such vectors can also be used with direct DNA delivery, such as with the particle gun or by direct DNA uptakes into protoplasts.

For eucaryotes, gene expression is the result of complex mechanisms comprising: DNA transcription in the nucleus; maturation of RNA transcripts and transportation to the cytoplasm; translation of mRNA to produce a protein; and possible post-translational modification of the protein. Transcription is regulated by sequences called promoters (p). There are many types of promoters, which are expressed under different conditions; we currently use p35S, isolated from cauliflower mosaic virus; p35S is referred to as constitutive because it is expressed in many types of cells throughout the growth of the plant. It has been proven that leader sequences (upstream of the coding region) are implicated in the translational phase of gene expression (Kozak 1988). With the aim of increasing the level of gene expression, we have modified, by mutagenesis, the leader of the ACMV CP gene in order to obtain the consensus sequence of the plant gene leaders.

Construction of chimaeric genes with the coat protein of ACMV

ACMV is a geminivirus with a genome composed of two molecules of single-stranded DNA of approximatively 2.7 kb (Bock and Harrison 1985). The sequence of the genome was elucidated in 1983 (Stanley and Gay 1983), and the CP open reading frame located on the positive strand of DNA A segment was isolated. Several constructs have been made using the CP coding region: with different promoters (p35S, pE35S, p35S+4xOCS); with two viral leader length sequences; with two different termination sequences; and in a sense and antisense orientation. Lastly, as noted above, a construct was also developed that included an improved translational consensus sequence.

Construction of chimaeric genes with the coat protein of CCMV

CCMV is a potexvirus, and its genome is composed of one molecule of single-stranded RNA of approximately 6.3 kb (Costa and Kitajima 1972). Partial cDNA cloning of the virus has been achieved, and the CP coding region located near the viral 3' end has been sequenced. The CP of CCMV has a molecular weight of 25 kd and comparisons of its amino acid sequence with six other potexviruses showed a 47-62% homology amongst them (Fauquet et al. 1990). In the case of CCMV, we developed constructs with only the enhanced promoter of the cauliflower mosaic virus (pE35S) (Kay et al. 1987), but we varied the viral leader sequences and the viral termination sequence in order to test their effect on the level of expression of the CP gene in transgenic tobacco plants.

Transgenic tobacco plants with ACMV and CCMV coat proteins

For each construct described above, *N. benthamiana* was transformed with *Agrobacterium* using the leaf-disc transformation technique (Horsch et al. 1985) and 20 independent lines of transformed tobacco plants were regenerated. The plasmid used for transformation contained one of the CP genes and also a gene conferring kanamycin resistance (*Npt*II gene). All the master plants regenerated were checked for the expression of the *Npt*II gene by ELISA, as well as for the expression of the CP genes of ACMV and CCMV. Because the *Npt*II gene and the CP gene are positioned very close to each other on the plasmid vector, there is a high probability that they will both be integrated at the same site in the plant genome and consequently the expression of *Npt*II gene can be used as a marker gene for selecting transgenic plants harbouring the CP genes. This is also expressed in the F_1 (and succeeding) generation; most of the plants contain one copy of the genes and are segregated in a Mendelian ratio (of 1:3:1). Assuming that the expression of the *Npt*II gene is strongly correlated to the expression of the viral gene, the plants that are saved for F_2 seed production are selected from among the highest expressors of *Npt*II. In most cases, this technique leads to plants homozygous for the viral CP and the *Npt*II gene.

Expression of coat protein genes of ACMV and CCMV in tobacco

The presence of the CP genes of ACMV and CCMV in the plant genome is verified by extracting the DNA from transgenic plants and hybridizing this DNA with DNA complementary to the coding region of the genes. The number of inserted genes can be evaluated by the number of bands and/or the intensity of the hybridizing bands compared with a known amount of DNA. Most of the transgenic tobacco plants had one gene or at least one site of integration of several genes. The transgenic plants are then checked for level of mRNA expressed from the CP gene with a similar hybridization technique but using poly-adenylated RNA instead of DNA. The ACMV transgenic *N. benthamiana* plants tested thus far have very low amounts of mRNA corresponding to the ACMV CP gene, while the CCMV plants have higher levels of mRNA. The final assay is to determine the amount of CP accumulated in the transgenic plants by western immunoblotting. As predicted from the mRNA analysis, the amount of CP produced by the ACMV gene is very low; the maximum concentration registered so far is 0.01% of the total soluble protein. The amount of CCMV CP in the transgenic tobaccos is extremely high in nearly all the transgenic plants, reaching as much as 4% of the total soluble protein. The ELISA technique has also been used for rapid evaluation of protein levels on a large number of plants.

Resistance of transgenic tobacco plants to ACMV and CCMV

After the production of F_2 transgenic tobacco plants, which are normally homozygous for the CP gene, the transgenic plants were challenged with the viruses to evaluate their level of resistance. In the case of ACMV, a limited number of F_1 plants have been tested, for a limited number of constructs; some of them showed a trend of resistance. This resistance was expressed as a short delay (7 days maximum) in the expression of infection in plants and/or a lower percentage of infected plants (for example, 30% instead of 90%). These encouraging preliminary results need to be confirmed in the F_2 generation to be sure that the CP strategy will control ACMV. A larger number of F_1 lines expressing the CCMV CP gene have been tested, and a few F_2 lines have also been tested. Most of these lines showed some degree of resistance and some were apparently highly resistant, nearly immune. In the case of the best resistant transgenic line, the percentage of infected plants was between 0 and 5% at 50 days after inoculation, using an inoculum concentration of CCMV of 1 µg/mL. This concentration of CCMV normally kills *N. benthamiana* in 30 days. The few plants that are infected developed very mild symptoms compared to the control plants, that is, a mosaic or a few necrotic spots instead of complete necrosis.

Transient Gene Expression Assays with Cassava

Principle and choice of promoters

Transient assays allow a rapid evaluation of gene constructs without the inconvenience of transformation and regeneration. Furthermore, such assays are important when transformation has not been achieved, as is the case with cassava. The principle is very simple: a reporter gene under the control of a chosen promoter is introduced into a cell, there is no integration of the gene into the plant genome, and expression of the gene is measured after a short period of time (24-48 hours). The reporter gene that we chose was the ß-glucuronidase (*Gus*) gene, and it was introduced into epidermal and mesophyll cells of intact leaflets of cassava with the help of a particle gun.

Cassava belongs to the Euphorbiaceae family and until now the expression of a foreign gene has not been studied in a member of this family. The first task was to determine if the available promoters were active in such a plant and, if so, whether it was possible to boost this expression by altering the promoter. Although p35S seems to be a very ubiquitous promoter in general, it needed to be tested in an euphorbiaceous host before being used for cassava transformation. As mentioned above, several promoters or modifications of the p35S promoter were tested, including a regulatory element of the promoter of the octopine synthase gene of *Agrobacterium* that was shown to enhance 200-fold the activity of heterologous promoters (Ellis et al. 1987). The promoter isolated from the ubiquitin gene of *Arabidopsis* species was also tested.

Optimization of the particle gun

The first particle gun was described in 1987 (Klein et al. 1987). The principle is as follows: DNA is coated onto tungsten particles 1.2 µm in diameter, which are then accelerated so that they can penetrate several layers of cells. Once in the cell, the DNA is freed and introduced genes are expressed. After optimization of our air-powered particle gun, 100-500 cassava cells expressed the *Gus* gene in each experiment (Franche et al. 1990).

Activity of *Gus* gene constructs in cassava leaves

Beta-glucuronidase is an enzyme that reacts with 5-bromo-4-fluoro-3-indolyl glucuronid to produce a blue precipitate, and the intensity of the reaction is correlated with the expression of the gene, and reflects the transcriptional efficiency of the promoter. We have demonstrated that all the promoters tested are active in cassava cells; p35S is slightly more active than the ubiquitin promoter, and pE35S is slightly more active than p35S+4OCS, and four to five times more active than p35S. This activity was evaluated by the intensity of blue spots on the cassava leaves, and by the activity of enzyme extracted from the leaves and assayed spectrophotometrically (Bogusz et al. 1990; Franche et al. 1990).

Regeneration of Cassava

The ICTP's objective of producing virus-resistant transgenic cassava plants can be achieved only if we are able to regenerate transformed cassava plants. In previous studies, many tissues of the cassava plant have been tested for regeneration but apparently only young leaves, meristems, and cotyledons are able to produce embryos that can be later regenerated into plants (Stamp 1984; Stamp and Henshaw 1986). This technique has been confirmed at CIAT with several cultivars which have been cultivated in the field (Szabados et al. 1987). Nevertheless, the technique remains difficult and is restricted to a few South American cultivars. The number of plants regenerated is generally very low, compared with the number of initial explants taken to produce embryos. We have been able to reproduce these results with two cassava cultivars from Colombia, MCol 22 and MCol 1505 (Schöpke et al. 1990). A third cultivar from Colombia, CMC76, produces many embryos, which can be used to produce secondary and tertiary embryos. This culture grows rapidly and is very convenient for transformation experiments, but the regeneration is unfortunately very difficult. Somatic embryogenesis is restricted to the tissue along the veins of very young leaflets of cassava (2-5 mm long) and there is a tendency for callus cells to overgrow the embryogenic cells. Finally, cassava embryos grow very slowly in culture and at least 8 weeks are needed to complete one experiment.

Transformation of Cassava

Transformation by *Agrobacterium*

Transformation by *Agrobacterium* requires a wound, but if there is a wound the explant has a tendency to produce callus tissue that will not regenerate into a plant. We nevertheless decided to transform callus tissue, in order to prove that we can transform cassava cells and detect the expression of the CPs. To do this, we used the leaf-disc transformation technique commonly used for tobacco transformation (discussed earlier), and selected transgenic cells on 100 μg/mL of kanamycin. The transformed cells were selected through several cycles of growth onto kanamycin and checked for their expression. We integrated the work with *Gus* expression and totally blue calluses have been obtained, showing that the transformation and the selection were effective. The same experiments with the CPs of ACMV and CCMV are currently in progress and the transformation will be confirmed by northern and western blotting analyses.

Transformation by particle gun

Different cassava tissues have been shot with DNA via the particle gun to study the ability of cassava tissues to express the *Gus* reporter gene. Transient expression of this gene has been demonstrated in young leaflets, in leaf embryogenic tissues, and in somatic embryos. The expression of the *Gus* gene has also been observed in different cell types, including the epiderm, the mesophyll, and associated with the phloem cells. These experiments showed the potential of the method and the ability of these different cell types to express a foreign gene and to be transformed. Stable transformation of embryogenic tissue has also been observed 4 weeks after gene introduction, and 2 weeks after selection on kanamycin. Clumps of several hundreds of transformed cells suggest that the transformation occurred and that multiplication of these cells also occurred, but this needs to be proven (Schöpke et al. 1990).

Conclusion

Although the ICTP has not yet produced cassava plants resistant to ACMV and CCMV, it has nevertheless thrown light on many previously unanswered questions. A very strong resistance to CCMV has been achieved in many different lines of *N. benthamiana*, and lines tolerant to ACMV have been obtained, showing that the CP strategy can be effective for resistance to cassava viruses, at least in tobacco. The weak resistance against ACMV can be related to the low mRNA and CP content in the transgenic plants, and further studies of mRNA stability will be needed to improve the level of resistance. It is not certain that a result in a tobacco model can be extrapolated to cassava, but we currently have no alternative. As yet, there is no known example where resistance achieved in a tobacco has not been confirmed or improved in the natural host of the virus under consideration. We have also demonstrated that cassava is able to express each of the gene constructs that we have produced and that the promoters used are efficient in different types of cassava cells. This could be a very important fact because ACMV is naturally transmitted by whiteflies, which inject the virus into cells associated with phloem cells, prior to the invasion of other cell types after the first cycles of viral replication. Cassava callus cells can be transformed and selected, and we anticipate that transformed calluses will express the CP genes of ACMV and CCMV. Lastly, we have been able to regenerate cassava plants from somatic embryogenetic tissues in a limited number of cassava cultivars, and we are currently using this route of regeneration to transform cassava by *Agrobacterium* and the particle gun. It appears that all the pieces of the puzzle have been gathered and that the production of genetically engineered virus-resistant cassava plants is now a matter of assembling the puzzle.

Discussion

QUESTIONS: Are the tomato plants growing in the field resistant to some antibiotics? What is the amount, expressed as % of total protein, of the TMV? What kind of promoters were used to express the TMV gene, and is it possible to drive the expression of these genes only when requested and in the right organ?

FAUQUET: The first generations of transgenic plants were resistant to kanamycin, but the next generations will use herbicide-resistant genes. With regard to the second question, the amount of CP expressed in transgenic plants depends on the nature of the virus, the plant, and the generated line. It can range from 0.001% to 4% of the total amount of protein. The correlation between the CP expression and the resistance obtained is not clear. It seems to be linear in the case of tobamoviruses, and not related in the case of potyviruses. In answer to the third question, so far we are using only p35S from CaMV or modified versions of that promoter. But other promoters, tissue specific or inducible, are under study.

HAMILTON: With regard to the 54 kd protein described by Dr Zaitlin, might conserved sequence in their protein be used to transform tobacco for resistance to several other tobamoviruses?

FAUQUET: I don't know if there is a conserved region in the 54 kd protein of TMV which can be used for transforming plants. But these transgenic plants were resistant to a strain which in fact was obtained by mutation from the original strain used, and not at all resistant for the most closely related tobamoviruses.

MURDOCK: Are you confident that coat protein gene-mediated resistance, once deployed over wide areas, will be durable — that is, not break down?

FAUQUET: Yes, for two reasons. First, plant viruses are very stable in time and space and it seems difficult to break resistance, even vertical resistance. Second, experiments in the laboratory proved that cycles of virus infection through transgenic plants have been unable to break this coat protein-mediated resistance. This resistance is, in most cases, a monogenic resistance, which is easy to transfer to other varieties but may be easier to break. Even in that event, it is easy to come back to the original coat protein, modify it, and produce a new transformant. Furthermore, it has been demonstrated that the spectrum of specificity with a virus group is fairly wide; viruses having at least 60% of homology with this coat protein are protected, leaving room for a lot of mutations.

ROBERTSON: Please do not regard this as a hostile question. Could you explain the present status of the patents for the coat protein approach and of the genes involved?

FAUQUET: A patent has been deposited in the USA by Roger Beachy from Washington University and Monsanto Company, covering the whole concept of coat protein-mediated resistance. If accepted, this patent will cover all the transgenic plants using the coat protein strategy. But at least in the case of cassava viruses, we have got a written agreement from Monsanto allowing us to freely use this patent. Such an agreement might be extended to other examples for the developing countries.

SINGH: Monsanto Company has been working on transgenic varieties for several years, and you showed beautiful pictures of transgenic tomato and tobacco varieties showing significant differences. Some of these date back to 1987. Have some of these transgenic varieties been released for general cultivation? If not, can you please enlighten us on the procedures and problems involved?

FAUQUET: Only field trials for research purposes have been done and Monsanto is forecasting that it will release transgenic plants by 1993-95, provided that rules, regulations and patents about transgenic plants are in place.

References

Beachy, R.N., S. Loesch-Fries, and N.E. Tumer. 1990. Coat protein-mediated resistance against virus infection. *Annual Review of Phytopathology* 28: 451-474.

Bock, K.R., and B.D. Harrison. 1985. *African Cassava Mosaic Virus*. AAB/Description of Plant Viruses, No 217.

Bogusz, D., C. Franche, C. Schöpke, C. Fauquet, and R.N. Beachy. 1990. Transient expression of chimaeric genes in cassava using high velocity micro-projectiles. In *Abstracts, Meeting, Molecular Strategies for Crop Improvement, UCLA, 16-22 April 1990, Keystone, Colorado, USA.*

Costa, A.S., and E.W. Kitajima. 1972. *Cassava Common Mosaic Virus*. AAB/Description of plant viruses, No 90.

Ellis, J.G., D.J. Llewellyn, J.C. Walkers, E.S. Dennis, and W.J. Peacock. 1987. The OCS element. *EMBO Journa*l 6: 3203-3208.

Fauquet, C., and R.N. Beachy. 1989. *International Cassava-Trans Project: Cassava Viruses and Genetic Engineering*. CTA/ORSTOM, Wageningen, the Netherlands.

Fauquet, C., L. Calvert, D. Bogusz, C. Franche, C. Schöpke, and R.N. Beachy. 1990. Sequence comparison of the coat protein gene of cassava common mosaic virus with other potexviruses shows common features and suggests important sequences for the coat protein gene expression in transgenic plants. In *Abstracts, Meeting, Molecular Strategies for Crop Improvement, UCLA, 16-22 April 1990, Keystone, Colorado, USA.*

Franche, C., D. Bogusz, C. Fauquet, C. Schöpke, C. Jobe, J. Pease, and R.N. Beachy. 1990. A la frontière de la balistique et de la biologie végétale: Le canon à particules. *ORSTOM Actualités* 29: 8-10.

Horsch, R.B., J.E. Fry, N.L. Hoffmann, D. Eichholtz, S.G. Rogers, and R.T. Fraley. 1985. A simple and general method for transferring genes into plants. *Science* 227: 1229-1231.

Kay, R., A. Chan, M. Daly, and J. McPherson. 1987. Duplication of CaMV 35S promoter sequences creates a strong enhancer for plant genes. *Science* 236: 1299-1302.

Klein, T.M., E.D. Wolf, R. Wu, and J.C. Sanford. 1987. High velocity microprojectiles for delivering nucleic acids into living cells. *Nature (London)* 327: 70-73.

Kozak, M. 1988. Leader length and secondary structure modulate mRNA function under conditions of stress. *Molecular Cell Biology* 6: 2737-2744.

Powell, A.P., R.S. Nelson, B. De, N. Hoffman, S.G. Rogers, R.T. Fraley, and R.N. Beachy. 1986. Delay of disease development in transgenic plants that express the tobacco mosaic virus coat protein gene. *Science* 232: 738-743.

Schöpke, C., C. Franche, D. Bogusz, C. Fauquet, and R.N. Beachy. 1990. Progress towards the transformation of cassava with a particle gun. In *Abstracts, Seventh Congress on Plant Tissue and Cell Culture, 24-29 June 1990, Amsterdam, The Netherlands.*

Stamp, J.A. 1984. In vitro plant regeneration studies with cassava (*Manihot esculenta* Crantz). PhD thesis, University of Birmingham, UK.

Stamp, J.A. and G.G. Henshaw. 1986. Adventitious regeneration in cassava. Pages 149-157 in *Plant Tissue Culture and Its Agricultural Applications.* Butterworths, London, UK.

Stanley, J., and M.R. Gay. 1983. Nucleotide sequence of cassava latent virus DNA. *Nature (London)* 301: 260-262.

Szabados, L., R. Hoyos, and W. Roca. 1987. In vitro somatic embryogenesis and plant regeneration of cassava. *Plant Cell Reports* 6: 248-251.

Acknowledgements

Work on this report was made possible through the support of ORSTOM, Washington University, and the Rockefeller Foundation. The research reported here was supported by grants from ORSTOM, the Rockefeller Foundation, GTZ, and USAID.

6.3

The use of monoclonal antibodies and cDNA for detection of plant viruses

R.I. Hamilton

Agriculture Canada Research Station, 6660 NW Marine Drive, Vancouver, British Columbia V6T 1X2, Canada

Abstract

Rapid and reliable methods of virus detection are essential for the proper management of virus diseases. Two developments in the past decade, monoclonal antibodies and complementary DNA (cDNA), now make it possible to design or select probes that are specific for all known isolates of a virus, that can readily distinguish virus strains, or that can detect all known members of a specific virus group. These probes thus have great potential for application in disease surveys, screening germplasm for exotic viruses, screening breeding lines for virus resistance, and in post-entry detection for quarantine purposes. The principles underlying the development of monoclonal antibodies directed against structural or non-structural virus proteins, and cDNA probes, which are complementary to specific regions of viral genomes, are discussed in this paper and examples of their application in the detection of plant viruses are presented.

The effective management of plant disease requires the use of rapid and reliable methods of pathogen detection. This is especially so with diseases caused by viruses, because symptoms are not an infallible indicator of the identity of the causal virus(es). Serological methods, using polyclonal antisera, have been widely applied for many years for the rapid detection and identification of plant viruses. Two developments in the past decade, monoclonal antibodies and complementary DNA (cDNA), have radically changed our capacity for the rapid detection of viruses, to the point where it is now possible to design or select probes that are specific for all known isolates of a virus, that can readily distinguish virus strains, or that can detect all known members of a specific virus group.

These probes have wide application in disease surveys, screening germplasm for exotic viruses, screening breeding lines for virus resistance, and in post-entry detection for quarantine purposes. This paper outlines the preparation of these diagnostic probes and illustrates their application in disease management.

Monoclonal Antibodies

By definition, a preparation of monoclonal antibodies represents a homogeneous class of antibodies produced by a single hybridoma, which consists of a lymphocyte from an immunized animal fused with a myeloma cell. These antibodies will combine with a specific antigen under a specified protocol (such as pH, salt concentration, and temperature) in a reproducible and defined manner. The traditionally used polyclonal antiserum is basically a mixture of antibodies produced by a multitude of lymphocytes, each antibody species being the unique product of a specific lymphocyte. The development of the monoclonal antibody technique (for reviews, see Sander and Dietzgen 1984; Halk and DeBoer 1985; Miller and Martin 1988) allows for the selection of specific populations of identical antibodies (monoclonal), which can be used for detection of a specific epitope. Monoclonal antibodies to about 50 distinct plant viruses are available, and production of many others is in progress.The main steps in the production of monoclonal antibodies can be summarized as follows (Regenmortel 1986):

1. A mouse (or rat) is immunized by a series of injections of viral antigen (10-100 μg).
2. The immunized mouse is sacrificed, the spleen is removed by aseptic technique and disrupted.
3. Lymphocytes are fused with a population of freshly cultured mouse myeloma cells by addition of a defined amount of polyethylene glycol to a mixture of these cells.
4. The resulting fused cells (hybridomas) are distributed to, and allowed to grow in, wells of a microtitre plate containing a nutrient solution that does not allow growth of cells that have not fused.
5. After removal of the old culture fluids and replacement with fresh fluid, supernatants of the new culture fluids are screened after 3-4 days for the presence of antibodies.
6. The hybridoma cell lines are subcloned by limiting dilution, to ensure that each cell line is monoclonal.
7. The hybridomas are injected into mice to produce ascitic fluid, which contains copious amounts of the antibodies, or grown in batch culture.

Although the coat protein has been the traditional immunogen for production of both polyclonal antisera and monoclonal antibodies, several non-structural viral proteins (such as viral replicase), whose amino acid sequences are highly conserved, are being investigated for the production of broad-spectrum monoclonal antibodies that may be useful for the detection of a conserved, virus group-specific amino acid sequence.

The enzyme-linked immunosorbent assay (ELISA) is usually the method of choice in screening culture supernatant fluids for evaluation of monoclonal antibodies (Regenmortel 1986). The most common reporter enzymes used are alkaline phosphatase and horseradish peroxidase, but penicillinase (Sudarshana and Reddy 1989) has been reported to yield equivalent results, and its lower cost is advantageous, especially in laboratories and field stations in developing countries. The screening procedure is critical because if the antibodies are to be used for detection of viruses in plant saps, rather than for characterization of highly purified virus, the screening procedure should be similar to the eventual test procedure (Al-Moudallal et al. 1984). A given monoclonal antibody does not necessarily react identically to the same antigen in all ELISA formats, and this reinforces the need to carefully select monoclonal antibodies for specific assay formats. For example, if a monoclonal antibody is used as a coating antibody, its conformation may be rearranged upon binding to the plastic, thus changing its avidity for antigens (Altschuh et al. 1985). Coupling of enzyme to the antibodies occasionally results in poor avidity of the antibody for the virus (Torrance and Pead 1986), and this may be obviated by the use of smaller reporter groups such as biotin (Zrein et al. 1986), or by a labelled antibody from a different animal species, which recognizes the second antibody, but not the coating antibody (for example, rabbit anti-mouse antibody).

Some, but not all, monoclonal antibodies that are very reactive in ELISA are useful in other types of solid-phase assay, such as immunosorbent electron microscopy (Hsu and Lawson 1985; Diaco et al. 1986). The application of monoclonal antibodies in gel diffusion has been successful with African cassava mosaic virus (Torrance and Pead 1986) and ilarviruses (Halk et al. 1984), but not with barley yellow dwarf virus (Torrance and Pead 1986). Dot-blot immunoassays (Sherwood et al. 1989) are equivalent to ELISA in some cases, and they are a useful and practical alternative in laboratories and field stations that are not equipped for ELISA. Monoclonal antibodies have been applied in the latex agglutination test for detection of rice grassy stunt virus, as an alternative to the more expensive ELISA (Hsu et al. 1990). A simple alternative to the preparation of plant saps for immunoassays is a variant of the dot-blot immunoassay, in which the cut surface of plant tissue is blotted on nitrocellulose membranes which are then treated with labelled monoclonal antibodies; this assay has been used successfully to detect tomato spotted wilt virus, and cucumo-, luteo-, potex-, and potyviruses in infected plants (Lin et al. 1990).

Monoclonal antibodies are generally superior to polyclonal antisera, especially in reaction with antigens in plant sap, because background absorbance due to reaction with plant proteins has been eliminated (Rose et al. 1987). The hybridomas can be stored for long periods at low temperature and are essentially "immortal" and thus can be regenerated to produce more of the same antibody when needed. They are being used to detect an increasing number of viruses in infected plants and are particularly useful in virus epidemiological studies for the identification of specific strains of viruses in their insect vectors (Cohen et al. 1989). For use in certification and quarantine schemes, it is essential to demonstrate that the monoclonal antibody to be used recognizes a wide range of virus isolates. Strain-specific antibodies are not suitable for such applications.

cDNA Hybridization

Nucleic acid hybridization, which involves the pairing of specific complementary nucleotide bases, has been employed as a rapid and powerful means of detecting viral nucleic acids in plant tissue extracts. In practice, a cDNA probe is synthesized in vitro by reverse transcription, using the viral RNA as a template for an RNA-dependent DNA polymerase; the cDNA is usually labelled with ^{32}P and the hybridization of the cDNA and the target RNA is detected by appropriate technology, usually autoradiography.

Nucleic acid probes (RNA or DNA) offer a significant advantage over serological methods because, theoretically, specific regions over the entire viral genome can be used as a template to synthesize the probe, whereas serological probes are almost exclusively generated from the coat protein gene product, which usually represents less than 10% of the total viral genomic information. The most useful and convenient probes are those made from cloned cDNA. Like synthetic probes, these can be directed at particular regions of the genome to give the desired selectivity or specificity. cRNA probes, made by in vitro transcription of cloned cDNA, have been used (Varveri et al. 1988); RNA:RNA hybrids are inherently more stable than DNA:RNA or DNA:DNA hybrids, but the probes are sensitive to RNase (Hames and Higgins 1985). For practical purposes, a randomly primed cDNA probe will contain sequences which cover most of the genome (Gould and Symons 1983), but production by chemical synthesis of probes complementary to specific regions of the genome has potential for the development of probes with very narrow or very broad specificities (Bar-Joseph et al. 1986). Cloned double-stranded DNA probes of viral RNA offer an uninterrupted supply of a specific probe, which can be labelled with isotopes or other reporter groups by nick translation or the simpler random-primed labelling ("oligolabelling") technique. These probes are analogous to monoclonal antibodies in the sense that there are no non-viral sequences (other than vector sequences) present in the probe, which might confound interpretation of hybridization assays, especially those of sap extracts (Gould and Symons 1983). In circumstances where viral RNA is difficult to purify, or where no virus-like particles can be associated with disease, viral dsRNA representing the replicative RNA of ssRNA viruses can be used as a template for synthesis of cDNA probes (Jelkmann et al. 1989) or end-labelled and used as probes (Jordan and Dodds 1983). One should be aware, however, that some high molecular weight dsRNAs are of host origin, and their function in plants is unknown (Wakarchuk and Hamilton 1990).

The most common assay procedure is the dot-blot assay (Hull 1986), which may be summarized as follows:

1. Extraction of a small volume of sap from infected plants
2. Deposition of a drop of the sap to a solid, nucleic acid-binding matrix, usually a nitrocellulose or nylon membrane
3. Binding of the nucleic acid molecules to the membrane by baking (at 80°C for 2 hours).
4. Application of a suspension of protein (bovine serum albumin or milk protein) and nucleic acid (calf-thymus or salmon-sperm DNA) to the membrane to block any free nucleic acid-binding regions
5. Binding of a labelled cDNA probe to target nucleic acid on the membrane
6. Removal of excess probe by washing
7. Detection of the labelled DNA (and thus the hybrid) by autradiography (for isotopically labelled probes) and by colorimetry for enzyme-labelled probes

For some dsRNA or dsDNA viruses, the nucleic acid must be de-natured prior to its binding to the membrane.

Dot-blot assays have been applied in numerous circumstances to detect viruses in plant and vector tissues (Czosnek et al. 1988; Rice et al. 1990; Xie et al. 1989). They have been used to discriminate between specific strains of virus and to detect most members of the same virus group (Linthorst and Bol 1986).

A serious limitation to the use of isotopically labelled probes in Africa is access to a reliable supply of short half-life radioisotopes. Another is disposal of radioactive waste, but storage for several months will reduce the activity to manageable levels. An alternative is the use of cDNA probes labelled with biotin (Eweida et al. 1989) or other reporter molecules. Such probes have great potential in field applications of nucleic acid hybridization for virus detection.

In some circumstances, especially if the virus concentration is below the limits of detection by the cDNA probe, or the incidence of infected samples in a batch-testing procedure is very low, the use of cDNA probes may be unreliable. Amplification of extremely low numbers of viral RNA molecules in such samples, and their subsequent detection, is possible by the polymerase chain reaction (PCR). In this method, viral RNA species are reverse-transcribed to DNA in the customary manner, and the DNA is then amplified by a specific DNA-dependent polymerase (Taq polymerase) to a high copy number and detected by a conventional cDNA probe. Reverse transcription of the viral RNA species depends on prior knowledge of its nucleotide sequence and the use of selected DNA primers for first-strand synthesis of cDNA. The PCR method has been applied recently to the detection of viroids and satellite viral RNAs (Hadidi and Yang 1990), and it will certainly be used for the detection of viruses. The method appears to be particularly suited to the detection of viruses in valuable germplasm accessions.

Conclusion

It is clear that biotechnological techniques will enhance research on tropical crops in Africa. With respect to crop protection research, monoclonal antibodies and cDNA probes to plant viruses have facilitated virus detection, and their introduction and application in Africa will assist in providing increased security of food supply and alleviating poverty. They have the potential to accelerate resistance breeding programmes by clearly identifying strains for which resistance is sought; they can be used in disease surveys to provide accurate data about virus epidemiology and the ecology of viruses and their vectors; and they will be useful in screening germplasm, thus facilitating the international exchange of breeding material.

References

Al-Moudallal, Z., D. Altschuh, J.B. Briand, and M.H.V. van Regenmortel. 1984. Comparative reactivity of different ELISA procedures for detecting monoclonal antibodies. *Journal of Immunological Methods* 68: 35-43.

Altschuh, D., Z. Al-Moudallal, J.P. Briand, and M.H.V. van Regenmortel. 1985. Immunological studies of tobacco mosaic virus. VI. Attempts to localize viral epitopes by monoclonal antibodies. *Molecular Immunology* 22: 329-337.

Bar-Joseph, M., S. Segerd, W. Blickle, V. Yesodi, A. Franck, and A. Rosner. 1986. Application of synthetic DNA probes for the detection of viroids and viruses. Pages 13-23 in *Proceedings, Conference on Developments and Applications in Virus Testing, 10-12 April 1985, Cambridge, England* edited by R.A.C. Jones and L. Torrance, Association of Applied Biologists, c/o National Vegetable Research Station, Wellesbourne, UK.

Cohen, S., J.E. Duffus, and H.T. Lin. 1989. Acquisition, interference and retention of cucurbit leaf curl viruses in whiteflies. *Phytopathology* 79: 109-113.

Czosnek, H., R. Ber, N. Navot, and D. Zamir. 1988. Detection of tomato leaf curl virus in lysates of plants and insects by hybridization with a viral DNA probe. *Plant Disease* 72: 949-951.

Diaco, R., R.M. Lister, J.H. Hill, and D.P. Durand. 1986. Detection of homologous barley yellow dwarf virus isolates with monoclonal antibodies in serologically specific electron microscopy. *Phytopathology* 76: 225-230.

Eweida, M., T.L. Sit, S. Sira, and M.G. AbouHaidar. 1989. Highly sensitive and specific non-radioactive biotinylated probes for dot-blot, southern and colony hybridizations. *Journal of Virological Methods* 26: 35-44.

Gould, R.G., and R.H. Symons. 1983. A molecular biological approach to relationships among viruses. *Annual Review of Phytopathology* 21: 179-199.

Hadidi, A., and X. Yang. 1990. Detection of some fruit viroids by enzymatic cDNA amplification. *Journal of Virological Methods* 30: 261-270.

Halk, E.L., and S.H. DeBoer, 1985. Monoclonal antibodies in plant disease research. *Annual Review of Phytopathology* 23: 321-350.

Halk, E.L., H.T. Hsu, J. Abeig, and J. Franke. 1984. Production of monoclonal antibodies against three ilarviruses and alfalfa mosaic virus and their use in serotyping. *Phytopathology* 74: 367-372.

Hames, B.D., and S.J. Higgins (eds). 1985. *Nucleic Acid Hybridization, A Practical Approach.* IRL Press, Oxford, UK.

Hsu, H.T., and R.H. Lawson. 1985. Comparison of mouse monoclonal antibodies and polyclonal antisera of chicken egg yolk and rabbit for assay of carnation etched ring virus. *Phytopathology* 75: 778-783.

Hsu, H.T., H. Bibino, and P.Q. Cabauatan. 1990. Development of serological procedures for rapid, sensitive and reliable detection of rice grassy stunt virus. *Plant Disease* 74: 695-698.

Hull, R. 1986. The potential for using dot-blot hybridisation in the detection of plant viruses. Pages 3-12 in *Proceedings, Conference on Developments and Applications in Virus Testing, 10-12 April 1985, Cambridge, England* edited by R.A.C. Jones and L. Torrance, Association of Applied Biologists, c/o National Vegetable Research Station, Wellesbourne, UK.

Jelkmann, W., R.R. Martin, and E. Maiss. 1989. Cloning of four plant viruses from small quantities of double-stranded RNA. *Phytopathology* 79: 1250-1253.

Jordan, R.L., and J.A. Dodds. 1983. Hybridization of 5'-end-labeled RNA to plant viral RNA in agarose and acrylamide gels. *Plant Molecular Biology Reporter* 1: 31-37.

Lin, N.S., Y.H. Hsu, and H.T. Hsu. 1990. Immunological detection of plant viruses and a mycoplasma-like organism by direct tissue blotting on nitrocellulose membranes. *Phytopathology* 80: 824-828.

Linthorst, H.J.M., and J.F. Bol. 1986. cDNA hybridization as a means of detection of tobacco rattle virus in potato and tulip. Pages 25-39 in *Proceedings, Conference on Developments and Applications in Virus Testing, 10-12 April 1985, Cambridge, England* edited by R.A.C. Jones and L. Torrance, Association of Applied Biologists, c/o National Vegetable Research Station, Wellesbourne, UK.

Miller, S.A., and R.R. Martin. 1988. Molecular diagnosis of plant disease. *Annual Review of Phytopathology* 26: 409-432.

Regenmortel, M.H.V. van. 1986. The potential for using monoclonal antibodies in the detection of plant viruses. Pages 89-101 in *Proceedings, Conference on Developments and Applications in Virus Testing, 10-12 April 1985, Cambridge, England* edited by R.A.C. Jones and L. Torrance, Association of Applied Biologists, c/o National Vegetable Research Station, Wellesbourne, UK.

Rice, D.J., T.L. German, R.F.L. Mau, and R.M. Fujimoto. 1990. Dot-blot detection of tomato spotted wilt virus RNA in plants and thrips tissues by cDNA clones. *Plant Disease* 74: 274-276.

Rose, D.G., S. McCarra, and D.H. Mitchell. 1987. Diagnosis of PVY^{N}: A comparison between polyclonal antisera and monoclonal antibodies and a biology assay. *Plant Pathology* 36: 95-99.

Sander, E., and R.G. Dietzgen. 1984. Monoclonal antibodies against plant viruses. *Advances in Virus Research* 29: 131-168.

Sherwood, J.L., M.R. Sanborn, G.C. Keyser, and L.D. Myers. 1989. Use of monoclonal antibodies in detection of tomato spotted wilt virus. *Phytopathology* 79: 61-64.

Sudarshana, M.R., and D.V.R. Reddy, 1989. Penicillinase-based ELISA for the detection of plant viruses. *Journal of Virological Methods* 26: 45-52.

Torrance, L., and M.T. Pead. 1986. The application of monoclonal antibodies to routine tests for two plant viruses. Pages 103-118 in *Proceedings, Conference on Developments and Applications in Virus Testing, 10-12 April 1985, Cambridge, England* edited by R.A.C. Jones and L. Torrance, Association of Applied Biologists, c/o National Vegetable Research Station, Wellesbourne, UK.

Varveri, C., T. Candresse, M. Ravelonandro, G. Macquaire, and J. Dunez. 1988. Use of in vivo transcribed RNAs for plum pox virus detection in a dot-blot hybridization assay. *Acta Horticulturae* 235: 319-324.

Wakarchuk, D.A., and R.I. Hamilton. 1990. Partial nucleotide sequence from enigmatic dsRNAs in *Phaseolus vulgaris*. *Plant Molecular Biology* 14: 637-639.

Xie, W., J.F. Antoniw, and R.F. White. 1989. Detection of beet cryptic viruses 1 and 2 in a wide range of beet plants using cDNA probes. *Plant Pathology* 38: 527-533.

Zrein, M., J. Burckard, and M.H.V. van Regenmortel. 1986. Use of the biotin-avidin system for detecting a broad range of serologically related plant viruses. *Journal of Virological Methods* 13: 121-128.

6.4

Broad spectrum virus resistance in *Solanum brevidens* and its introduction by protoplast fusion to cultivated potato

R.W. Gibson[1], J. Valkonen[1], E. Pehu[1], M.G.K. Jones[1] and A. Karp[2]

Natural Resources Institute, Central Avenue, Chatham Maritime, Kent ME4 4TB, UK[1]; IACR-Rothamsted Experimental Station, Harpenden, Herts AL5 2JQ, UK[2]

Abstract

In the study reported in this paper, *Solanum brevidens* developed no obvious disease symptoms when inoculated with potato virus Y (PVY) and potato virus X (PVX), and both viruses reached only a low virus titre, apparently because relatively few cells became infected. However, protoplasts were readily infected in vitro.

Solanum brevidens is a wild plant species which is distantly related to the cultivated potato, *S. tuberosum*. It does not produce tubers but has been identified as a potential source of resistance to the late blight fungus, *Phytophthora infestans* (Helgeson et al. 1986), to a soft rot bacterium, *Erwinia carotovora* (Austin et al. 1988), to potato leafroll virus (PLRV) (Jones 1979), to potato virus Y (PVY) (Gibson et al. 1988), and to potato virus X (PVX) (Gibson et al. 1990).

Plants of *S. brevidens* inoculated with PLRV, PVY, and PVX develop no obvious symptoms, despite being infected. Experiments using enzyme-linked immunosorbent assay (ELISA), electron microscopy, and a local lesion assay show that both PVY and PVX attain only low virus titres in inoculated plants (Gibson et al. 1990). Among the protoplasts prepared from leaves of plants of *S. brevidens* inoculated with either PVY or PVX, few could be identifed as infected, whereas many were infected in preparations from inoculated plants of the cultivated potato. However, when protoplasts of either *S. brevidens* or the cultivated potato were inoculated in vitro with either PVY or PVX, protoplasts of both species were found to be similarly susceptible to infection with the two viruses (Valkonen et al. 1991). These results suggest that resistance to both viruses is associated with slow cell-to-cell spread in *S. brevidens*.

S. brevidens cannot be hybridized with the cultivated potato by cross-pollination, but has been hybridized directly by protoplast fusion to produce somatic hybrid plants, using dihaploid cultivated potato so that the tetraploid level could be attained (Austin et al. 1985; Fish et al. 1987). The hybrids exhibit a range of attributes, differing in chromosome number, cytoplasm, phenotype, and resistance to virus infection (Gibson et al. 1988; Pehu et al. 1990). Tetraploids inoculated with PLRV, PVY or PVX developed a range of virus titres, whereas hexaploids comprising two genomes of *S. brevidens* were susceptible. Among aneuploid hybrids, resistance (as measured by virus titre) to PVY and PVX was positively correlated, confirming the possibility of a common resistance mechanism.

References

Austin, S., M.A. Baer, and J.P. Helgeson. 1985. Transfer of resistance to potato leafroll virus from *S. brevidens* into *S. tuberosum* by somatic fusion. *Plant Science* 39: 75-82.

Austin, S., E. Lojkowcka, K.M. Ehlenfeldt, A. Kelman, and J.P. Helgeson. 1988. Fertile interspecific somatic hybrids of *Solanum*: A novel form of resistance to *Erwinia* soft rot. *Phytopathology* 78: 1216-1220.

Fish, N., A. Karp, and M.G.K. Jones. 1987. Improved isolation of dihaploid *S. tuberosum* protoplasts and the production of somatic hybrids between dihaploid *S. tuberosum* and *S. brevidens*. *In Vitro* 23: 575-580.

Gibson, R.W., M.G.K. Jones, and N. Fish. 1988. Resistance to potato leafroll virus and potato virus Y in somatic hybrids between dihaploid *Solanum tuberosum* and *S. brevidens*. *Theoretical and Applied Genetics* 76: 113-117.

Gibson R.W., E. Pehu, R.D. Woods, and M.G.K. Jones. 1990. Resistance to potato virus Y and potato virus X in *Solanum brevidens*. *Annals of Applied Biology* 116: 151-156.

Helgeson, J.P., G.J. Hunt, G.T. Haberlach, and S. Austin. 1986. Somatic hybrids between *Solanum brevidens* and *Solanum tuberosum*: Expression of a late blight resistance gene and potato leafroll resistance. *Plant Cell Reports* 3: 212-214.

Jones, R.A.C. 1979. Resistance to potato leafroll virus in *Solanum brevidens*. *Potato Research* 22: 149-152.

Pehu, E., R.W. Gibson, M.G.K. Jones, and A. Karp. 1990. Studies on the genetic basis of resistance to potato leafroll virus, potato virus Y, and potato virus X in *Solanum brevidens* using somatic hybrids of *Solanum brevidens* and *Solanum tuberosum*. *Plant Science* 69: 95-101.

Valkonen, J.P.T., E. Pehu, M.G.K. Jones, and R.W. Gibson. 1991. Resistance in *Solanum brevidens* to both potato virus Y and potato virus X may be associated with slow cell-to-cell spread. *Journal of General Virology* 72(2): 231-236.

6.5

Proteinase inhibitor gene transfer for improving insect resistance in plants

J. Narvaez and C.A. Ryan

Institute of Biological Chemistry, Washington State University, Pullman, Washington 99164-6340, USA

Abstract

Tobacco plants transformed with a tomato trypsin Inhibitor I cDNA from a wild tomato species (*Lycopersicon peruvianum*), under regulation by the CaMV 35S promoter, accumulated Inhibitor I proteins in leaves to levels of about 100 μg/g tissue. When fed to *Manduca sexta* larvae, the leaves reduced growth of the insects by 50%, compared to control leaves. In similar experiments, tobacco transformed with a tomato chymotrypsin Inhibitor I gene from a modern tomato species (*L. esculentum*) had little effect on growth of the larvae when present in leaves at levels of over 200 μg/g tissue. Thus the inhibitory specificities of the two Inhibitor I family members were responsible for their differential effects on the *M. sexta* larvae. When the CaMV 35S promoter construct-tomato chymotrypsin Inhibitor I was expressed in tomato, tobacco, nightshade, and alfalfa plants to compare levels of transgenic Inhibitor I protein accumulation in leaves of different plant species, the differences were striking. The highest accumulation of tomato Inhibitor I was in leaves of the homologous host tomato, and the lowest in the most distantly related plant, alfalfa. The synthesis of the Inhibitor I is reflected in the Inhibitor I mRNA levels, indicating that this foreign gene is transcribed at different rates in different plant species. These experiments indicate that the CaMV promoter may have limited usefulness in achieving high levels of expression of proteinase inhibitor genes in certain crop plants, such as alfalfa.

Studies have shown that tobacco plants transformed with proteinase inhibitor genes coding for inhibitor proteins having trypsin inhibitor specificities provided increased resistance toward lepidopteran insects (Hilder et al. 1987; Johnson et al. 1989). Hilder et al. (1987) first transformed tobacco plants with a Bowman-Birk family trypsin inhibitor gene regulated by the CaMV 35S promoter that produced about 1% of the leaf proteins as the inhibitor. The transformed plants were more resistant to feeding by larvae of *Heliothis virescens* than untransformed control plants or transformed plants that did not express the gene.

The cowpea inhibitor is an antinutrient agent against a wide range of insects (Hilder et al. 1987), including the larvae of *Heliothis*, *Spodoptera*, *Diabrotica*, and *Tribolium*, all agronomically important insect pests. More recently, Johnson et al. (1989) transformed tobacco plants with genes coding for tomato and potato Inhibitor II proteins (having both chymotrypsin and trypsin inhibitory activities) and a tomato Inhibitor I protein (having only chymotrypsin inhibitory activity), all under the control of the CaMV 35S promoter. The leaves of plants expressing the Inhibitor II proteins at levels of 50 µg/g tissue or above, when supplied to larvae of *Manduca sexta* as their only food source, caused inhibition of growth compared to the effect of larvae feeding on untransformed plants. At higher levels of inhibitors (100 µg/g tissue), larvae growth was even more inhibited, and some died (Johnson et al. 1989). Larvae feeding on tobacco leaves transformed with the CaMV 35S-Inhibitor I gene were relatively unaffected by the presence of the chymotrypsin inhibitory activity.

Effects of Tomato Inhibitor I Specificity on *Manduca sexta* Larval Growth

In order to further understand the roles of the trypsin inhibitory activity in reducing growth of *M. sexta* larvae on tobacco leaves, a tomato Inhibitor I gene, coding for an Inhibitor I protein with an inhibitory reactive site specific for trypsin (P_1-$P_{1'}$= Lys-Asp) (Wingate et al. 1989) and regulated by the CaMV 35S promoter, was used to transform tobacco plants (Narvaez and Ryan, unpubl.). This gene was constructed with a trypsin Inhibitor I cDNA obtained from a wild species of tomato, *Lycopersicon peruvianum* (Wingate et al. 1989). The plants expressed the trypsin Inhibitor protein in leaves at levels near 100 µg/g tissue(s). In insect feeding studies, the leaves of these plants were compared with leaves of tobacco plants previously reported to have levels of a tomato chymotrypsin Inhibitor I protein of over 200 µg/g tissue (Johnson et al. 1989).

Figure 1 shows the growth of *M. sexta* larvae on the leaves of tobacco plants which have been transformed with the wild *L. peruvianum* trypsin Inhibitor I gene, expressed in different plants to levels between 60 and 170 mg/g tissue. The data represent the larval weight over a period of feeding on the leaves, as previously described (Johnson et al. 1989). In the inset in Figure 1, the data show the growth of larvae feeding on the intact leaves of transformed tobacco plants expressing 160 µg trypsin Inhibitor I/g tissue. These results are in sharp contrast to the growth data of the *M. sexta* larvae on tobacco leaves containing the chymotrypsin Inhibitor I protein. The effects of larvae growing on intact leaves is more evident than on detached leaves (Johnson et al. 1989). It is likely that the presence of the trypsin inhibitory activity is the causal effect of the reduced larval growth. The two Inhibitor I proteins are very similar in their primary structures, having over 74% identity between them. The two inhibitors are similar in virtually every property except their specificities against proteinases.

The limited studies to date (feeding larvae transgenic plants expressing foreign proteinase inhibitor genes) have firmly established that at least three different families of trypsin inhibitors in leaves can provide resistance to insect pests. The cowpea inhibitor (Hilder et al. 1987) and the tomato trypsin Inhibitor I are both strictly trypsin inhibitors. The tomato and potato Inhibitor II proteins both possess potent trypsin inhibitor activities but

Figure 1 **Growth of *Manduca sexta* larvae when fed detached leaves of transgenic tobacco plants containing increasing levels of the foreign tomato trypsin Inhibitor I protein**

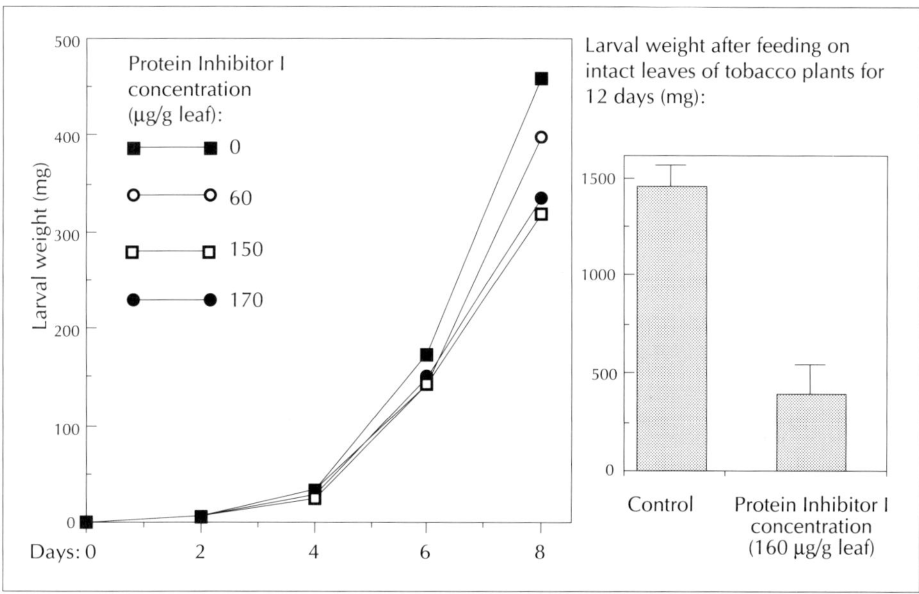

contain strong chymotrypsin inhibitor activities as well (Bryant et al. 1976; Plunkett et al. 1982). Thus, although the chymotrypsin inhibitor activity, together with its trypsin inhibitor activity, may have contributed to the antinutrient effects, it is likely that the trypsin inhibitory activity is mainly responsible for the effects in the lepidopteran species assayed. In many plant tissues, several different inhibitors are often found with varying specificity (Ryan 1990). It is possible that an array of protease inhibitors, including inhibitors with a spectrum of specificities, acting in concert, may add to the defence potential of the plant tissues. Genes coding for some of these inhibitors have already been isolated and characterized (Ryan 1990), and they may be of use in transforming plants with broad arrays of protease inhibitors for improved natural defences against insects and pathogens.

Protease inhibitors have been of particular interest in plant recombinant DNA research because they are single gene products and are therefore more easily manipulated and tested than the more complex defensive chemicals that are products of biosynthetic pathways. Amylase inhibitors and lectins are also natural plant defensive proteins, coded by single genes that are of considerable interest in the transformation of plants in order to increase insect resistance (Ryan 1990). With time, it is anticipated that plants will be transformed with cassettes with several defensive genes that will express a spectrum of proteins and enzymes. The successful transformation of crop plants with such cassettes could have important beneficial effects, not only for plant defence but also for increasing the nutritional properties of the transgenic plants. Many of the inhibitors are rich in essential amino acids such as lysine and cysteine, and when these proteins are cooked they are inactivated (denatured) and can be excellent food proteins (Huang et al. 1981).

Species Specificity of the CaMV-35S Inhibitor I Gene Fusion Construct

To add to the complexity of isolating and selecting the most effective genes for plant defence, the problems associated with the expression of foreign genes are very important. Promoters must be available to express foreign genes in the most desirable tissues. The development of transgenic plants with high levels of expression of foreign genes in all tissues may not be cost-effective, and it could negatively affect the productivity that the transgenic plants were developed to enhance. Thus, tissue specific or temporal expression of defensive proteins and/or enzymes may be necessary or desirable. Promoters conferring such specificities are being developed, but whether any given promoter will be effective in various tissues of different species of plants is poorly understood.

As a first step in investigating whether a specific construct, containing a proteinase inhibitor gene driven by a well-known constitutive promoter, the CaMV 35S promoter, would be similarly expressed in different species of plants, we transformed three species of Solanaceae and one species of Fabaceae with a fused CaMV 35S-tomato Inhibitor I gene and monitored levels of tissue-specific expression (Narvaez and Ryan, unpubl.). About 20-30 transformants of each species that expressed the tomato Inhibitor I gene and accumulated Inhibitor I protein were selected. Figure 2 shows a comparison of the accumulation of mRNA Inhibitor I and protein in the leaves of transformed nightshade,

Figure 2 Comparisons of the accumulation of the foreign tomato Inhibitor I mRNA and protein in leaves of different plants*

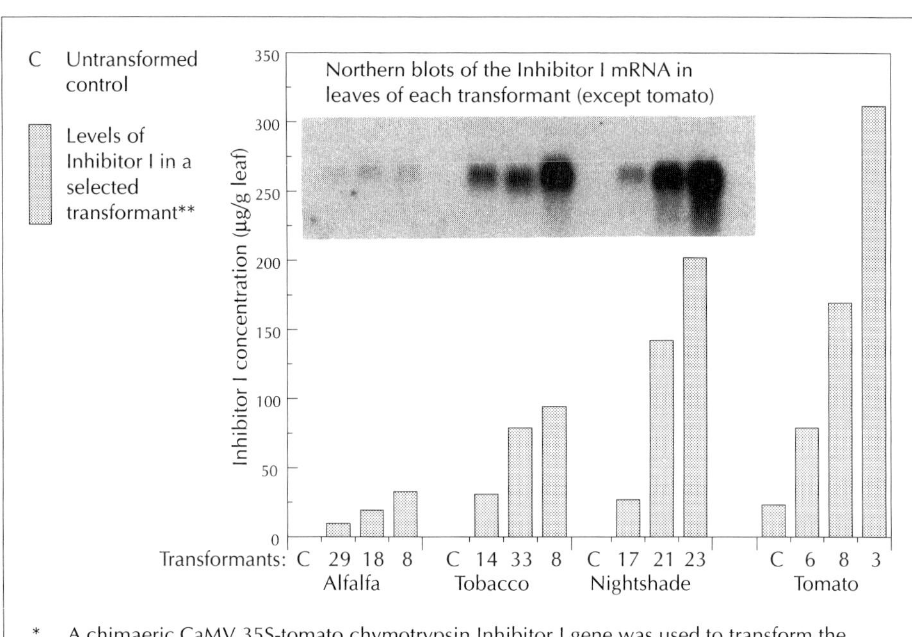

* A chimaeric CaMV 35S-tomato chymotrypsin Inhibitor I gene was used to transform the plants; over 20 transformants were recovered.

** For each plant the lowest and highest expressing transformants are presented, as well as a transformant near the average of each group.

tobacco, alfalfa, and tomato. In tomato, mRNA levels had previously been shown to reflect the levels of Inhibitor I protein (Graham et al. 1986); mRNA levels are not shown for this species. It is clear from Figure 2 that the CaMV 35S promoter is differentially effective in the four plant species when driving the tomato Inhibitor I gene expression. The tomato Inhibitor I protein accumulated to the highest levels in tomato leaves. The accumulation of the inhibitor protein in nightshade leaves was lower than in tomato leaves, but higher than in tobacco. Nightshade is also more closely related to tomato than tobacco. The alfalfa, a member of the Fabaceae family, showed only low levels of the tomato Inhibitor I protein.

To target any given proteinase inhibitor in a plant at high levels in order to provide effective protection against insect pests, it will be crucial to know something of the nature of the gut proteolytic enzymes used by the insect to digest its food, and the role of these proteinases in the regulation of the overall physiology of digestive processes of the insect. The general principles for digestion will probably be similar among genera, or perhaps throughout families of insects; however, there may be differences, depending upon the evolutionary pressures placed on the members of that family or species. With this knowledge, various proteinase inhibitors with the appropriate specificities can be selected and assayed with the isolated insect enzymes, or fed to the insects in defined diets, to determine which inhibitor might be most effective. This strategy will probably also be necessary to assess the effectiveness of other defensive chemicals, such as amylase inhibitors, lectins and thionines. It is a considerable undertaking that cannot be done easily or cheaply. At present, only limited data on insect digestive enzymes are available. More fundamental research is needed to establish a good information base. The application of such knowledge to agriculture, particularly in developing countries, could be very useful, since the aim is to endow the plants with better protective systems and to help increase productivity without the use of expensive and environment-damaging chemical pesticides.

Discussion

FAUQUET: What is the range of specificity of action of the trypsin inhibitor? What would be the effect of trypsin inhibitors on human and animal consumption?

RYAN: The specificity of any trypsin inhibitor against a broad spectrum of insects has not been established. Some trypsin inhibitors may (and do) not inhibit all trypsins. We are working in an area where little is known. The precedent of having a high level of protein inhibitors in food has already been established in nature. Potatoes, wheat, beans, rice and sweet potatoes all have high (5% and over) protein inhibitor levels.

BOSQUE-PEREZ: What insects are you trying to protect cassava from (using proteinase inhibitors)?

RYAN: Cassava hornworm. It is possible that other insects might be affected, such as nematodes. Additionally, plant pathogens induce the inhibitor genes in some plants. Perhaps some resistances against pathogens may show up. We are in a situation where we are looking for any improvements.

OGBE: What is the prospect of biotechnology for traits that are polygenic, with small effects that are additive?

RYAN: I heard Dr Rob Fraley speak last month and I thought he mentioned that at least five genes can be expressed from a cassette on a single plasmid.

References

Bryant, J., T.R. Green, T. Gurusaddaiah, and C.A. Ryan. 1976. Proteinase inhibitor II from potatoes: Isolation and characterization of the isoinhibitor subunits. *Biochemistry* 15: 3418-3424.

Graham, J.S., G. Hall, G. Pearce, and C.A. Ryan. 1986. Regulation of synthesis of proteinase inhibitors I and II mRNAs in leaves of wounded tomato plants. *Planta* 160: 399-405.

Hilder, V.A., A.M.R. Gatehouse, S.E. Sheerman, R.F. Barker, and D. Boulter. 1987. A novel mechanism of insect resistance engineered into tobacco. *Nature* 330: 160-163.

Huang, D.Y., B.G. Swanson, and C.A. Ryan. 1981. Stability of proteinase inhibitors in potato tubers during cooking. *Journal of Food Science* 46: 287-290.

Johnson, R., J. Narvaez, G. An, and C.A. Ryan. 1989. Expression of proteinase inhibitors I and II in transgenic tobacco plants: Effects on natural defense against *Manduca sexta* larvae. *Proceedings of the National Academy of Sciences, USA* 86: 9871-9875.

Plunkett, G., D.F. Senear, G. Zuroske, and C.A. Ryan. 1982. Proteinase inhibitor I and II from leaves of wounded tomato plants: Purification and properties. *Archives of Biochemistry and Biophysics* 213: 463-472.

Ryan, C.A. 1990. Protease inhibitors in plants: genes for improving defenses against insects and pathogens. *Annual Review of Phytopathology* 28: 425-449.

Wingate, V.P.M., R.M. Broadway, and C.A. Ryan. 1989. Isolation and characterization of a novel, developmentally regulated trypsin inhibitor I protein and cDNA from the fruit of a wild species of tomato. *Journal of Biological Chemistry* 264: 17734-17738.

Acknowledgements

This research was supported in part by the Washington State University College of Agriculture and Home Economics, Project No. 1791, and by grants from the National Science Foundation, USA.

6.6

Improving insect resistance in cowpea through biotechnology: Initiatives at Purdue University, USA

L.L. Murdock

Department of Entomology, Purdue University, West Lafayette, Indiana 47905, USA

Abstract

Several research initiatives aimed at laying the foundation for introducing insect resistance into cowpea are being pursued at Purdue University, USA. This work is being done in collaboration with scientists in Italian institutes and at the International Institute of Tropical Agriculture (IITA). Ongoing studies at Purdue focus on: developing improved bioassay systems to use in finding and testing specific insect resistance genes; identifying specific genes that confer resistance on specific post-flowering pests; attempting to make interspecific crosses between wild, insect-resistant *Vigna* species and cultivated *V. unguiculata;* and the genetic transformation of cowpea, using particle-mediated and *Agrobacterium*-mediated gene transfer.

Cowpea is an important food crop for the peoples of many developing countries in sub-Saharan Africa and Latin America. Its yield potential is very high (1.5-3.0 t/ha), but the actual yields on small-scale farms in much of the developing world are far less, averaging 0.2-0.4 t/ha. Insects are one of the major causes of these large yield deficits. Not only do they damage cowpea in the field, they also cause extensive losses when the crop is stored after harvest. Among the international agricultural research centres (IARCs) in the Consultative Group on International Agricultural Research (CGIAR) system, the International Institute of Tropical Agriculture (IITA) has the global mandate for cowpea improvement through research. IITA's Grain Legume Improvement Program (GLIP) conducts the Institute's research and training programmes on cowpea and other grain legumes.

Breeding/Screening Approach Developing Insect-Resistant Cowpea

Over the millennia, crop species have been continuously attacked by insect pests. Here and there, resistant plants have developed through processes of mutation, recombination, natural selection, and farmer selection. Genes for resistance have thus accumulated in the genepool of the crop species. Since there are many insect pests, some local and sporadic, and since the process of accumulating resistance genes has involved vast stretches of time, the resistance genes that have arisen are scattered throughout the crop genome. It is the task of the scientist to find these genes and make use of them. Insect management through classical plant resistance to insects involves systematically screening the world's germplasm collections of the crop species using insect bioassays; that is, the insect pest is used to tell the scientist that he has a resistant line in hand. Once a resistant accession is found, the breeders go to work, using crossing techniques to bring the resistance genes into cultivars suited to various crop ecologies and farmer and consumer preferences (Painter 1951).

IITA scientists have had some marked successes using this approach. They have discovered cowpea lines with high degrees of resistance to aphids (Bata et al. 1987), for example, and this resistance has been widely incorporated into the cowpea lines released to national programmes. Similarly, they have found cowpeas with moderate resistance to the cowpea weevil, *Callosobruchus maculatus*, the major post-harvest pest, and have built this resistance into their advanced lines.

Despite these and other successes, several important cowpea insect pests have not yielded to the breeding/screening approach. The cowpea pod borer, *Maruca testulalis*, is an example; the larvae of this species eat cowpea flowers as well as the delicate shoots and peduncles of cowpea, reducing yields greatly (Singh and Jackai 1985). Another serious pest group are the complex of pod-sucking bugs, which probe into developing seeds within the green pods. They inject a toxic saliva and suck out the digested seed contents, destroying the seed and leaving an ugly scar on the pod. Pod-sucking bugs are particularly difficult pests because they are highly mobile and because there are several species of them; a cowpea line resistant to one species may not be resistant to another. As mentioned above, the GLIP scientists have found a source of resistance to the cowpea weevil. Unfortunately, the resistance is only moderate; better as well as different sources of resistance are needed. Furthermore, this single known source may be lost if the weevils themselves are able to develop resistance to it. Among other insect pests of cowpea which still have not yielded to systematic breeding and screening for resistance are cowpea thrips, which can cause massive flower drop.

One constraint to breeding/screening efforts to improve cowpea for insect resistance is the available genepool: it is a rather poor source of resistance genes for cowpea pests. GLIP's efforts over the years to discover new sources of resistance to cowpea weevil have involved screening more than 12 000 accessions from the world germplasm collection; only a single source turned up. Similarly, substantial efforts have been made to discover sources of resistance to cowpea pod borer and pod bugs, but success has been limited.

Biotechnological Tools for Plant Resistance to Insects

The host plant resistance approach to insect management is sustainable, potentially very effective, and widely accepted by small farmers in developing countries. Given this, there needs to be some way out of the impasse blocking wider use of pest-resistant varieties for cowpea insect pest management. In fact, there are two possible ways, each of which involves elements of advanced plant and molecular science.

First, there is the route of interspecific hybridization. Over the years, scientists have noted that many of the wild relatives of cowpea are highly resistant to the insect pests which plague cultivated cowpea. In principle, if hybrids could be made between these wild species and the cultivated species, the insect resistance traits could then be brought into the cultivated cowpea by systematic breeding. The difficulty is in making the hybrids. Sophisticated laboratory techniques, such as embryo rescue, anther culture, and protoplast fusion, may have to be brought to bear.

Second, there is the route of genetic engineering. In principle, if genes coding for resistance to one or more of the cowpea insect pests could be identified, cloned, transmitted into cowpea, and expressed at the right time and place in the plant and at the right concentration, we could transform a susceptible cowpea plant into a resistant plant. The resistance genes could come from any species, not only from plants but even from bacteria and fungi.

Neither interspecific hybridization nor genetic engineering are approaches that the IARCs can follow easily without collaboration with advanced research institutes. Research in biotechnology is moving extremely rapidly. The most advanced laboratories have unhindered and continuous access to the latest information, theory, and techniques, and to a full and complex set of molecular technology, reagents, kits, and so on. Because biotechnology is driven by new basic discoveries and is multidisciplinary, large university groups, which can draw on expertise from many fields,have become the centres of origin of much biotechnology. Two such university groups have developed, over the years, strong and continuing commitments to cowpea improvement, and have evolved collaborative partnerships with GLIP at IITA as well as with each another. These are Purdue University, USA and a consortium of Italian research institutes led by L. Monti of the University of Naples. The activities of the Italian scientists in biotechnology are described elsewhere in this volume. What follows here is a brief summary of cowpea research activities going on at Purdue University.

Interspecific Hybridization Studies

Because of its promising resistance to insects, *Vigna vexillata* has received much attention from scientists in the past few years. Many attempts have been made to cross *V. unguiculata* with *V. vexillata* at IITA, at Purdue, and elsewhere, none of them successful. In an attempt to understand the barriers to hybridization between these species, detailed histological studies on materials from IITA and Purdue were carried out by F. Saccardo and colleagues at the University of Naples, Portici. They observed evidence of partial pollen stigma incompatibility, but in some cases pollen tubes were able to reach ovaries and ovules. Generally, pod retention was limited to 8-15 days. Embryonic development was generally blocked at the globular embryonic stage. This was followed by vacuolization in each embryonic cell and collapse of the endosperm, which can lead to embryo abortion.

Work carried out at Purdue by S.R. Schnapp, under the direction of P.M. Hasegawa, led to the successful crossing of *V. luteola* and *V. oblongifolia*. Both these species have seed resistance to the cowpea weevil and resistance to certain other post-flowering pests of cowpea. The *oblongifolia* x *luteola* hybrids, which are self-fertile, showed morphological traits intermediate between the two species. The cross was confirmed using DNA fingerprinting (Schnapp et al. 1990). Since both of these species exhibit resistance to multiple insect pests of cowpea, it may be possible to use the cross as a bridge to bring resistance into cultivated cowpea. Ongoing work is seeking to cross the hybrid with *V. unguiculata*. In addition, many crosses between *V. oblongifolia* and *V. unguiculata* have been attempted but, so far, without success.

Bases of Gene Transfer for Insect Resistance

Genes for resistance

Over the past several years, the Purdue team has focused on: developing hypotheses about the types of genes which could be transferred into plants to confer insect resistance; studying vulnerable physiological and biochemical systems in insects which ought to be susceptible to specific protein gene products; developing methodologies for discovering anti-insect genes; and testing these hypotheses and applying these methodologies to identify and clone promising genes for specific insect pests. This work at Purdue has led us to the view that we ought to use genes which code for protein products and which can themselves serve as the agents of plant protection, that is, biologically active proteins, such as protease inhibitors, a-amylase inhibitors, and lectins (Foard et al. 1983). If possible, these genes (and their protein products) should already be present in human and animal foods, so that any potenital hazard to the ultimate consumer would be low or even negligible. The problem is then reduced to one of how one finds useful genes for specific insect pests. In the following summary of activities and progress toward finding genes with the potential for controlling post-flowering pests of cowpeas, it needs to be emphasized that there are no magic gene bullets which will control all insect pests for all time. On the contrary, if the genetic transformation of crop plants to confer insect resistance is to become a widely accepted and sustained approach, adding to our present repertoire of insect management tools, then a system will necessarily have to be developed which guarantees a novel supply of resistance genes.

Bioassay systems

One of the keys to the process of finding useful genes through evaluation of protein gene product (often given insufficient attention) is a reliable insect bioassay system. The artificial seed bioassay system for bruchids developed at Purdue University (Shade et al. 1986) has been essential for the successful identification of protein gene products whose genes have the potential to control the cowpea weevil. The development of other bioassay systems, such as the artificial seed to test the impact of specific chemicals on pod bugs (work involving L.E.N. Jackai of IITA and R.E. Shade and L.L. Murdock at Purdue) is another example.

A further example is the availability of an artificial diet for cowpea pod borer and techniques for getting eggs from this difficult insect, developed by Jackai at IITA. Still another example is the recent development of techniques for rearing cowpea pod borer on cowpea callus, work performed by Shade and Schnapp at Purdue University. This bioassay system will enable the biotechnology group to actually test individual genes for effectiveness against a major target pest on transformed callus.

Candidate genes for post-flowering pests of cowpea

Bacillus thuringiensis d-endotoxin

The gene coding for the insect-toxic protein found in parasporal inclusions of *Bacillus thuringiensis* has received more attention than any other gene from genetic engineers interested in conferring insect resistance on crop plants. The toxin acts in two ways: it inhibits feeding, and it causes mortality by disrupting the physiological integrity of the midgut epithelium. The *B. thuringiensis* protoxin has a limited range of activity, however, being most effective against lepidopteran species. P.E. Dunn, Professor of Insect Biochemistry at Purdue, and colleagues, are exploring the impact of *B. thuringiensis* protoxin on cowpea pod borer. Protoxin preparations were incorporated into the cowpea pod borer diet (Jackai and Raulston 1988) and larvae allowed to feed on them. Most of the 10 forms of the protoxin tested were effective against cowpea pod borer, causing greatly stunted growth. Strain HD 73 showed good growth-suppressing effects, although other strains such as alesti, sotto, and berliner appeared to be more effective. The Purdue group has in hand a number of clones of *B. thuringiensis* protoxin genes which could be used to introduce cowpea pod borer resistance into cowpea.

Protease inhibitors

These are widely present in human foods, and they have received attention as possible plant protectants. They act by blocking the digestion of the insect and so limiting growth, development, and reproduction.

Work at Purdue University has focused on one class of proteinase inhibitors — the cysteine proteinase inhibitors. These inhibitors are of particular interest because many insect species, including pest bugs and beetles, use cysteine proteases (Murdock et al. 1988). Working with the bioassay unit, S.S. Nielsen at Purdue is leading studies on characterizing cysteine proteinase inhibitors from plants and assessing their impact on insects. Several cysteine proteinase inhibitors of plant origin are known (for example, rice, pineapple stem, soybean, and maize) but have been little studied. Experiments were carried out at Purdue using the microbial inhibitor E-64 with the cowpea weevil and the common bean weevil. E-64 dramatically retarded the growth and development of the two species of weevils (Murdock et al., 1987). Adding free amino acids to the diets containing E-64 restored normal growth and development. This work established that the cysteine proteases in the guts of these species are crucial for normal growth and development and that, if a gene for an active cysteine proteinase inhibitor can be found and used to transform cowpea, this could result in bruchid-resistant plants.

Work at Purdue University is currently directed towards isolating and purifying a cysteine proteinase inhibitor from soybean. Once sufficient inhibitor is in hand, it will be bioassayed against the main insect pests of cowpea.

a-Amylase inhibitors

Working under the direction of Nielsen and Murdock, J. Parker is giving attention to isolating and testing a-amylase inhibitors for effects against cowpea weevil. Studies conducted by Shade, Murdock, and J.E. Huesing confirmed reports in the literature that common bean a-amylase inhibitor is effective against the cowpea weevil when incorprated into the weevil's diet (Gatehouse et al. 1986; Ishimoto and Kitamura 1989). Using purified inhibitor provided by M. Chrispeels of the University of California, San Diego, they demonstrated that levels of inhibitor in the order of 0.2-0.6% (w/w) severely retards the bruchid's growth and development. Since the inhibitor accumulates to about this level in common bean, it should be relatively straightforward to obtain the gene, transmit it into cowpea, and get it expressed to the desired level, thereby conferring a new source of weevil resistance on cowpea. Fortunately, Chrispeels has already cloned the gene in question and made it available to the cowpea biotechnology group for use in cowpea transformation. He has also consented to provide a promotor sequence, which will ensure good expression of the gene in seeds.

Because it is desirable to have a repertoire of genes from which to chose, Parker is focusing on additional a-amylase inhibitors. Her hope is to find forms of the inhibitor even more effective than that from common bean. She will give most of her attention to cereals, which are known to be rich in these inhibitors.

Lectins

Another class of genes that have potential as insect control genes are those coding for lectins — carbohydrate-binding proteins. The Purdue team has previously shown that many different lectins are not effective against the cowpea weevil when fed in artificial seeds. However, one class of lectins (those which bind N-acetylglucosamine) had marked toxic and growth-retarding effects.

This is particularly interesting because N-acetylglucosamine is a component of chitin, which is a key component of the insect body. The most effective N-acetylglucosamine specific lectin studied to date is wheat germ agglutinin (Murdock et al. 1990). All the single isolectins of wheat germ agglutinin are equally effective against the cowpea weevil (Huesing et al. 1990). Several other N-acetylglucosamine-specific lectins were nearly as effective as wheat germ agglutinin. In principle, genes coding for wheat germ agglutinin might be used to transform cowpea and so impart resistance to cowpea weevil.

An observation made by Jackai that pod-sucking bugs will feed readily on dry seeds was taken advantage of in order to begin exploring the vulnerability of pod-sucking bugs to specific inhibitors. Using the artificial seeds which had been developed at Purdue University for cowpea weevil, Jackai and Shade recognized that they could bioassay specific chemicals for effects on pod-sucking bugs. The initial indications of this research are that both the cysteine proteinase inhibitor E-64 as well as wheat germ agglutinin have detrimental effects on the pod-sucking bug *Clavigralla tomentosicollis* (Jackai and Shade, unpubl.). This indicates that this group of insects may be vulnerable to both lectins and cysteine proteinase inhibitors. The efforts to engineer genes coding for these defensive genes into cowpea for bruchid resistance might therefore have an unexpected benefit for another group of insects attacking cowpea.

Research on the Genetic Transformation of Cowpea

Several *Agrobacterium tumefaciens* strains and Ti plasmids have been used in attempts to transform cowpea embryo and bud meristems by infection. These include LBA4404/pLBA4404; C58/pBiBo542; EHA101/pBiBo542; and EHA105/pBiBo542. The C58 and EHA strains also contained GUS marker genes on either pB1121 or pK1W1 plasmids. Over 5000 seedlings derived from 436 parent plants, grown after meristem infections, were tested for marker gene expression. So far, no positively expressing R2 plants have been found.

We have attempted to transform embryo and bud meristems with GUS marker genes by particle bombardment. Many examples of transient GUS expression in meristem cells have been found. Also, apparently permanent expression (2 weeks after transformation) of GUS has been found in the plantlets resulting from growth of the bombarded meristems.

Chimaeric gene constructs of the a-amylase inhibitor gene from common bean have been transferred to *Agrobacterium* and used to infect cowpea hypocotyls. We are currently isolating a-amylase inhibitor-expressing callus from these infections.

References

Bata, H.D., B.B. Singh, S.R. Singh, and T.A.O. Ladeinde. 1987. Inheritance of resistance to aphid in cowpea. *Crop Science* 27: 892-894.

Foard, D.E., L.L. Murdock, and P.E. Dunn. 1983. Engineering of crop plants with resistance to herbivores and pathogens: An approach using primary gene products. Pages 223-233 in *Plant Molecular Biology* edited by R. Goldberg. Alan R. Liss, New York, USA.

Gatehouse, A.M.R., K.A. Fenton, I. Jepson, and D.J. Pavey. 1986. The effects of a-amylase inhibitors on insect storage pests: Inhibition of a-amylase in vitro and effects on development *in vivo*. *Journal of the Science of Food and Agriculture* 37: 727-734.

Huesing, J.E., L.L. Murdock and R.E. Shade. 1990. Effects of wheat germ isolectins on development of cowpea weevil. *Phytochemistry* 30: 785-788.

Ishimoto, M., and K. Kitamura. 1989. Growth inhibitory effects of an a-amylase inhibitor from the kidney bean (*Phaseolus vulgaris* L.). *Japanese Journal of Breeding* 38: 367-370.

Jackai, L.E.N., and J.R. Raulston. 1988. Rearing the legume pod borer, *Maruca testulalis* Geyer (Lepidoptera: Pyralidae) on artificial diet. *Tropical Pest Management* 34: 168-172.

Murdock, L.L., G. Brookhart, P.E. Dunn, D.E. Foard, S. Kelley, L.W. Kitch, R.E. Shade, R.H. Shukle, and J.L. Wolfson. 1987. Cystein digestive proteinases in Coleoptera. *Comparative Biochemistry and Physiology* 87B: 783-787.

Murdock, L.L., R.E. Shade, and M.A. Pomeroy. 1988. Effects of E-64, a cysteine proteinase inhibitor, on cowpea weevil growth, development, and fecundity. *Environmental Entomology* 17: 467-469.

Murdock, L.L., J. E. Huesing, S. S. Nielsen, R. Pratt, and R.E. Shade. 1990. Biological effects of plant lectins on the cowpea weevil, *Callosobruchus maculatus* (L.). *Phytochemistry* 29: 81-85.

Painter, R.H. 1951. *Insect Resistance in Crop Plants*. University of Kansas Press, Lawrence, Kansas. USA.

Schnapp, S.R., R.E. Shade, L.W. Kitch, L.L. Murdock, R.A. Bressan, and P.M. Hasegawa. 1990. Utilization of in vitro culture methods to facilitate introgression of insect resistance into cowpea (*Vigna unguiculata*). Page 200 in *Abstracts, Seventh International Conference on Plant Tissue and Cell Culture, 24-29 June 1990, Amsterdam, The Netherlands.*

Shade, R.E., L.L. Murdock, D.E. Foard, and M.A. Pomeroy. 1986. Artificial seed system for bioassay of cowpea weevil (Coleoptera: Bruchidae) growth and development. *Environmental Entomology* 15: 1286-1291.

Singh, S.R., and L.E.N. Jackai. 1985. Insect pests of cowpea in Africa: Their life cycle, economic importance and potential for control. Pages 217-232 in *Cowpea Research, Production and Utilization* edited by S.R. Singh and K.O. Rachie. John Wiley, Chichester, UK.

Acknowledgements

The author is grateful to all the scientists in the Purdue University team whose work has been summarized here and to colleagues at IITA, the University of Naples, the Italian Germplasm Institute and other Italian colleagues, who have provided stimulation and encouragement to the Purdue team. Special thanks go to K. Ibrahim and R. Bressan for their assistance in preparing the manuscript.

6.7

Insect-resistant plants with improved horticultural traits from interspecific potato hybrids grown in vitro

E.D. Earle, Z. Lentini, R. Hopper, L.A. Batts, Z. Ganga and R.L. Plaisted

Department of Plant Breeding and Biometry, Cornell University, Ithaca, New York 14853-1902, USA

Abstract

In vitro culture was used to facilitate the introgression of trichome-mediated insect resistance from *Solanum berthaultii* into potato (*S. tuberosum*). Plants were regenerated from calluses of advanced tetraploid hybrids between the two species and backcrossed to potato. Some regenerates produced many more backcross progeny with desirable trichome traits than non-regenerated controls. Several culture-derived lines show promise in the Cornell potato breeding programme. Similar studies using diploid F_1 hybrids are now in progress. Many of the diploid explants produced tetraploid regenerates that facilitate further crosses with potato. Results from this project suggest that somaclonal variation may aid introgression of desirable traits from wild species into crop plants.

Many attempts to use somaclonal variation as an aid to plant breeding have sought to recover novel traits not present in the original plant material. Our work at Cornell University has a more limited and perhaps more realistic goal: to generate a reassortment of genes already present in the original material.

The plant materials used were interspecific hybrids of potato, *Solanum tuberosum* L., and *S. berthaultii*, an insect-resistant wild species from South America. Glandular trichomes on the foliage of *S. berthaultii* confer resistance on insects, including aphids (*Myzus persicae, Macrosiphum euphorbiae*), colorado potato beetle (*Leptinotarsa decemlineata*), leafhoppers (*Empoasca fabae*), mites (*Tetranychus urticae*), and thrips (*Thrips tabaci*). Two types of trichomes are involved in the resistance: type A (short-stalked, with a gland on the tip that can exude a phenolic substance) and type B (longer, with a viscous sticky droplet consisting mainly of sucrose esters).

Earlier studies in the Cornell potato breeding programme showed that backcrosses of *S. tuberosum* x *S. berthaultii* hybrids to *S. tuberosum* gave predominantly the cultivated phenotype without insect resistance (Kalazich 1989). There appeared to be restricted recombination between the wild species and the cultivated genomes, which made it difficult to combine the horticultural characteristics of potato with the resistance of the wild species. Because phenotypic and chromosomal alterations have often been observed after tissue culture of potato (Karp et al. 1989), we have tested in vitro culture of the interspecific hybrids as a strategy for facilitating the introgression of insect resistance into potato (Lentini et al. 1990).

Studies with Tetraploid Hybrid Materials

Plant materials

Advanced tetraploid hybrids of *S. tuberosum* x *S. berthaultii* from the Cornell potato breeding programme were used in most of the work. Hybrid F743-4 was 50% *S. tuberosum* (not an F_1 but a line obtained after a number of intercrosses) and hybrid J114-1 and five others tested were 75% *S. tuberosum*. Callus was initiated from petioles of in vitro grown virus-free plantlets of these hybrids using a Murashige-Skoog (MS) medium with 3% sucrose and 0.2 mg/L a-naphthalene acetic acid (NAA), 2.3 mg/L 6-benzyl-aminopurine (BAP), and 10.0 mg/L gibberellic acid (GA3). Regeneration was evaluated after transfer of callus to the medium with different growth regulators (no NAA, 2.3 mg/L BAP, 5.0 mg/L GA3).

Most petioles from *S. tuberosum* controls formed callus; about 25% of them regenerated one or more plants. *S. berthaultii* petioles produced less callus and almost no plants. The seven hybrid lines tested varied substantially in callus formation and in plant regeneration. Hybrid J114-1 produced 56 plants per 10 petioles; F743-4 produced only 5 plants per 10 petioles. Plants regenerated from J114-1 and F743-4 after short (2 weeks) or long (up to 14 weeks) periods of callus growth were used to obtain clonal or sexual progeny for evaluation in the greenhouse and field.

Evaluation of clonal progeny of regenerated plants

Clonal progeny were evaluated in replicated trials conducted in 1987 and 1988, using two populations of plants propagated by nodal cuttings from the original regenerates maintained in vitro. Progeny of two of the 58 R_0 plants recovered from calluses of hybrid J114-1 showed stable and heritable differences from the original hybrid over two cycles of evaluations in the field. One regenerate showed significantly higher marketable yield (about twofold) than the J114-1 hybrid and retained trichome characteristics comparable to the hybrid. The other regenerate had higher levels of phenolic exudate in the A type of trichome. This second regenerate also showed a heritable mutation in flower colour, from blue to white with blue flecks.

Evaluation of sexual progeny of regenerated plants

Sexual progeny of some regenerates showed desirable changes which were not evident in their clonal progeny. A backcross of hybrid F743-4 to *S. tuberosum* gave very few (0-3%) progeny with droplets on trichome B, and only 18% with browning in trichome A. In contrast, seven of 14 regenerates from hybrid F743-4 had more backcross progeny (up to 15 times as many) with these desirable trichome traits than the original hybrid. One of the seven had been regenerated from 2-week-old callus, suggesting that even a very brief period in culture may give rise to genetic changes. In field studies, these culture-derived plants showed better resistance to leafhoppers than controls, and some had good vine type and yield.

Progeny from a second backcross of at least one regenerated plant again showed a high percentage of plants with droplets on type B trichomes and high browning of type A trichomes. Further backcrosses and evaluation of these culture-derived materials are in progress in the Cornell potato breeding programme.

We do not know the genetic basis of the altered phenotypes, but they are not associated with changes in ploidy; the regenerates whose progeny show higher levels of insect resistance are tetraploid, like the original hybrid lines.

Studies with Diploid F_1 Hybrids

The studies described above used tetraploid advanced hybrids because diploid F_1 hybrids were not available. We are now working with diploid F_1 hybrids which were derived from a cross between a dihaploid *S. tuberosum* line and *S. berthaultii*. Petiole explants from 14 different F_1 hybrid plantlets showed great differences in callus growth and regeneration. We have regenerated over 150 vigorous plants from five of the F_1 hybrids, some after more than 7 months of callus culture. Plants are currently being evaluated in the greenhouse for trichome characters. Selected plants will be used for selfs, intercrosses, or backcrosses with *S. tuberosum*.

It will be possible to do the backcrosses either at the diploid or tetraploid level because both diploid and tetraploid regenerates (R_0 plants) were recovered. The ploidy of small regenerated plantlets was determined by a convenient procedure: nuclei were isolated from small amounts (~50 mg) of chopped leaves, stained with a fluorescent DNA-binding stain, and passed through a flow cytometer for analysis of DNA content (Arumuganathan and Earle 1991). Many of the R_0 plants (~50%) were tetraploids, rather than diploids. Tetraploid plants were recovered from some cultures maintained as callus for only 1 month; other calluses still produced diploid plants even after 7 months of unorganized growth. Some stunted octoploid plants were also obtained.

The backcross progeny of R_0 plants from the diploid F_1 hybrid explants will be compared to the backcross progeny from the non-regenerated F_1 hybrids. The objectives of this work are to determine whether changes which are similar to those observed in the earlier work with tetraploid advanced hybrids occur, whether work at the diploid level or tetraploid level and/or with F_1 or more advanced hybrids gives more useful materials, and whether a longer period of callus culture before plant regeneration is helpful. These studies should help determine which type of explants and culture procedures should be used in further work.

Conclusion

The results of this study suggest that callus culture of interspecific hybrids, followed by plant regeneration, may facilitate the introgression of desirable traits from wild species into crop plants. The callus phase also facilitates chromosome doubling that may be helpful in further breeding procedures. Our work has been limited to potato, but the same approach may also be useful for other crops.

References

Arumuganathan, K., and E.D. Earle. 1991. Estimation of nuclear DNA content of plants by flow cytometry. *Plant Molecular Biology Reporter* 9(3): 229-241.

Kalazich, J. 1989. Introgression of the trichome characteristics of *Solanum berthaultii* (Hawkes) into *S. tuberosum* (L.). PhD thesis, Cornell University, Ithaca, New York, USA.

Karp, A., M.G.K. Jones, D. Foulger, N. Fish, and S.W.J. Bright. 1989. Variability in potato tissue culture. *American Potato Journal* 66: 669-684.

Lentini, Z., E.D. Earle, and R.L. Plaisted. 1990. Insect-resistant plants with improved horticultural traits from interspecific potato hybrids grown in vitro. *Theoretical and Applied Genetics* 80: 95-104.

Acknowledgements

This work was supported by the United States Agency for International Development (USAID), the International Potato Center, Hatch Project 422 (EDE), the New York State College of Agriculture and Life Sciences, and the New York State Integrated Pest Management Program.

6.8

Biotechnological tools for improving the quality of seed legume crops[1]

R. Rao[2], P. Spigno[2] and A. Costa[3]

Department of Agronomy and Plant Genetics, University of Naples, Via Universita 100, 80055 Portici, Italy[2]; Research Centre for Vegetable Breeding, National Research Council, Via Universita 133, 80055 Portici, Italy[3]

Abstract

This paper discusses the structure of legume seed protein in relation to the poor nutritional quality of legume seeds. Various biotechnological applications to improve seed nutritional quality are reviewed, such as the construction of modified and hybrid genes containing new methionine codons or the identification and utilization of plant genes coding for protein of high nutritional value.

Plant proteins are an important component of the human diet and in some countries they represent the major part of food protein available to the population. Legume seeds contain a high level of protein, which unfortunately has an unbalanced amino acid composition. Although they are rich in lysine and other essential amino acids, sulphurated amino acids, such as cystine and methionine, are present in very low amounts, thus affecting the biological value of the seeds. It is known that seed amino acid content is related to the structure of seed proteins, which represent the major constituents of the seed.

Legume Seed Proteins

Legume seed proteins consist of a storage fraction (globulins), which is the most abundant fraction, and a non-storage fraction (albumin). Both these fractions are developmentally

1 Contribution No. 67 from the Research Centre for Vegetable Breeding, National Research Council, Via Università 133, 80055 Portici, Italy.

regulated and thus they accumulate inside the seed at different stages (Gatehouse et al. 1986).

The albumin fraction is composed of a few polypeptides showing macro- and micro-heterogereity, with structures that differ within species (Rao et al. 1987). They are important in determining the nutritional quality of seed proteins, having a well-balanced amino acid composition and providing a significant amount of the sulphurated amino acids present in the seed (Shroeder 1984).

Globulins belong to two majör protein classes 7S and 11S. Both classes contain several polypeptides and share many features of size, subunit composition, and processing within different legume species. Data are available on the biochemistry, genetic localization, and number and inheritance of the genes coding for these proteins in the genera *Pisum, Phaseolus, Glycine* and *Vigna*, and in *Vicia faba* (Casey et al. 1986; Pedalino et al. 1991, Tucci et al. 1991). The 7S family is represented mainly by trimeric proteins formed by three subunits of 50-70 kD, beside several low molecular weight subunits derived from post-translational processing of specific precursor polypeptides. The 50-70 kD subunits are coded by a gene family composed of about 7-8 genes (Domoney and Casey 1985); they seem to lack cystine and methionine almost completely, and therefore have a very low nutritional value. The 11S family is represented by hexameric proteins, formed by six subunits of about 60 kD, which are coded by a gene family of about 10-11 genes (Domoney and Casey 1985). Each subunit is composed of a pair of polypeptides (a-ß pair) having a better amino acid composition than those involved in the 7S protein family.

There are both quantitative and qualitative genetic variants that could possibly be used to improve the nutritional quality of legume seeds (reviewed in Casey et al. 1986), which is a difficult task because of the polygenic control of the synthesis and deposition of seed storage proteins (Monti and Leone 1990). Moreover, some legumes, such as lentils and chickpeas, have a limited genepool controlling protein content and composition (Muehlbauer et al. 1985; Smithson et al. 1985), thus preventing the improvement of nutritional quality using a classical approach.

The abundance and tissue specificity of the seed storage proteins suggest a specialized role in the provision of amino acid, ammonia, and carbon skeletons to the developing seedling. Undoubtedly, the amino acid composition of storage proteins has evolved for efficient nutrition of the developing embryo. Any molecular modification of storage proteins for improved animal nutrition should be performed within the constraints of their role in seed physiology.

Biotechnological Applications

One possible strategy is to increase selectively the high-methionine globulin components. As these are naturally occuring components, their enrichment or exclusive synthesis should not create problems in protein body deposition or in their mobilization upon germination. Elevation of high-methionine components could be coupled with 'silencing' of the low-methionine globulin genes. A possible way of selectively altering the expression of globulin components is to target regulatory mechanisms specific for methionine-rich or methionine-poor globulins. This strategy should be available soon, as both naturally occurring and induced mutations have begun to identify potential regulatory loci in several species (Goldberg et al. 1989). Another potential application of genetic engineering of storage proteins is the improvement of the nutritional balance of essential amino acids within the major storage proteins, thereby correcting deficiencies in the total protein complement.

The recent isolation of several seed protein genes and promoter regions and their utilization in gene manipulation and transformation experiments promise the possibility of new strategies to improve legumes.

An elegant experiment was performed by Saalbach et al. (1988) who changed the codon composition of a gene coding for one of the 11S faba bean subunits, legumin B, in favour of increased nutritional quality of the corresponding protein (Figure 1). The target sequence of this two-step in vitro mutation was the last exon of the gene, coding for the C-terminal part of the ß-polypeptide chain.

Figure 1 Scheme of the two-step mutation of the faba bean legumin B gene, generating 4 methionine codons in the originally methionine codon-free gene and changing the 3'-terminal sequence of exon 3

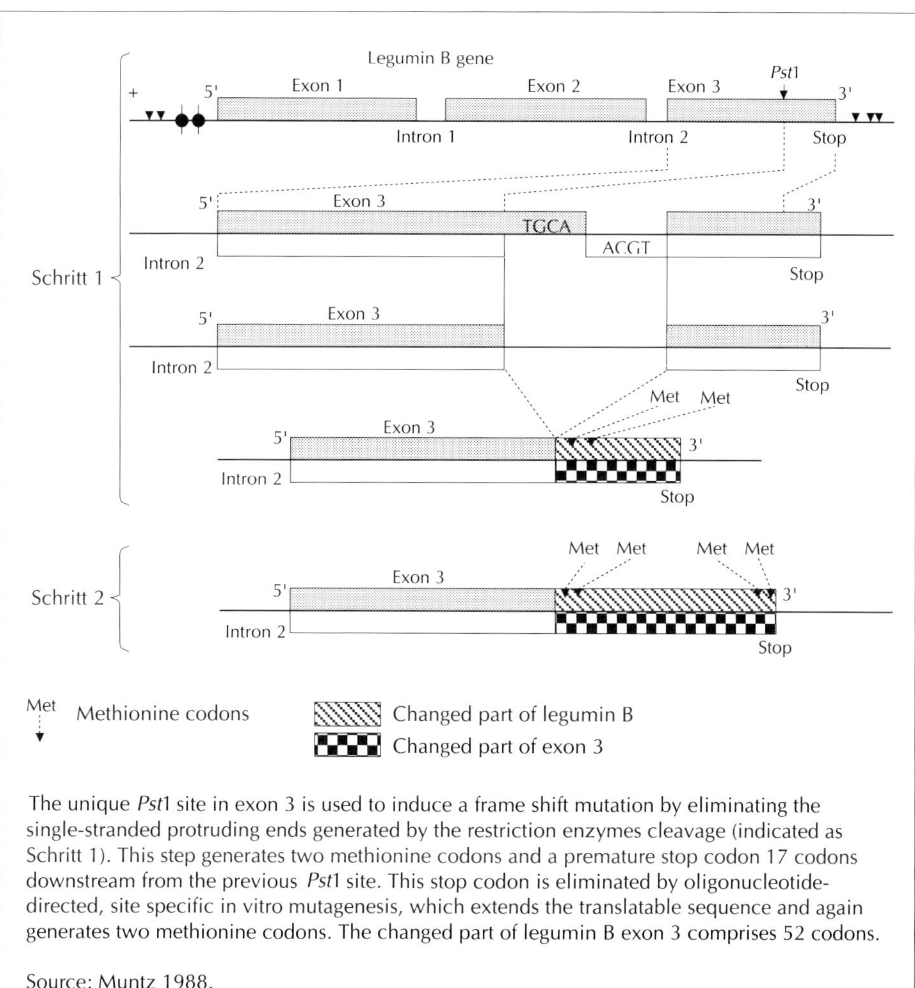

The unique *Pst*1 site in exon 3 is used to induce a frame shift mutation by eliminating the single-stranded protruding ends generated by the restriction enzymes cleavage (indicated as Schritt 1). This step generates two methionine codons and a premature stop codon 17 codons downstream from the previous *Pst*1 site. This stop codon is eliminated by oligonucleotide-directed, site specific in vitro mutagenesis, which extends the translatable sequence and again generates two methionine codons. The changed part of legumin B exon 3 comprises 52 codons.

Source: Muntz 1988.

The unique *Pst*I site of exon 3 was used to induce a frame-shift mutation with subsequent transformation of the stop codon, in order to extend the sequence of 36 additional triplets terminated by a new stop codon. This construction originates in four methionine codons. The protein coded by the modified gene appears to have normal stability. The gene modified (as described) was also ligated to a gene isolated from soybean and coding for one of the 11S storage proteins, glicin, to produce the so-called glicigumin gene, which contains 11 methionine codons. This hybrid gene contains the 5' region, including the promoter, of glycinin, while the 3' end contains the legumin sequence.

Both genes have been used to transform *Nicotiana* species, but their level of expression was unsatisfactory. However, more encouraging examples have been described regarding the transformation for genes conferring high nutritional quality similar to that of the Brazil nut, a methionine-rich protein which accumulates to a high level, up to 8% of total seed proteins (Altenbach et al. 1987; Altenbach and Simpson 1990).

It is worth noting that the manipulation of genes controlling protein in plant seeds could serve as a model for engineering changes in other constituents of seeds.

Regarding seed quality, it should benoted that the nutritional quality of leguminous seeds is sometimes limited also by the presence of toxic proteins such as lectins, or anti-nutritional compounds which are thermo-tolerant or thermo-resistant, the former being more difficult to control. Some naturally occurring soybean mutants presenting insertion sequences at the lectin locus suggest the possibility of using transposons to switch off these undesirable genes (Goldberg et al. 1983). However, lectins are sometimes involved in conferring insect resistance (Ming-Li and Sun 1990; Chrispeels and Raikhel 1991), which makes unnecessary the use of insecticides where residues may also affect seed quality.

According to statistics produced by the Food and Agriculture Organization of the United Nations (FAO), about 10% of the world's population — some 450 million people — are on the edge of starvation or suffer severe malnutrition. On the other hand, only 10% have their nutritional needs adequately met and another 10% are only moderately well nourished (Briskey et al. 1983). A major long-term problem will be crop quality, and biotechnology could make its most valuable contribution through the application of new techniques to traditional plant breeding.

References

Altenbach, S.B., K. Pearson, F.W. Leung, and S.S.M. Sun. 1987. Cloning and sequence analysis of a cDNA encoding a Brazil nut protein exceptionally rich in methionine. *Plant Molecular Biology* 8: 239-246.

Altenbach, S.B., and R.B. Simpson. 1990. Manipulation of methionine-rich protein genes in plant seeds. *Trends in Biotechnology* 8: 156-160.

Briskey, E.J., U.W. Hays, and R.L. Mitchell. 1983. Future approaches for meeting nutritional needs. Pages 118-143 in *Agriculture in the Twenty-First Century* edited by S.W. Rosenblum. John Wiley, New York, USA.

Casey, R., C. Domoney, and N. Ellis. 1986. Legume storage proteins and their genes. Pages 1-95 in *Plant Molecular and Cell Biology* (Vol. 3) edited by B.J. Miflin. Oxford University Press, Oxford, UK.

Chrispeels, M.J., and N.V. Raikhel. 1991. Lectins, lectin genes, and their role in plant defense. *The Plant Cell* 3: 1-9.

Domoney, C., and R. Casey. 1985. Measurement of gene number for seed storage proteins in *Pisum*. *Nucleic Acid Research* 13: 687-699.

Gatehouse, J.A., I.M. Evans, R.R.D. Gray, and D. Boulter. 1986. Differential expressions of genes during legume seed development. *Philosophical Transactions of the Royal Society of London* Series B 314: 367-384.

Goldberg, R.B., G. Hoschek, and L.O. Vodkin. 1983. An insertion sequence blocks the expression of a soybean lectin gene. *Cell* 33: 465-475.

Goldberg, R.B., J. Barker, and L. Perez-Grau. 1989. Regulation of gene expression during plant embryogenesis. *Cell* 56: 149-160.

Ming-Li, W., and S.S.M. Sun. 1990. Molecular analysis of bean arcelin genes encoding an insect resistant protein. *Supplement to Plant Physiology* 92: 14.

Monti, L., and A. Leone. 1990. Miglioramento genetico delle leguminose da granella per gli aspetti qualitativi. *Rivista di Agronomia* 24: 132-145.

Muehlbauer, F.J., J.I. Cubero, and R.J. Summerfield. 1985. Lentils (*Lens culinaris medic)*. Pages 266-311 in *Grain Legume Crops* edited by R.J. Summerfield and E.H. Roberts. Collins, London, UK.

Muntz, K. 1988. *Genetic Engineering of Plant Storage Proteins: Present Research and Future Prospects*. Italian Association for International Development, Santa Margherita Ligure, Italy.

Pedalino, M., M. Paino D'Urzo, A. Costa, S. Grillo, and R. Rao. 1991. Biochemical characterization of cowpea seed protein. Pages 81-89 in *Cowpea Genetic Resources* edited by N.Q. Ng and L.M. Monti. IITA, Ibadan, Nigeria.

Rao, R., S. Grillo, and M. Tucci. 1987. The use of electrophoretic techniques in genetic studies of legume seed proteins. Pages 45-57 in *Agriculture: Protein Evaluation in Cereals and Legumes* edited by V. Pattakou. Cereals Institute, Thessaloniki, Greece.

Saalbach, G., R. Jung, I. Saalbach, and K. Muntz. 1988. Construction of storage protein genes with increased number of methionine codons and their use in transformation experiments. *Biochemie and Physiologie der Pflanzen* 183: 211-218.

Schroeder, H.E. 1984. Major albumins of *Pisum* cotyledons. *Journal of the Science of Food and Agriculture* 35: 191-198.

Smithson, J.B., J.A. Thompson, and R.J. Summerfield. 1985. Chickpea (*Cicer arietinum L*). Pages 312-390 in *Grain Legume Crops* edited by R.J. Summerfield and E.H. Roberts. Collins, London, UK.

Tucci, M., R. Capperelli, A. Costa, and R. Rao. 1991. Molecular heterogeneity and genetics of *Vicia faba* seed storage protein. *Theoretical and Applied Genetics* 81: 50-58.

7.1

Scientific, social, and economic implications of biotechnology for developing countries[1]

R. Barker

Department of Agricultural Economics, Cornell University, Ithaca, New York, USA

Abstract

This paper addresses a range of questions: How does biotechnology differ from traditional agricultural technologies? What are the ways in which biotechnology could have an impact on agriculture? What socio-economic impact will biotechnology have on agriculture in developing countries? What are the ethical and regulatory issues associated with biotechnology? What are the issues related to the protection of intellectual property rights? What steps should be taken to improve the access of developing countries to biotechnology? Biotechnology will be of benefit to developed and developing countries alike, but a major concern is that the gap between "favourable" and "unfavourable" farms, regions, and nations is likely to grow. The continued protectionist policies for agricultural products in the developed countries will ensure that the highly developed agricultural research systems will gain, relative to the rest. This fact notwithstanding, it is in the best interests of developing countries to adopt a strategy which will facilitate access to new biotechnologies.

Productivity growth in agriculture and industry in the period after World War II depended in large measure on advances in the petrochemical and transportation industries. With the maturing of these industries, the potential for further growth in productivity from these sources has diminished. Furthermore, the rapid rise in energy consumption and heavy dependency on fossil fuels has led to growing concerns about resource depletion and environmental pollution.

1 This paper draws heavily on an earlier paper: Barker, R. 1990. Socioeconomic impact. Pages 299-310 in *Agricultural Biotechnology: Opportunities for International Development* edited by G.J. Persley. CAB International, Wallingford, UK.

Biotechnology and microelectronics (or information technology) are the new "core technologies" which offer prospects for further growth in productivity, saving raw materials, reducing energy consumption, and mitigating pollution problems. Over time, these technologies are becoming increasingly interlinked. Biotechnology is knowledge intensive and advances in biotechnology require considerable scientific manpower and investment. The main impact will not be realized for decades to come. The challenge for the developing countries is to generate the capability to utilize this knowledge and adapt biotechnologies to suit local conditions.

Several questions concerning the new biotechnology arise, including the following: Will the introduction of biotechnology result in a continued rapid growth in agricultural productivity? To what degree will the development of the new technology be based on public as opposed to private research and development (R&D) investments, and what are the implications? How will technology transfer take place between and among the developed and developing countries? What impact will these technologies have on trade between developed and developing countries?

The purpose of this paper is to identify and discuss briefly the broad range of technical, socio-economic, and ethical issues related to biotechnology and its potential impact on developing countries.

How does Biotechnology Differ from Traditional Agricultural Technologies?

Biotechnology differs from more traditional forms of technology in three important respects. First, the creation of biotechnologies involves input from basic scientists, such as molecular and cell biologists and cytogeneticists, who are not a part of the traditional agricultural research extension network. In fact, many of these scientists engaged in agriculture-related research are not in colleges of agriculture or agricultural research stations, and they may have little, if any, familiarity with agriculture. This poses a problem in establishing research priorities that reflect the needs of farmers and consumers.

Second, the processes and/or products of biotechnology are often patentable. This has greatly stimulated private investment in agricultural research. Whereas traditionally the bulk of agricultural research was conducted in the public sector, today private sector investment in biotechnology-related research exceeds public sector investment. A division of labour is developing between the two sectors, with the public sector focusing on basic science and the private sector on technology development.

Third, research in biotechnology is expensive. Simple forms of biotechnology research, employing techniques such as tissue culture to propagate a disease-free perennial, may cost less than US $1 million and take 3-6 years. However, genetic engineering or gene transfer may require between US $50 million and US $500 million and one to four decades to achieve the goal.

The implications for developing countries are clear. Basic biotechnology research is too costly and too demanding of scientific skills for the limited resources of most of these countries. Access to biotechnologies may be further restricted by the private sector. However, in order to gain access to biotechnologies and to adapt these technologies to local conditions, developing countries should first and foremost develop a research establishment strongly grounded in the traditional agricultural disciplines such as plant breeding, agronomy, plant pathology, and entomology.

In what Ways could Biotechnology have an Impact on Agriculture?

Biotechnology could have an impact on agriculture through:

- Increased yields and productivity
- Improved quality and shelf life of products
- Improved resistance to pests and diseases
- Substitution of one crop for another or, alternatively, a synthetic for an agricultural commodity
- Substitution of renewable (agricultural) for non-renewable (fossil fuel) energy sources
- Adaption of existing crops and livestock to different environments
- Shift in marketed surplus to farmers, regions, and nations with high productivity

Of course, even before the advent of the new biotechnology, agricultural research has always affected all the areas listed above. However, the growing dependence on science-based agriculture and associated advances in biotechnology are likely to speed up the process.

The list of areas for biotechnology research is long. It includes pest and pathogen resistance, quality enhancement, manipulation of growth regulators, improved tolerance to stresses (including temperature, moisture, and soil conditions such as salinity), enhancement of photosynthetic activity, and achievement of nitrogen fixation in non-leguminous crops. Thus, there is the prospect of lowering production costs, reducing dependency on energy from non-renewable resources, increasing productivity of agriculture in hitherto disadvantaged environments, and mitigating environmental and pollution problems. Major advances in almost any of these areas would seem to offer significant benefits for developing countries. All this sounds too good to be true, and indeed there is a major gap between the potential and the reality.

The reality is that biotechnology applications must be specifically designed for target organisms, and it may require intensive research to perfect the many separate techniques required by a process. Furthermore, technology once perfected must be adapted to local conditions; this has been a major obstacle in the dissemination of traditional technology. However, biotechnology could facilitate this process. For example, tissue culture is being widely used, even in developing countries, to accelerate the process of screening large numbers of cultivars for desired agronomic traits.

The agricultural applications of biotechnology will be more rapid in those crops where cloning is possible, where genetically manipulated organisms or enzymes can be used to produce or displace agricultural products, and in animal applications, where there has been considerably more research to date, some of it a spillover from research on human health. Progress will be slower in the major cereal and feed grains, which are of importance to the developing world. This is because in these monocots much work needs to be done in gene mapping and perfection of recombinant DNA techniques before the important genes can be identified and transferred.

In summary, biotechnology will vastly expand the range of choices in research and it offers the prospect for advances in a wide range of areas that would be of benefit to society. However, with limited budgets, choices have to be made. Which line of research to pursue, or which gene to transfer, involves both technical and socioeconomic considerations and often even an ethical consideration. The establishment of research priorities will determine who benefits in the long run. In the next three sections, we examine some of the socio-economic and ethical issues in biotechnology research as they relate to the developing countries.

What Socioeconomic Impact will Biotechnology have on Agriculture in Developing Countries?

Although biotechnology will be of benefit to developed and developing countries alike, the gap between "favourable" and "unfavourable" farms, regions, and nations is likely to grow, for the following reasons:

- The difficulty in transferring knowledge-intensive technology to under-developed areas
- Limited access because of the increasing privatization of research
- Loss of export markets for developing country commodities through substitution in end products

Much of the developing world today depends on public sector investments in agricultural research and extension, but over the past decade budgets for research and the quality of national research institutions have declined in many developing countries. Support from developed countries and donor agencies is declining with the growing pressures for assistance to eastern Europe. Also, there has been a tendency to spread investments in research and extension over a large number of institutions, rather than developing a few quality institutions. The result has been a weakened capacity for technology transfer, and hence reduced ability to adapt biotechnologies to local conditions.

There are exceptions to this trend. For example, large countries such as Brazil, China, and India and a number of the smaller more advanced developing countries do have considerable domestic research capacity. The gap between these countries and the less advanced developing countries is likely to grow.

The private sector now represents more than half of the total investment in biotechnology research. Increasingly, the research agenda in developed countries is being driven by the priorities of the private sector. The private sector will invest in R&D only in those areas where they see a profit. Many of the commodities and research problems important to the developing world (for example, in the staple crops of Africa such as cassava, plantain, and millets) are likely to receive too little attention without continued strong support for public sector research.

Lastly, biotechnology offers the potential for further substitution in end products, which could reduce the demand for exports from developing countries. The failure of the General Agreement on Trade and Tariffs (GATT) negotiations to reach agreement on reducing agricultural protection in the developed countries creates an additional incentive to produce substitutes. For example, liquid maize sweetener has replaced about 50% of the sugar consumed in the USA and accounts for 15% of the world's sweetener market. The high price support for domestic sugar in the USA (designed to protect the domestic sugar beet industry) provided the incentive for developing liquid maize sweetener. This resulted in a sharp drop in demand for sugar cane exports from the developing countries. Biotechnology facilitates the focus on the use of end products such as starches, proteins, cooking oils, or sweeteners and thus makes it easier for scientists to respond to distorted prices by creating cheaper substitutes. The poorer African countries that rely largely on exporting tropical products and importing food are the most vulnerable.

What are the Ethical and Regulatory Issues Associated with Biotechnology?

The following questions relate to some of the ethical and regulatory issues associated with biotechnology:

- How can regulatory regimes be established that will ensure health, safety, and the protection of the environment without unduly dampening the incentives for research?
- What scientific, technical, economic, and psychological information is needed to rationalize regulatory regimes?
- How should we deal with the often false public perceptions of the risks involved?
- What position should be taken on the private sector conduct of research, testing, and dissemination of technologies in developing countries which have no regulations?

Another way to state the issue is as follows: Can it be anticipated that environmental pressures are likely to stimulate the development and diffusion of biotechnologies for sustainable agricultural production systems? Or will the regulatory processes, problems related to the protection of intellectual property rights, and public fears about the new technologies in food and agriculture inhibit their development?

What are the Issues Related to the Protection of Intellectual Property Rights?

The proponents of intellectual property protection argue that it will:

- Encourage private investment in biotechnology R&D and plant breeding
- Make available modern varieties which are beneficial to agricultural producers

The opponents argue that intellectual property protection will:

- Hamper developing country access to plant genetic resources
- Increase the burden on a farmer's income by raising the price of seeds or other patented inputs
- Result in a "brain drain" or loss of public sector scientists to the private sector

If the developing countries are to abide by laws governing intellectual property rights, they will see a clear economic advantage in compliance. Most do not. But where farmers grow crops for the export market, they will have difficulty exporting their products if they are protected by patent in the importing country.

While developed countries argue for intellectual property protection, the countervailing strategy of developing countries is to promote farmers' rights. Those advocating farmers' righers argue that farming communities in areas of diverse plant genetic resources should participate in the benefits derived from the conservation and utilization of these resources.

What Steps should be Taken to Improve Developing Country Access to Biotechnologies?

It is in the interest of the developed countries to encourage agricultural development and economic growth in the developing countries. These countries represent the greatest potential source of increased demand for exports from developed countries. However, it is very difficult to predict the impact of biotechnology on both developed and developing countries. Two very different scenarios can be envisaged, one pessimistic and the other optimistic, with respect to the impact of biotechnology on development in the developing world.

If existing protectionist policies for agricultural products continue, the debt problem faced by developing countries remains unresolved, and too little attention is given to the development of agricultural research capacity in these countries, the slowdown of economic growth and agricultural trade witnessed in the past few years will become a chronic problem. In this environment, advances in biotechnology will tend to increase inequities, not only between developed and developing countries, but also among the more and less advanced developing countries. Those countries with the most highly developed agricultural research systems will gain, relative to the rest.

On the other hand, if the structural maladjustments of the 1990s are reduced or removed, and if through technical assistance and other measures the applied agricultural research capacity is strengthened throughout the developing world, the benefits from advances in biotechnology are likely to be more evenly distributed.

Many participants at the IITA conference expressed concern that the first scenario seemed already to be a reality in Africa. However, regardless of which scenario prevails, the optimum strategy for helping developing countries to gain access to new biotechnologies will be the same. This strategy can be outlined as follows:

- Strengthen conventional agricultural research/extension capacity in areas such as plant breeding and plant protection.
- Provide more support to biotechnology research laboratories in the developed countries to tackle problems that are important for the developing countries (for example, the Rockefeller Foundation's international programme on rice biotechnology).
- Strengthen links between public and private sector research institutions in the developed and developing world where a conflict of interest can be avoided.
- In this context, encourage the international agricultural research centres to play a key role in: identifying and transferring high priority technologies or end products from biotechnology to developing countries; exploring opportunities for establishing links with private sector companies; and promoting biosafety.
- In the long term, strengthen basic science research and training capacity in the tropical world.

The success of this strategy will require close cooperation between national and international organizations, and between the public and private sector.

7.2

International agricultural research centres and the private sector

G.J. Persley

Agriculture and Rural Development Department, World Bank, 1818 H Street, Washington DC 20433, USA

Abstract

The major change in agricultural research funding in industrial countries in the past decade has been the greatly increased role of the private sector, mainly in funding research and development in modern biotechnology. This paper examines the implications of such a change for the international agricultural research centres (IARCs) and their collaborators in the national agricultural research systems. The IARCs could foster the greater use of modern biotechnology on the major tropical commodities. They could also foster greater private sector involvement in the applications of biotechnology to these commodities. A major issue affecting such applications is the lack of protection of intellectual property in most developing countries. The IARCs and the national research systems need to clarify their policies and practices in regard to intellectual property management. The primary purpose of intellectual property management would be to facilitate greater access to new developments in biotechnology by the IARCs and the national research systems.

The major change during the past decade in the way agricultural research in industrial countries is funded has been the substantially increased role of the private sector, largely in funding research in modern biotechnology. In 1985, it was estimated that approximately US $4 billion was spent on research and development (R&D) activities in modern biotechnology worldwide. Of this, US $2.7 billion (67%) came from the private sector. It was estimated that US $900 million (22%) was spent on agriculture-related research, and US $3100 million on other forms of modern biotechnology, mainly for human health care. Within the total estimated agricultural biotechnology expenditure of US $900 million in 1985, US $600 million (66%) was spent on seeds and US $300 million (34%) on agricultural microbiology. An estimated US $550 million (60%) of the total expenditure on agricultural R&D came from the private sector (Persley 1990).

Since private companies are able to appropriate many of the benefits of research investments in modern biotechnology, they are far more willing to invest in it than in more traditional agricultural research, where it is more difficult to obtain proprietary rights to the technology generated. The exceptions are agrochemicals and hybrid seed, where there have been substantial private sector R&D investments because both these types of products can be protected.

Structure of the Biotechnology Industry

The companies investing in modern biotechnology in agriculture are: new biotechnology firms; major agrochemical and seed companies; and other major companies, particularly food companies. Several invest at least US $5 million per year on R&D in biotechnology. In 1986 there were about 134 companies in the USA involved in agricultural biotechnology, of which 70 were new biotechnology firms (OTA 1986). The commercial applications of biotechnology are currently dominated by American and European companies, with increasing interest from Japanese companies.

Several large agrochemical companies are developing their markets for new products by purchasing seed companies. This allows them to integrate modern biotechnology approaches into conventional plant breeding programmes. There are now a series of mergers and acquisitions in progress as R&D investments are rationalized in the light of more realistic market forecasts and better estimates of the likely time frames for the availability of novel products.

A new feature of modern biotechnology has been public/private sector collaboration, with private companies often contracting specific research projects with public sector institutions and, in return, having some proprietary rights to the technology generated. Continued public sector investment in biotechnology and creative partnerships between public and private sector interests are critical for a country attempting to establish a competitive strategy in biotechnology.

A new pattern of agricultural research funding is emerging from the links from biotechnology to both academia and commerce, with increased research financing coming from sources beyond traditional government appropriations. In several industrial countries, many public research institutions are now required to raise a substantial part of their research budget from non-government sources through contractual research, licensing agreements, and royalties. The result is a new pattern of agricultural research activity, one in which traditional methods of open interaction and communication in research are being rapidly altered, and where the potential availability of intellectual property rights for new technologies influences the extent of private investment.

Intellectual Property Management

Private sector companies have been investing substantially in R&D for modern biotechnology, largely because they perceive that there are opportunities for the proprietary protection of the new products and processes. Protection of intellectual property rights is demonstrably one of the most effective ways of stimulating private sector innovation in the development of biotechnology applications.

Patent protection in the USA has been granted under existing legislation for biotechnological processes (in the 1970s), and for novel microorganisms (1980), plants (1985), and animals (1988) produced using recombinant DNA technology (OTA 1986). Several other Organization of Economic Cooperation and Development (OECD) countries also offer some patent protection for biotechnological products and processes. Few non-OECD countries offer any patent protection.

The legal systems available for protecting innovations include: seed and breed certificates; plant patent and variety protection; invention patents; petty patents; inventors' certificates; industrial design patents; copyrights; and trade secrecy.

Many industrialized countries and transnational companies see what they regard as inadequate protection of intellectual property rights as a major disincentive to preparing novel products for use in the developing world. Conversely, many newly industrialized and developing countries consider that the current international patent regime works to the disadvantage of those countries with the ability to copy inventions and that these countries receive nothing in return for protecting inventions produced in industrialized countries. If no ways are found to build bridges between these divergent viewpoints, disagreements over intellectual property rights will be a major impediment to generating applications of biotechnology for agriculture in developing countries.

The issues of concern include: the limitations to be placed on the franchise of foreign firms to supply technology; the protection to be provided to the progeny of plants and animals covered by an intellectual property rights agreement; the breadth of distribution of germplasm originating from the international agricultural research centres (IARCs) with the Consultative Group on International Agricultural Research (CGIAR) system; and the role of petty patents in stimulating local adaptive research. These issues are examined in more detail by Evenson and Putnam (1990).

The current international debate on 'patenting life' stems from: the economic debate as to what is an equitable international system for recognizing the contribution of an invention, while not stifling the development of inventive capacity elsewhere, particularly in the developing world; and the ethical and legal debates as to whether living organisms produced by new biotechnologies are patentable subject matter or part of the common heritage of mankind. Many countries specifically exclude agricultural inventions from patent protection on the grounds that they provide basic needs for food and health care for the population (OTA 1989).

There is a need to facilitate greater access by individual countries and the IARCs to biotechnological products and processes, while protecting the legitimate interests of their inventors. This could be done by licensing arrangements and other contractual agreements.

The issues involved intellectual property management, and their possible resolution, may be summarized as follows:

- *Model patent systems.* Individual countries need to consider the benefits and costs of intellectual property rights in relation to biotechnology and, where the benefits outweigh the costs, design patent systems tailored to their requirements.
- *Trade negotiations.* Individual countries need to consider their position on negotiating intellectual property rights in exchange for improved trade opportunities.
- *Patent policy information.* Information and advice should be made available to staff of national agricultural research systems, IARCs, and international development agencies on the issues involved in the management of intellectual property in biotechnology. This

would facilitate the development of greater public/private sector collaboration in biotechnology.

- *Negotiating skills.* Improved skills in IARCs, national agricultural research systems, and international development agencies are needed in order to negotiate with private companies on behalf of consortia of developing countries to gain access to technologies likely to be useful when adapted to commodities important in developing countries.
- *Patenting IARC inventions.* The IARCs need to identify the advantages and disadvantages of patenting their significant inventions in biotechnology. The ability of IARCs to identify such advantages will be critical for the negotiation of substantive collaboration with private companies in biotechnology.

The IARCs could patent their significant inventions and then license these (free if appropriate) for use by national research systems and other collaborators. This action could be defensive patenting, to protect the IARCs against third parties patenting their inventions. It could also be offensive patenting, in terms of earning royalties for IARC inventions and using these to fund further research and development in collaboration with national programmes. Patenting IARC inventions would also facilitate the negotiation of substantive collaborative activities with private sector companies.

International development agencies and IARCs could also negotiate access to new technologies on behalf of consortia of countries, particularly the small technology importers with limited capacity to adapt technology themselves. This situation exists particularly in Africa, where there are many small countries which have little local adaptive capacity at present and would benefit from greater access to new technologies.

Emerging Role of the International Agricultural Research Centres

The IARCs can play an important role in adapting the tools of modern biotechnology to the needs of agriculture in the developing world. Most IARCs are already involved in biotechnology to varying degrees; their current programmes are described in detail by Plucknett et al. (1990) and in a recent World Bank technical report (World Bank 1991). In order for them to play their role to its full potential, there will need to be further shifts in research strategies and, in some centres, a reallocation of resources and personnel.

The major clients of the centres' efforts in biotechnology are likely to be the many smaller and poorer countries (70 or more), rather than the technologically more advanced countries (10 or less). The latter are developing their own capacity in biotechnology and are likely to require little assistance from the IARCs, except in an occasional advisory role.

In order to be a serious player in biotechnology, an international centre needs to have:

- A critical mass of in-house scientific expertise to monitor, choose, and use new products and processes, and adapt them to the limiting factors in agricultural production in the developing world
- Skills to acquire new technologies from the public or private sector in industrial countries, under suitable licensing and/or royalty arrangements if necessary
- Scientific and managerial skills to develop well-chosen and effective collaborative research programmes with the best public and private sector laboratories in industrial and newly industrialized countries. This will require legal and financial skills not

commonly available at the centres, but which could perhaps be provided on a CGIAR system-wide basis
- Research capacity to integrate new technologies into existing R&D programmes, particularly plant breeding
- Support for the necessary complementary research in such disciplines as physiology, pathology, entomology, and cytogenetics to provide the scientific basis for future genetic improvements
- Adequate biosafety guidelines, compatible with the regulatory environment of the host country

To enable the IARCs to adapt the tools of biotechnology to the demands of agriculture in the developing world, new methods for acquiring technologies need to be considered. The centres need to be able to negotiate in a creative fashion with the private sector on their own behalf and, if requested, on behalf of consortia of their client countries. This may involve licensing a particular technology for use by a centre, in a manner similar to the purchase of proprietary computer software. Identifying and understanding the relative strengths of the IARCs and the private sector is the first step in defining working relationships between the two parties (USAID 1989; James and Persley 1990).

To acquire new technologies, especially in an area that is changing as rapidly as biotechnology, the centres need the capability to observe, choose, and use new concepts that might prove useful in crop or livestock improvement. International centres located in countries remote from new developments may wish to consider innovative staffing arrangements to facilitate their access to new scientific developments. For example, a centre staff member could be located in an advanced institute in an industrialized country. His/her brief would be to monitor new developments in modern biotechnology and establish collaborative links with the R&D programmes of the IARC. Outposting staff in this way may prove more cost effective than trying to ensure that a molecular biologist located at the centres headquarters keeps abreast of a rapidly changing field.

The acquisition of desirable new technologies necessitates access to new expertise by the international centres. Decisions must be made on the type of technology needed, choices made between the various options available, and estimates made of the costs associated with each technology and of the expected time required to obtain practical applications for agriculture. Once this decision-making process is in place and a technology selected, the means of its acquisition must be decided (Plucknett et al. 1990).

Acquiring new technologies should not be limited to those of the public domain alone. Recent advances in the private sector, coupled with their interest and ability in working with the IARCs, should be considered as an option. Collaborative relationships with advanced laboratories can be effective ways of acquiring and developing these new technologies. Technology acquisition, development, and application by the IARCs and the national research systems could benefit greatly from expanded links between the centres, national institutions, and biotechnology practitioners elsewhere, both public and private.

The different features of modern biotechnology, relative to more conventional agricultural research, suggest that the IARCs need to consider:

- Distributing improved germplasm through the private sector, as well as through national agricultural research systems
- Moving their research upstream towards more strategic programmes

- Relying on patents or other property rights for processes and products, and obtaining such protection of proprietary lines prior to use
- Integrating molecular and cellular biology with conventional breeding to make new advances
- Complying with national and international regulatory standards, especially those involving genetic engineering
- Maintaining and enhancing strategic alliances for the collaborative research required to augment centre-oriented research
- Undertaking more direct commercial involvement, as national programmes and IARCs collaborate with the private sector
- Depending upon new germplasm to achieve genetic gains

Additional support should be provided to the IARCs to enable expansion of the following types of activities (Persley 1990; Plucknett et al. 1990):

- *Scientific expertise at the IARCs.* Scientific interchange, including university and industrial post-doctoral research; development of interdisciplinary teams to handle new technologies and apply them to specific constraints; posting centre staff to advanced laboratories to monitor new developments and identify potentially useful technologies.
- *Technology acquisition.* Acquisition of new technologies, including ones from the private sector, through purchase, licensing, or other agreements on royalties; collaborative research designed to serve the interests of both parties; market analysis; technology appraisal.
- *Verification.* Testing of model systems in contained experiments prior to wider adoption.
- *Application.* Encouragement of participation by national programmes and commercial concerns in the application of biotechnology to agricultural development.
- *Outreach.* Provision of opportunities for training, extension, and adaptive research with national agricultural research systems.
- *Policy and management issues.* Collaboration with national agricultural research systems in defining national policies on biotechnology and preparing national programmes appropriate to the size and scientific infrastructure of a particular country.

Discussion

BOKANGA: I would like to suggest one approach to fight back the loss of export market. First, let's not always emphasize export of raw materials but rather encourage the transformation of raw materials into processed products with high added-value. Two examples will illustrate this point. First, thaumatin, a protein with a sweetening power 3000 times greater than that of sucrose, is extracted from a wild berry plant growing in West and Central Africa. Its use in soft drinks could be a great source of revenue for Africa. Japan is going to produce thaumatin by genetic engineering. Second, papain, a protease from papaya, is used widely in the food industry. It could also be produced in bacteria through genetic engineering. However, in the early 1880s a company in Zaire set up a efficient purification of papain and took over about 60% of the world market. The question is:

Will there be assistance from donor countries or IARCs to help developing countries increase their capacity to transform raw materials into products with high added value, into products of good quality which could compete favourably on the world market?

ENONUYA: While the developed countries seek more public sector investments in biotechnology, the need in Africa is for more private sector investment. However, since the bulk of the investible funds in Africa are in the hands of the governments, the issue of developing biotechnology is these parts boils down to government policy objectives, and commitment to actual implementation of rational, logical solutions, rather than simply ceremonial pronouncements about them. Developing African countries need some help and training in technology, priority identification, decision making, planning, and implementation. The question that arises is what could the Africa Plant Biotechnology Network (APBNET) and individual scientists do to influence government policies in the right development-oriented direction in these matters?

THOTTAPPILLY: We had a planning meeting in August 1988 to discuss the priorities in cassava, yam, and plantain. Another planning meeting was held in February 1989 to discuss the constraints in cowpea. These meetings have guided IITA in setting up its priorities. Significant progress has been made here since we met last. Also, collaborative research projects were initiated after these meetings.

FAUQUET: The patent is a false problem in biotechnology in that if patents are not accepted, private sector research will stop. If they accepted, we have to distinguish between industrial and food crops. For industrial crops, the concept is to share the benefits between the farmer and the private companies, and any industrial crop can support biotechnology as it was supporting pesticides. For food crops, if the private companies can have a return with industrial crops they will be able to support the development of biotechnologies for food crops. On this last point, I think that the public sector should not invest in the private companies; on the contrary, private companies should spend a little part of their investment on tropical research, and this is a political decision.

FAUQUET: About the APBNET, I want to reinforce the arguments of Dr Barker and Dr Persley. To have an active and efficient network, we need money, objectives, priorities and connections. It is not possible to take care of 50 different crops and all the targets for each crop. Choices about priorities have to be made in order to concentrate efforts. Funds need to be found to run the network, to initiate programmes, and to train people. Connections need to be developed with other networks and particularly with advanced laboratories in developed countries. After some time, the APBNET should handle collaborative programmes between advanced laboratories and African laboratories. IITA should be integrated in the APBNET to play important roles as leader and catalyst.

References

Evenson, R.E., and J. Putnam. 1990. Intellectual property management. Pages 332-336 in *Agricultural Biotechnology: Opportunities for International Development* edited by G.J. Persley. CAB International, Wallingford, UK.

James, C., and G.J. Persley. 1990. Role of the private sector. Pages 367-377 in *Agricultural Biotechnology: Opportunities for International Development* edited by G.J. Persley. CAB International, Wallingford, UK.

OTA. 1986. *Technology, Public Policy and the Changing Structure of American Agriculture.* OTA-F-285. US Government Printing Office, Washington DC, USA.

OTA. 1989. *New Developments in Biotechnology: Patenting Life.* Special Report, OTA-BA-370. US Government Printing Office, Washington DC, USA.

Persley, G.J. 1990. *Beyond Mendel's Garden: Biotechnology in the Service of World Agriculture.* CAB International, Wallingford, UK.

Plucknett, D., J.I. Cohen, and M. Horn. 1990. Role of the international agricultural research centres. Pages 400-414 in *Agricultural Biotechnology: Opportunities for International Development* edited by G.J. Persley. CAB International, Wallingford, UK.

USAID. 1989. *Proceedings, Workshop on Strengthening Collaboration in Biotechnology: International Agricultural Research and the Private Sector, 17-21 April 1988.* Science and Technology Bureau, USAID, Washington DC, USA.

World Bank. 1991. *Agricultural Biotechnology Study: The Next Green Revolution?* World Bank Technical Report No. 133. World Bank, Washington DC, USA.

7.3

Biosafety in agricultural technology

G.J. Persley

Agriculture and Rural Development Department, World Bank, 1818 H Street, Washington DC 20433, USA

Abstract

One of the major issues that will affect successful applications of biotechnology to agriculture is the regulatory climate governing the release of new products. This paper describes the effect that biotechnology is having on regulatory systems in several countries. In terms of field releases, there have been over 200 small-scale releases in 17 countries, with no untoward effects. The first commercial releases of new, genetically engineered plant varieties are expected within the next 2 years. The national agricultural research systems are establishing biosafety systems, generally based on guidelines similar to those which have been endorsed by the Organization for Economic Cooperation and Development (OECD). There is increasing interest in regional cooperation in preparing guidelines suitable for various geographic and ecological regions.

A major issue that will affect successful applications of biotechnology to agriculture is the regulatory climate governing the release of new products. A safe and efficient regulatory process, able to ensure public health and environmental safety, is in itself a comparative advantage in biotechnology.

In many countries, the existing legislation is adequate to regulate the use of most agricultural products which are likely to be produced using biotechnology. The new requirement is for guidelines to cover the handling of genetically engineered organisms at the experimental stage, and for the assessment of any risk associated with the commercial release of new organisms into the environment. The guidelines should be framed so as to complement and support existing regulatory agencies. Several recent studies report that the benefits of the use of the new technologies outweigh the risks, and that risk assessments can be undertaken to guard against the release of any potentially damaging organisms. Guidelines based on those which have been endorsed by the Organization for Economic

Cooperation and Development (OECD) are now in use in several countries to allow such an assessment.

There have been at least 200 small-scale releases of genetically modified plants in 17 countries (up to mid-1990), with no untoward effects. Current work in several countries is concerned with developing procedures for handling large-scale, commercial releases of new products. This will enable experimental material to be carried forward into new products for wide-scale use. In agriculture, these products will be primarily new plant varieties with novel characteristics.

Regulatory Systems

Four definitive statements on regulating environmental applications of genetically engineered organisms have been published in USA:

- US National Academy of Science (NAS 1987)
- Office of Technology Assessment of the United States Congress (OTA 1988)
- Ecological Society of America (Tiedje et al. 1989)
- US National Academy of Science (NRC 1989)

These studies share common judgements regarding the appropriate approaches for regulatory authorities to take. All are cautiously optimistic that the benefits of the use of new technologies outweigh the risks and that mechanisms exist to allow regulatory authorities to assess the instances of low, medium, and high levels of risk, and to enforce appropriate safeguards.

Statements on evironmental release have also been made by the OECD (1986) and the Australian Recombinant DNA Monitoring Committee (Millis 1990). The UK Royal Commission on Environmental Pollution has published its report on the release of genetically engineered organisms into the environment (Anon. 1989). The OECD report provides internationally endorsed recommendations concerning safety considerations when releasing genetically engineered organisms. The Australian guidelines for risk assessment are further developed in the recent report of the Ecological Society of America (Tiedje et al. 1989).

Many OECD countries already have regulations concerning the release of new biological products from any source. These regulations usually provide sufficient protection for the approval of the commercial sale of the products of biotechnology because legislation governing the release of agrochemicals, biological control agents, and new plant varieties is as relevant to the products of modern biotechnology as it is to more conventionally produced products.

In countries which do not have satisfactory regulatory procedures in agriculture, the position is less satisfactory, as there is no suitable regulatory framework into which biotechnology products and processes can fit. Several United Nations organizations, including the Food and Agriculture Organization (FAO), the United Nations Environment Programme (UNEP), the United Nations Industrial Development Organization (UNIDO) and the World Health Organization (WHO), are reviewing the needs of their member countries in order to advise on the development of safe and efficient regulatory processes in all countries. Environmentally safe application of biotechnology is also a major concern of the United Nations Conference on the Environment and Development (UNCED) to be held in Brazil in 1992.

Guidelines for Release of Genetically Modified Organisms

The need for statutory regulations on the conditions governing release are not embraced enthusiastically when they are excessively rigid, expensive, and time consuming. Such regulations may be self-defeating in that they encourage unauthorized experimentation. Guidelines established by a technical committee rather than legislative regulations are the preferred choice in some countries, as they have the advantage over statutory regulations of flexibility and rapid change in a field where the expansion of knowledge is rapid (Millis 1990).

In implementing appropriate guidelines, the key step is the formation of a hierarchy of biosafety committees, at the institutional and national level, so that they can carry tiered responsibility in the following areas:

- Certifying the security of the facilities used for genetic engineering work, according to the category of risk
- Approving proposals at the institutional level for work of the lowest category of risk and forwarding to the national committee for review any proposals where the institutional committee is uncertain of the risk, or where the genetic construct is one of perceptibly high risk
- Ensuring that advice from the national biosafety committee is followed, including the appropriate training of workers
- Inspecting regularly the containment facilities
- Providing the national biosafety committee with regular listings of all current work involving genetic manipulation

All countries need to establish functioning national review bodies and institutional biosafety committees, and to develop guidelines on monitoring and regulating their applications of biotechnology. Possible terms of reference for a national biosafety committee would need to be listed. Table 1 (*overleaf*) provides an example of such a list, derived from the Australian experience.

Risk Assessment Criteria

The Ecological Society of America produced a list of criteria which should be taken into comsideration with respect to the safety of planned introductions (Tiedje et al. 1989). It develops further the risk assessment criteria established in Australia, and shows how such criteria might be linked together in a flexible review scheme that should be of great interest and value to regulators.

The US National Academy of Science (NAS) and its National Research Council (NRC) have released a report on the establishment of a framework for decision making procedures regarding the introduction of genetically modified microorganisms and plants into the environment (NRC 1989). The study was charged with identifying the criteria for defining

Table 1 Terms of reference for an advisory committee on genetic manipulation

Having regard to the Government's wish for a voluntary monitoring system for recombinant DNA technology and for advice to be provided to the Minister on the continuing assessment of the hazards associated with the production and/or application of material incorporating recombinant DNA molecules unlikely to occur in nature, the Committee shall:

1. Establish and review as necessary guidelines for both physical and biological containment and/or control procedures appropriate to the level of assessed risk involved in relevant research, development, and application activities.

2. Review relevant proposals, except those that relate to research performed under laboratory containment conditions, and recommend any conditions under which this work should be carried out, or that the work not be undertaken.

3. Consult with relevant Government agencies and other organizations as appropriate.

4. Report to the Minister at least annually. Report promptly to the Minister on breaches of the guidelines referred to in (1) above, and on other relevant matters referred to the Recombinant DNA Monitoring Committtee (RDMC) by the Minister.

5. Establish contact and maintain liaison with such monitoring bodies in other countries and with international organizations, as is appropriate.

6. As necessary, advise on the training of personnel with regard to safety procedures.

7. Collect and disseminate information relevant to the above, having due regard to the special circumstances relating to proprietary information.

8. Oversee the work of a scientific subcommittee whose terms of reference, in addition to (3),(5),(6), and (7) above, insofar as they relate to laboratory-contained research, are set out below.

The scientific subcommittee shall:

1. Enter into discussions directly with scientists and the institutions where they work, and with fund-granting bodies, in determining the conditions under which laboratory-contained research with recombinant DNA molecules should be carried out.

2. Review proposals for such research and recommend any conditions under which experiments should be carried out, or that work not be undertaken.

3. Provide technical advice to the Committee and contribute to the function of the Committee in relation to laboratory-contained research.

The Committee shall report within 5 years of its establishment on the need for continuing to monitor recombinant DNA activities.

Source: RDMC 1987.

risk categories and recommending ways to assess the potential risks associated with introducing modified organisms into the environment. The report reiterates the principle contained in the earlier document produced by the National Academy (NAS 1987) that safety assessment of a recombinant DNA-modified organism "should be based on the nature of the organism and the environment into which it will be introduced, not on the method by which it was modified. No conceptual distinction exists between genetic modification of plants and microorganisms by classical methods or by molecular methods that modify DNA and transfer genes."

The NRC report (1989) also recognizes that there is a long history of utility and safety in the use of plants and microorganisms. Society has benefited greatly from the use of genetically modified microorganisms and plants, and field testing is essential in order to increase our knowledge about the relative safety or risk of large-scale utilization of genetically modified organisms and to determine the potential ability of the modified organisms.

With regard to the field testing of genetically modified plants, the NRC (1989) report concludes that:

- Plants which are modified by classical genetic methods are judged to be safe for field testing on the basis of experience with hundreds of millions of genotypes field-tested over decades. The current means for making decisions about the introduction of classically bred plants are entirely appropriate and no additional oversight is needed or suggested.
- Crops modified by molecular and cellular methods should pose risks no different from those modified by classical genetic methods for similar traits. As the molecular methods are more specific, users of these methods will be more certain about the traits they introduce into the plants. Traits that are unfamiliar in the specific plant will require careful evaluation in small-scale field tests where plants exhibiting undesirable phenotypes can be destroyed.
- The potential for enhanced weediness is the major environmental risk perceived for introductions of genetically modified plants. The likelihood of enhanced weediness is low for genetically modified, highly domesticated crop plants, on the basis of our knowledge of their morphology, reproductive systems, growth requirements, and unsuitability for self-perpetuation without human intervention.
- Confinement is the primary condition for ensuring safety of field introduction for classically modified plants.
- Depending on the crop species, proven confinement options include biological, chemical, physical, spatial, environmental, and temporal isolation, as well as size of field plot.
- Plants grown within field confinement for experimental purposes rarely, if ever, escape to cause problems in the natural ecosystem.
- Established confinement options are as applicable to field introduction of plants modified by molecular and cellular methods as to introductions of plants by classical genetic methods.

The NRC has developed frameworks for the assessment of risk, based primarily on familiarity criteria. These frameworks are discussed in detail in Persley (1990).

Biosafety Procedures at the International Agricultural Research Centres

The IARCs have established their own institutional biosafety committees. These function in accordance with host country approval mechanisms, where available. When these regulations are not in place, the IARCs use internationally endorsed guidelines, such as those agreed by the OECD, to ensure that adequate safeguards are followed. The successful testing of organisms in countries with established regulatory processes provides firm justification for subsequent testing in other countries. Flexibility in establishing regulations will be required to accommodate rapidly evolving innovations from recombinant DNA research, while respecting environmental and ethical concerns (Plucknett et al. 1990).

Figure 1 outlines the possible relationships between national review bodies and the institutional biosafety committee of an IARC. Tables 2 summarizes the recommended biosafety procedures for IARCs and national agricultural research systems (Plucknett et al. 1990).

Figure 1 Potential relationships between institutional biosafety committees of international agricultural research centres (IARCs) and national agricultural research systems

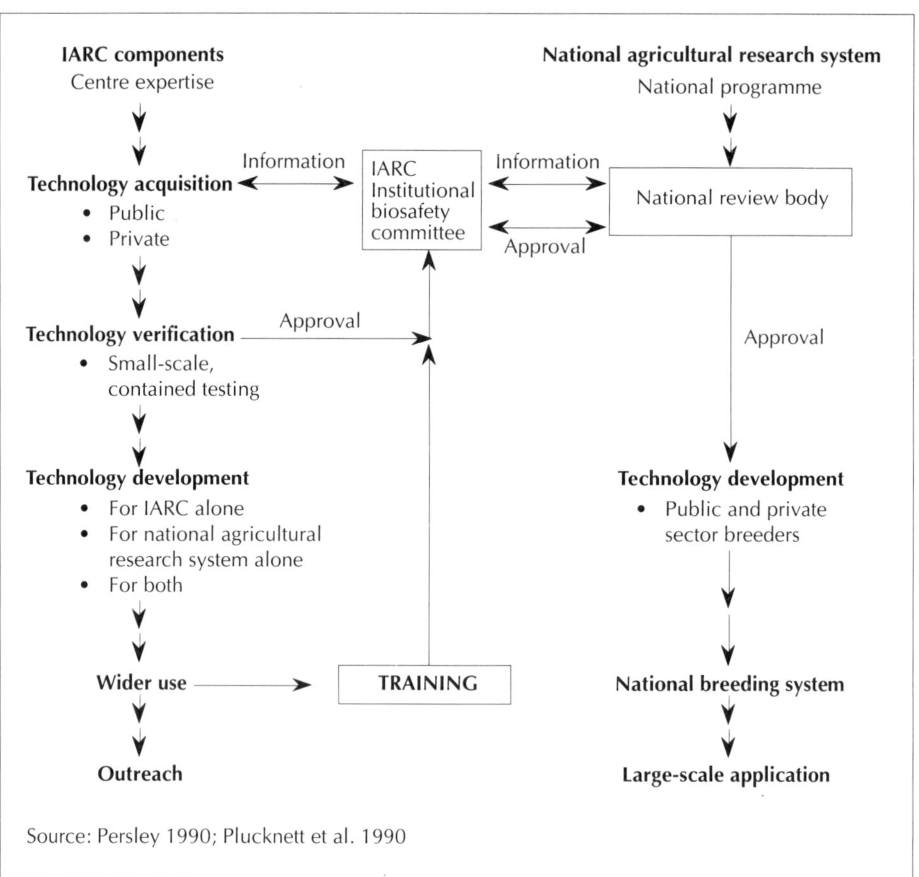

Source: Persley 1990; Plucknett et al. 1990

Table 2 **Biosafety procedures for national agricultural research systems and international agricultural research centres (IARCs)**

1. National review bodies

   There is a need for individual countries to establish functioning and well-informed national review bodies and institutional biosafety committees to monitor and regulate the release of genetically engineered organisms into the environment. The national body and its guidelines should complement and support existing regulatory agencies.

2. IARC institutional biosafety committees

   There is a need for the IARCs to establish international biosafety committees, where these do not exist, and ensure that these function in accordance with host country approval mechanisms.

3. Environmental risk assessment procedures

   National review bodies and the IARCs need ready access to regular assessments of the experience and current policies of countries with functioning national review committees, particularly in relation to environmental risk assessment procedures, so that they can benefit rapidly from the experience being gained in other countries and by international organizations.

4. Project biosafety reviews

   International development agencies need to ensure that biosafety reviews, using functioning national guidelines, are conducted prior to the release of genetically engineered organisms in any sponsored projects.

Source: Persley 1990.

References

Anon. 1989. *Royal Commission on Environmental Pollution 13th Report. The Release of Genetically Engineered Organisms into the Environment.* HMS0, London, UK.

Millis, N.F. 1990. Regulating release of organisms. Pages 322-331 in *Agricultural Biotechnology: Opportunities for International Development* edited by G.J. Persley. CAB International, Wallingford, UK.

NAS. 1987. *Introduction of Recombinant DNA-Engineered Organisms into the Environment: Key Issues.* National Academy Press, Washington DC, USA.

NRC. 1989. *Field Testing Genetically Modified Organisms: Framework for Decisions.* National Academy Press, Washington DC, USA.

Persley, G.J. 1990. *Beyond Mendel's Garden: Biotechnology in the Service of World Agriculture.* CAB International, Wallingford, UK.

Plucknett, D., J.I. Cohen, and M. Horne. 1990. Role of the international agricultural research centres. Pages 400-414 in *Agricultural Biotechnology: Opportunities for International Development* edited by G.J. Persley. CAB International, Wallingford, UK.

OECD. 1986. *Recombinant DNA Safety Considerations.* OECD, Paris, France.

OTA. 1988. *Field Testing of Engineered Organisms: Genetic and Ecological Issues.* OTA-BA-350. US Government Printing Office, Washington DC, USA.

RDMC. 1987. *Procedures for the Assessment of the Planned Release of Recombinant DNA Organisms*. RDMC Publication No. 7. Department of Industry, Technology and Commerce, Canberra, Australia.

Tiedje, J.M., R.K. Colwell, Y.L. Grossman, R.E. Hodson, R.E. Linski, R.N. Mack, and P.J. Regal. 1989. The planned introduction of genetically engineered organisms: Ecological considerations and recommendations. *Ecology* 70: 298-315.

Recommendations

General Recommendations

1. Collaboration between researchers in different fields is essential for progress in biotechnology. Collaboration — the sharing of a common, important goal, and the sharing of hypotheses, techniques, concepts, work, and credit — will engender a more effective effort than can be generated by groups working in isolation, and it can prevent duplication. For major constraints, there is benefit in having two or three groups at work on the problem — a little competition among the collaborators can be quite beneficial.

2. The International Institute of Tropical Agriculture (IITA) can be a bridge between the advanced laboratories and the national programmes to mobilize the effort needed to break major constraints. It can do this by:

   (a) motivating advanced laboratories (where the basic research leading to new biotechnologies is being done) to become involved in addressing the problems of IITA's mandate crops
   (b) providing training (in methodologies, for example) for African scientists
   (c) ensuring that IITA's own scientists benefit by learning the latest ideas and techniques through training in advanced laboratories; this should be encouraged by IITA management via sabbaticals or specialized retraining programmes
   (d) promoting and fostering linkages between advanced laboratories and African research groups; the idea of a sister institution relationship catalyzed by IITA is an attractive one
   (e) fostering and stimulating the development and dissemination of appropriate biotechnological techniques (those most needed in African laboratories and which can be implemented there in the near future and mid-term)
   (f) identifying people to work at IITA to facilitate the transfer of technologies

3. The leadership of IITA should take an active role in contacting private biotechnology companies to explore the possibility of these companies making available, on a humanitarian basis, genes, promoters, hypervariable probes, and other methodologies.

4. Biosafety of the products of biotechnology must be of high concern. The IITA leadership should ensure that all appropriate measures be taken to ensure biosafety in the implementation of the products of biotechnology.

Wide Crosses

Wild species are known to be a source of useful traits, such as good-quality seed, and resistance to biotic and abiotic stresses. The transfer of these traits is often difficult, because of interspecific barriers. The *constraints encountered* in wide crosses are: (1) availability of germplasm; and (2) incompatibility (a) pre-fertilization and (b) post-fertilization. In order to *overcome the constraint*s, (1) the right type of germplasm must be found; and (2) crossing barriers need to be overcome, using bud pollination, hormone treatment, embryo culture, in vitro fertilization, and protoplast fusion. Following successful crossing, appropriate selection strategies are required by conventional techniques or by advanced techniques such as restiction fragment length polymorphisms (RFLPs) and DNA fingerprinting.

The working group identified the constraints to cassava, yam, plaintain, cowpea, and rice improvement and then made recommendations for interspecific hybridization to address such constraints. These are summarized here, followed by a summary of the recommendations made for germplasm collection and for training.

Cassava

No crossing barriers were identified; however, additional sources are required for resistance to: green mite; mealybugs; cassava mosaic virus; cassava bacterial blight; and anthracnose. Wild species, such as *Manihot glaziovii*, have been identified as good sources of resistance to these pests and diseases. The working group recommended that efforts should continue on assembling more wild *Manihot* germplasm for evaluation and on interspecific hybridization between cassava and wild *Manihot* to introduce useful genes to cassava.

Yam

There is a lack of basic knowledge on crossability between *Dioscorea* species.The working group recommended that experiments on interspecific hybridization and cytogenetic studies be conducted to exploit the diverse genepool of *Dioscorea* for yam improvement.

Plantain

The successful identification of sources of resistance to black sigatoka from wild diploid *Musa* and the incorporation of the resistant genes into plantain is a classic example of the utilization of a wild genepool for crop improvement. Such efforts should be continued. The working group recommended the collection of more wild *Musa* germplasm for use in the breeding programme; interspecific hybridization research should also be continued.

Cowpea

Cowpea production is constrained by many post-flowering pests, such as *Maruca* pod borer, legume pod-sucking bugs, flower thrips, and bruchids. Good sources of resistance have not been found in the cultivated germplasm. On the other hand, several good sources of resistance to those pests have been found in wild *Vigna* germplasm. The working group recommended that every effort should be made to utilize the wild *Vigna* germplasm, especially those species belonging to the primary genepool of cowpea. Other wild *Vigna* species have good potential and some have been shown to have better resistance to pests than the close relatives of cowpea. The genepool of these species should be exploited. However, there are many constraints related to crossing incompatibility which have to be solved. Interspecific hybridization studies on genetic affinity should be continued; with the help of cytogenetics and biotechnology, they will be able to identify appropriate species for intensive crossing to transfer the desirable genes from wild species to cowpea. Research on embryo rescue should also be strengthened.

Rice

Germplasm of *Oryza glaberrima* and lines of wild rice in Africa have good potential as sources of resistance to insect pests and diseases, such as rice yellow mottle virus and stem borers, and to physiological stress. The working group recommended that the genepool of these African rice species should be exploited for the improvement of *O. sativa*. Interspecific hybridization between *O. sativa* and these African rice species should be studied.

Germplasm collection and conservation

Since plant genetic resources constitute the base for crop improvement, and advances in biotechnology will enhance the use of these resources, including distantly related species or genera, the working group recommended that ongoing efforts to collect and conserve germplasm should be sustained. Further collection efforts should be made to fill gaps.

Training

The working group recommended that more personnel in national programmes should be trained in biosystematics, cytogenetics, and embryo and cell cultures, as well as in germplasm evaluation and management.

Major Problems and Available Biotechnological Solutions

An effort was made to generate a matrix of current crop problems and priorities, on the one side, and available biotechnological techniques, on the other, as this was expected to provide a clearer picture of what may be possible in the short term and what will remain to be done in the medium or long term (Table 1 *overleaf*). While ready or demonstrable techniques are available for application in some cases, the need for better tools to facilitate diagnostics was noted in several others.

Policy Issues

Policy issues were explored in this discussion group, primarily with a view to developing specific advice on the types of training needed and on enabling mechanisms at IITA to

Table 1 **Matrix of major problems and available tools/techniques of biotechnology, by crop (developed from discussion/consensus in a working group)**

Crop priorities		Genes	Trans./ regen.	Wide crosses	Embryo rescue	— RFLP maps — Status	Need
Cassava:	HCN	BN	NA/D	Yes	Yes	TB	Yes
	Mealybug	BN					
	Low protein	BN					
	Mosaic virus	A					
	Diagnostics						
	Grasshopper						
	Nematode						
	Anthracnose						
	Eating quality						
	Mites						
Yam:	Fungus complex	BN	NA/D	Yes	Yes	TB	Yes
	Nematodes	BN					
	Mosaic virus	A					
	Potyvirus	TB					
	Yam beetle	BN					
	Diagnostics						
Maize:	MSV-MMV	A, BN	NA/R	No	No	A	Yes
	Stem borer	BN					
	Sitophilos	BN					
	Striga	BN					
	Diagnostics						
	Drought, downy mildew, Helminthos						
Rice:	Pyriculariosis	BN	NA/D	Ongoing	Ongoing	A	Yes
	Nematodes	BN					
	Rice yellow mottle	A					
	Gall midge						
	Rice weevil						
	Eating quality						
	Salt stress						
Cowpea:	*Maruca*	A	NA/NA	Yes	Yes	TB	Yes
	Pod bugs	A, TB					
	Thrips	BN					
	Bruchids	A					
	Potyviruses	TB					
	Diagnostics						
	Striga						
Plantain + Banana:	Black sigatoka	BN	NA/D	Yes	Yes	TB	Yes
	Banana weevil	BN					
	Mosaic virus	A					
	Banana bunchy top	TB					
	Banana streak	A					
	Diagnostics						
	Lodging, root rot, Panama disease						

Note: trans. = transformation; regen. = regeneration; A = available; NA = not available; BN = basic need; D = demonstrated; R = routine; TB = to be developed.

promote research collaboration on biotechnology. In the process, a shopping list was developed of desirables among the biotechniques available at present. The groups' recommendations are listed here.

In training, IITA should

- Be crop neutral
- Utilize "spillover" technologies
- Disseminate (biotechnology) information
- Be based on specific technology/problem, not crop
- Provide opportunities for post-doctoral training
- Provide facilities for regional training courses
- Provide joint training with (African) medical and animal biotechnology groups
- Present biotechnology which is low-cost and suitable for national programmes and universities

In research, IITA should

- Play an intermediary or bridge role between developed countries and developing countries in Africa
- Develop low-cost biotechnology tools for use in Africa
- Devote adequate human resources to carry out research goals (for example, at least two principal scientists, four post-doctoral fellows, technicians, and visiting scientists)
- Establish links with African medical and animal biotechnology groups
- Maintain links with laboratories in developed countries

In policy, the following issues are important

- Biosafety
- Sustainable biotechnology products
- Links with private sector
- Patents/policy of the Consultative Group on International Agricultural Research (CGIAR)

Priority list for biotechniques (not necessarily in order)

- RFLPs/DNA fingerprinting
- Wide crosses technology
- Tissue culture
- Polymerase chain reaction/DNA amplification technology

Acronyms

ACIAR	Australia Centre for International Agricultural Research
ACMV	African cassava mosaic virus
ACP	Africa-Caribean-Pacific group of countries
AIDAB	Australian International Development Assistance Bureau
AlMV	alfalfa mosaic virus
APBNET	Africa Plant Biotechnology Network
ARS	Agricultural Research Service (USDA)
BBTV	banana bunchy-top virus
BCMV	bean common mosaic virus
BIOTASK	Task Force on Biotechnology (CGIAR)
BNYVV	beet necrotic yellow vein virus
BYDV	barley yellow dwarf virus
BYMV	bean yellow mosaic virus
CAbMV	cowpea aphid-borne mosaic virus
CaMV	cauliflower mosaic virus
CATIE	Centro Agronomico Tropical de Investigacion y Ensenanze (Costa Rica)
CBB	cassava bacterial blight
CCMV	cassava common mosaic virus
cDNA	complementary DNA
CGIAR	Consultative Group on International Agricultural Research
CGM	cassava green mite
CIAT	Centro Internacional de Agricultura Tropical
CIMMYT	Centro Internacional de Mejoramiento de Maíz y Trigo
CM	cassava mealybug
CMeV	cowpea mottle virus
CNR	Consiglio Nazionale delle Ricerche (National Research Council, Italy)
CP	coat protein
cpDNA	chloroplast DNA
CTA	Technical Centre for Agricultural and Rural Cooperation
CTCRI	Central Tuber Crops Research Institute (India)
CuMV	cucumber mosaic virus
CW	coconut water
CYMV	cowpea yellow mosaic virus
EEC	European Economic Community
ELISA	enzyme-linked immunosorbent assay
EMBRAPA	Empresa Brasileira de Pesquisa Agropecuária
FAO	Food and Agriculture Organization of the United Nations
FHIA	Fundacion Hondureana de Investigacio Agricola
GLIP	Grain Legume Improvement Program (IITA)
GTZ	Deutsche Gesellschaft für Technische Zusammenarbeit
HCN	hydrocyanic acid
IAEA	International Atomic Energy Agency
IAPTC	International Association of Plant Tissues Culture
IARCs	international agricultural research centres
IBPGR	International Board for Plant Genetic Resources
ICARDA	International Center for Agricultural Research in the Dry Areas
ICRISAT	International Crops Research Institute for the Semi-Arid Tropics

ICTP	International Cassava-Trans Project
IDRC	International Development Research Centre
IFAR	International Fund for Agricultural Research
IITA	International Institute of Tropical Agriculture
INIBAP	International Network for the Improvement of Bananas and Plantains
IRRI	International Rice Research Institute
ISEM	immunosorbent electron microscopy
ISNAR	International Service for National Agricultural Research
MCDV	maize chlorotic dwarf virus
MDMV	maize dwarf mosaic virus
MMV	maize mosaic virus
MSV	maize streak virus
NAS	National Academy of Science, USA
NRC	National Research Council (NAS)
ODA	Overseas Development Agency (UK)
OECD	Organization for Economic Cooperation and Development
ORSTOM	Institut Français de Recherche Scientifique pour le Développement en Coopération
ORSV	ontontoglossum ring spot virus
OTA	Office of Technology Assessment (US Congress)
OTU	operational taxonomic unit
PEBV	pea early browning virus
PEG	polyethylene glycol method
PLRV	potato leafroll virus
PMMV	pepper mild mosaic virus
PRSV	papaya ringspot virus
PVR	plant variety rights
PVX	potato virus X
PVY	potato virus Y
QTL	quantitative trait locus
R&D	research and development
RAPD	random amplified polymorphic DNA
RDMC	Rccombinant DNA Monitoring Committee (Australia)
RFLP	restiction fragment length polymorphism
SBMV	southern bean mosaic virus
SHMV	sunnhemp mosaic virus
SMV	soybean mosaic virus
STG	shoot-tip grafting
TEV	tobacco etch virus
TGMV	tomato golden mosaic virus
TMGMV	tomato mild green mosaic virus
TMV	tobacco mosaic virus
TobRSV	tobacco ring spot virus
ToMV	tomato mosaic virus
TRV	tobacco rattle virus
TSV	tobacco streak virus
TYMV	turnip yellow mosaic virus
UNDP	United Nations Development Programme
UNEP	United Nations Environment Programme
UNIDO	United Nations Industrial Development Organization
USAID	United States Agency for International Development
USDA	United States Department of Agriculture

WDMV	wheat dwarf mosaic virus
WHO	World Health Organization
WMV	watermelon mosaic virus
ZYMV	zucchini yellow mosaic virus

Index

This index lists plants, countries, institutions and organizations referred to in this volume; it is not intended as an index to the plant biotechnology techniques and concepts discussed in the volume.